Geographies of Developing Areas

Dividing the world into neat compartments – East and West, First and Third World, industrialized and rural – serves to code and classify our understandings of difference. Such definitions, however, provide a limiting and fractured perspective on how people live their lives. This significant new textbook questions traditional conceptions of Africa, Asia, Latin America and the Caribbean to provide a new understanding of the 'Global South'. Eschewing narrow perspectives of 'developing' and 'developed' countries, the book examines people and places within broader debates from economic, political, social and cultural geography, highlighting the rich diversity of regions that are usually only viewed in terms of their 'problems'.

Providing a positive but critical approach to a number of key issues affecting these important areas, the book is divided into four parts to:

- examine the powerful but problematic ways in which the Global South is represented, and the values which underpin these conceptions and images;
- explore how the South is shaping, and being shaped by, global economic, political and cultural processes;
- look at the impact these have on peoples' lives and identities, from consumption and home-making, to work and local politics;
- assess the possibilities and limitations of different development strategies, whether directed by the state, the market or NGOs.

Packed with insightful case studies from fieldwork conducted across a range of communities and nations, the book challenges common assumptions to stimulate lively debate. This is a timely assessment of the way global processes are perceived from the Global South, and the role local realities take in mediating these wider movements.

Including a full glossary of key terms, introductions to the wider literature in this field and over 60 colour photographs, this is a wonderful textbook for all students interested in Human Geography and Development Studies.

Dr Glyn Williams is a Senior Lecturer in the Department of Town and Regional Planning at The University of Sheffield, UK. His main research interests lie in the fields of poverty and participation; state power and political practices; and environmentalism and environmental governance. He has conducted most of his research in India.

Dr Paula Meth is a Lecturer in Town and Regional Planning at The University of Sheffield, UK. Her research focuses on social development, particularly in Southern Africa, with a special interest in processes of gendered marginalization.

Dr Katie Willis is a Reader in Development Geography and Director of the Centre for Developing Areas Research at Royal Holloway, University of London, UK. Her research focuses on two main areas: social development, particularly health, in urban Latin America, and transnationalism and migration.

Geographies of Developing Areas is a path breaking textbook. The text is exceptionally well presented, with a lively and accessible writing style. As it draws on the detailed contemporary research and personal field experiences of the authors, it easily engages the reader. It will be a great boost to geography teaching in many different contexts.

Jenny Robinson, Open University, UK

The text departs from standard development debates to probe contemporary social processes shaping people's lives in the Global South. It will become a standard reference for development geographers and deserves to be read by human geographers and anthropologists much more broadly; a terrific achievement.

Dr. Craig Jeffrey, Associate Professor in Geography and International Studies
University of Washington, US

In clear prose and with boxes, illustrations and supporting tables and figures, the authors liberate developing areas from the distorting shadow of development, imbuing their populations with agency and individuality.

Jonathan Rigg, Department of Geography, University of Durham, UK

Skilfully weaving the latest research into a structured and highly accessible point-by-point discussion, this book is a welcome addition to human geography resources and suitable for a range of undergraduate courses.

Sarah Radcliffe, University of Cambridge, UK

Geographies of Developing Areas

The Global South in a changing world

Glyn Williams, Paula Meth and Katie Willis

Routledge
Taylor & Francis Group

LONDON AND NEW YORK

First published 2009 by Routledge
2 Park Square, Milton Park, Abingdon, Oxon, OX14 4RN

Simultaneously published in the USA and Canada
by Routledge
270 Madison Ave, New York, NY 10016

Routledge is an imprint of the Taylor & Francis Group, an informa business

Typeset in Garamond by Pindar NZ, Auckland, New Zealand
Printed and bound in India by Replika Press Pvt, Ltd

British Library Cataloguing in Publication Data
A catalogue record for this book is available from the British Library

Library of Congress Cataloging-in-Publication Data
Williams, Glyn, 1970-
Geographies of developing areas: the Global South in a changing world / Glyn
Williams, Paula Meth, and Katie Willis.
    p. cm.
  Includes bibliographical references.
  ISBN 978-0-415-38123-9 (hardback)—ISBN 978-0-415-38122-2 (pbk.) 1.
Developing countries—Textbooks. 2. Globalization—Developing countries—
Textbooks. 3. Economic geography—Textbooks. I. Meth, Paula. II. Willis, Katie,
1968- III. Title.
  HC59.7.W533 2009
337.09172'4—dc22   2008036924

ISBN10: 0-415-38123-1 (hbk)
ISBN10: 0-415-38122-3 (pbk)
ISBN10: 0-203-08624-4 (ebk)

ISBN13: 978-0-415-38123-9 (hbk)
ISBN13: 978-0-415-38122-2 (pbk)
ISBN13: 978-0-203-08624-7 (ebk)

For
Anna and Jamie
Joel, James and Harry
and in memory of Diana Willis (1944–2005)

Contents

Part I: Representing the South

Part 4: Making a difference

Plates

Figures

Tables

Boxes

Concept boxes

Acknowledgements

As with most publications, this book hides most of the negotiations, tensions and general stresses involved in its production. After agreeing to write the book in 2004 the enormity of the task we had set ourselves gradually dawned, especially as we saw the reactions of colleagues when we outlined what the book would entail. We have persevered because we remain convinced of the importance of recognizing the diversity of the peoples and places of the Global South and that 'Development' is only one lens through which to consider them. In a book of a few hundred pages we clearly could not provide a comprehensive overview of how global and local processes affect and are influenced by people in the Global South. The aim has been to highlight key debates which can provide readers with some starting points in considering the geographies of the Global South.

A book like this draws heavily on our teaching experience, and we would like to thank our students and colleagues at Keele, King's College London, Sheffield Hallam, Sheffield, Liverpool and Royal Holloway who have helped us to develop some of the ideas it contains. In particular, the late Graham Drake's love of Geography and writing textbooks revealed to Paula what it meant to be a dedicated teacher. We have also ourselves been inspired by those who have taught us the value of constructive critical thinking and fired our geographical imaginations, and so thanks are due to Stuart Corbridge, Colin Clarke, David Harvey, Linda McDowell, Susan Owens, Judy Pallot, Di Scott and Alison Todes. Glyn would also like to make special mention of Elsbeth Robson and Liz Young who many years ago helped devise a proposal for a similar volume, which for various reasons never got written.

The team at Routledge has provided excellent support and encouragement throughout what has been a much longer process than any of us originally envisaged. They have been very understanding as life events (both joyful and tragic) have affected progress on the book and we are very grateful to them for their patience. In particular, we would like to thank Andrew Mould, the commissioning editor, who has stuck with us through thick and thin. In addition, Harriet Brinton,

Liz Dawn, Russell George, Jennifer Page and Zoe Kruze have all made valuable contributions and have used just the right balance of carrot and stick to help us finish the manuscript!

We would also like to thank the anonymous reviewers for their detailed and insightful reports, both on individual chapters and on the manuscript as a whole. We have not been able to address all their comments in the revisions, but we took them all seriously and they certainly helped restructure, clarify and develop many of our arguments.

A key element of the book is its use of images. Jenny Kynaston at Royal Holloway drew most of the figures, often from rather illegible originals provided by the authors. We are extremely grateful for her professionalism and constant good humour. Sue May also provided photographic assistance for a number of images and patiently and creatively produced quality illustrations.

Many friends, colleagues, students and family members have provided input into the production of this book. For ideas, explanations, clarifications, contacts, or acting as sounding boards for the various themes discussed within the book, we would like to thank, Dwaipayan Bhattacharyya, Catherine Campbell, Sarah Charlton, Steve Connelly, Bob Doherty, Margo Huxley, Tariq Jazeel, Colin Marx, Emma Mawdsley, Anna McCord, Colin McFarlane, Charles Meth, Di Scott, Zarina Patel and Liz Watson. The staff in the Foyle Reading Room at the Royal Geographical Society (with the Institute of British Geographers) provided valuable support in the research for Chapter 2. We would also like to extend our thanks to Chrissie Webb, archivist at Oxfam, for enabling us to access the Oxfam newspaper cuttings and posters collection.

Providing photographs for a book of this scope was a significant task in itself, and here we would like to thank Muyiwa Agunbiade (Plates 5.1, 5.6 and 8.7), Amita Baviskar (Plate 9.2), Katherine Brickell (Plate 8.5), Steve Connelly (Plate 5.9 and the Cairo picture at the start of Part 1), Claire Cowie (Plate 4.1), Vandana Desai (Plate 11.4), Madeleine Dobson (10.1), Margo Huxley (Plate 1.4), Brian Jones (Plate 1.3), Melanie Lombard (Plates 7.2 and 8.3), Andrew Marton (Plate 8.4), Colin McFarlane (Plate 11.5), Claire Mercer (Plates 4.5 and 5.4), Hassan Sani (Plates 6.5 and 7.1), SDCEA (Plate 6.8), David Simon (Plate 3.1), Katie Walsh (Plate 2.9), Nikky Wilson (Plates 6.2, 7.5 and image at the start of Part 3). Gerhard Anders, Susan Appe, Sarah Charlton, Tim Conway, James Fairhead, Llynne Jones, Charles Meth, Kirsten Mumford, Francie Lund, Ben Page, Sara Parker, Zarina Patel and Caroline Skinner provided additional support and advice in sourcing images.

In addition, we acknowledge the following for their permission to reprint images in this book: DFID (Figure 7.2); Divine Chocolate (Plate 10.2); Jamaica Trade and Invest (Plate 2.8); John Friedmann (Figure 7.1); MapAbility.com (globe icons for text boxes); Maps in Minutes (base map data for world maps); Oxfam (Plates 2.5, 2.6); PA Photos (Plate 7.4); the Franklin D. Roosevelt Library (Plate 9.1); David Rose (Plate 7.3); The Royal Geographical Society (Plates 2.2, 2.3, 6.1); Still Pictures (Plates 1.5, 2.7, 3.2, 4.4, 5.2, 5.7, 6.6, 7.6, 10.3); *The Sunday Times Travel*

ACKNOWLEDGEMENTS

Magazine (Plate 2.4); Tate Gallery (Plate 2.1); Suzanne van Hook, WIEGO (Plate 7.7); Martin Westlake (Plate 2.4). Every effort has been made to trace the owners of copyright material. If copyright material has been inadvertently reproduced, if notified the publishers will rectify omissions or errors in future editions.

Finally, on a personal note, Katie would like to extend particular thanks to the following for their friendship and support in relation to 'el libro' during her sabbatical term in Oaxaca City, Mexico: Gudrun Dohrmann at the Instituto Welte, David and Kerry Langley, Teresa Villareal and Lucia Dávila at Las Mariposas. She is also very grateful to Glyn, Paula, Anna and Jamie for their unstinting hospitality during book visits to Sheffield. Paula and Glyn would like to thank family and friends who have looked after us (and particularly our children) throughout the writing of this book: Cristina Cerulli, Steve Connelly, Natasha Erlank, Mike Evans, Liz Gagen, Katrina Hulse, Margo Huxley, Liz Johnston, Emma Mawdsley, Deanna Meth, Fiona Meth, Mark Parsons, Mitch Rose, Rebecca Scambler, Liz Watson, Greg Whitmore, Marian Williams, Ken Williams, Nikky Wilson, and all the staff at Beech Hill Nursery. We would particularly like to thank our children Anna and Jamie for their endless patience and putting up with bouts of neglect. Last, but by no means least, Glyn and Paula would like to thank Katie for her unfailing patience with us, and her energy and enthusiasm throughout.

London and Sheffield,
August 2008

Acronyms

ANC	African National Congress
APPO	Asamblea Popular de los Pueblos de Oaxaca (Popular Assembly of the Peoples of Oaxaca)
BMI	Body Mass Index
BRAC	Bangladesh Rural Advancement Committee
CARICOM	Caribbean Community
CARIFTA	Caribbean Free Trade Association
CBO	Community-based Organization
CCER	Civil Coordinator of Emergency and Reconstruction
CGIAR	Consultative Group for International Agricultural Research
CIDA	Canadian International Development Agency
CSR	Corporate Social Responsibility
DFID	Department for International Development
DRC	Democratic Republic of Congo
ECLA	Economic Commission for Latin America
EIA	Energy Information Administration
ELAM	Escuela Latin Americana de Medicina (Latin American School of Medicine)
EPZ	Export Processing Zone
EU	European Union
FAO	Food and Agriculture Organization
FDI	Foreign Direct Investment
GALZ	Gays and Lesbians of Zimbabwe
GATT	General Agreement on Tariffs and Trade
GDP	Gross Domestic Product
GEF	Global Environmental Facility
GM	Genetically Modified
GNI	Gross National Income
GNP	Gross National Product

HDI	Human Development Index
HIPC	Highly Indebted Poor Country
HIV/AIDS	Human Immunodeficiency Virus/Acquired Immune Deficiency Syndrome
HYV	High-yielding Variety
IBRD	International Bank for Reconstruction and Development
IFI	International Financial Institution
ILO	International Labour Organization
IMF	International Monetary Fund
ISI	Import-substitution Industrialization
KFC	Kentucky Fried Chicken
MCC	Millennium Challenge Corporation
MDGs	Millennium Development Goals
MNC	Multinational Corporation
MOSOP	Movement for the Survival of the Ogoni People
NAFTA	North American Free Trade Area
NAM	Non-Aligned Movement
NBA	Narmada Bachao Andolan (Save the Narmada Movement)
NGO	Non-governmental Organization
NIC	Newly Industrializing Country
NIDL	New International Division of Labour
NIE	Newly Industrializing Economy
NIMBY	Not-In-My-Back-Yard
NPM	New Public Management
ODA	Official Development Assistance
OECD	Organization for Economic Cooperation and Development
OLS	Operation Lifeline Sudan
OPEC	Organization of Petroleum Exporting Countries
PAP	People's Action Party (Singapore)
PLA	Participatory Learning and Action
PPP	Purchasing Power Parity
PRA	Participatory Rural Appraisal
PRSP	Poverty Reduction Strategy Paper
PT	Partido dos Trabalhadores (Brazilian Workers' Party)
RRA	Rapid Rural Appraisal
SACU	Southern African Customs Union
SAP	Structural Adjustment Policy/Programme
SAR	Special Administrative Region
SDCEA	South Durban Community Environmental Alliance (South Africa)
SDI	Shack-Slum Dwellers International
SEWA	Self Employed Women's Association (India)
SEWU	Self Employed Women's Union (South Africa)

SIDS	Small Island Developing States
SIPRI	Stockholm International Peace Research Institute
SMEs	Small and Medium Enterprises
SPLM	Sudanese People's Liberation Movement
SSP	Sardar Sarovar Project (Narmada Valley, India)
TB	Tuberculosis
TNC	Transnational Corporation
TVA	Tennessee Valley Authority
UAE	United Arab Emirates
UN	United Nations
UNCTAD	United Nations Conference on Trade and Development
UNDP	United Nations Development Programme
UNEP	United Nations Environment Programme
UNHCR	United Nations High Commissioner for Refugees
UNHSP	United Nations Human Settlements Programme
UNICEF	United Nations (International) Children's (Emergency) Fund
USAID	United States Agency for International Development
WHO	World Health Organization
WIEGO	Women in Informal Employment: Globalizing and Organizing
WTO	World Trade Organization

1 Introduction

DEFINING THE GLOBAL SOUTH: REAL AND IMAGINED DIVIDING LINES

This is a textbook about the Global South – the parts of Latin America, the Caribbean, Africa and Asia that are often referred to as 'the developing world'. That should be a simple enough opening sentence, but it hides some contentious definitions that go right to the heart of our reasons for writing this book. Had we been writing thirty years ago, these might have been brushed aside more easily: in 1980, ex-German Chancellor Willy Brandt chaired a Commission whose report, *North–South: A Programme for Survival* presented a clear dividing line between a rich and powerful North, and a poor and marginalized South. Positioned outside the capitalist 'First World' as well as the Soviet 'Second World', the Global South or 'Third World' was seen as something of a residual category, in need of sustained international assistance to ensure its development. From today's perspective, however, the world looks rather different. Rich and poor nations fall on either side of Brandt's Line (Figure 1.1): the old Soviet Union has dissolved, leaving new low income countries (such as Tajikistan and the Kyrgyz Republic) in its wake, economic development in some parts of the South has given countries unquestionable 'High Income' status (such as South Korea, Singapore and Kuwait), and still others (such as Brazil, India and China) have become regionally or even globally powerful in their own right.

But although Brandt's Line no longer neatly divides rich from poor and the powerful from the powerless (and never did, given the variation that exists within as well as between countries), North–South divides are still important in shaping the way in which the Global South is imagined, talked about and studied today. Far too often in social science research and teaching, the Global South remains a marginal, residual and generalized category. It is vastly under-represented in research: Jonathan Rigg notes that only one-eighth of recent papers in top Anglo-American Geography journals were directly concerned with the countries and

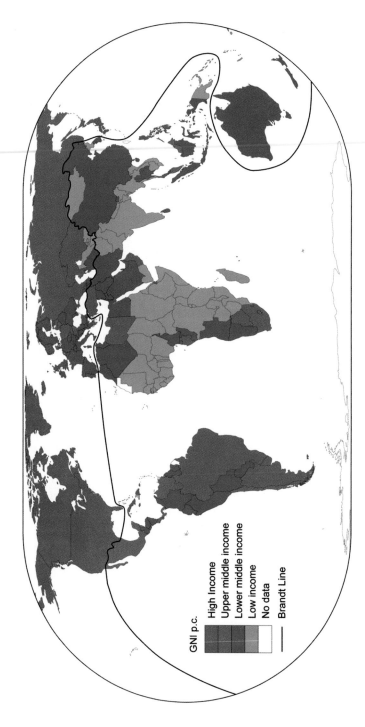

Figure 1.1 North–South divisions according to the Brandt Line.
Source: Map data © Maps in Minutes™ (1996).

GNI p.c.

High Income
Upper middle income
Lower middle income
Low income
No data
—— Brandt Line

conditions of the Global South, despite the fact that it contains over 80 per cent of the world's population (Rigg, 2007: 2). Not only this, but when students in the North are introduced to the Global South through courses, it is often via a set of *problems* – such as **poverty**, debt or environmental degradation – and the development theories and policies used to 'solve' these. This conveys some useful information, and links with many students' personal interests in issues of global inequality or ecological sustainability, but it does so with some considerable problems. One of these is that it reinforces negative stereotypes of developing areas, and represents the Global South as a collection of places and peoples in need of external (i.e. Northern) intervention. Equally importantly, it can help to reinforce academic divisions that place the study of the Global South as a 'specialism' separate from the 'mainstream' of social science disciplines. Debates in social, economic and political theory often implicitly take conditions in the Global North as their point of reference. Where the Global South contributes to theoretical debate at all, it is often through the separate category of 'development theory', a point we return to below.

One of our reasons for producing this book was that our own experiences of living and working in the Global South (see Boxes 1.1, 1.2 and 1.3) don't fit easily or comfortably within this academic framework. We all see ourselves as geographers, rather than 'development experts', and in thinking about our various research topics in Mexico, South Africa, India and elsewhere, we've drawn directly on debates ranging across the social sciences, and not from development theory alone. Even though our own work addresses social, economic and political concerns, we're each acutely aware that there's a lot more to life in the South than a list of problems – and one of our motivations for writing is to communicate something of the richness of its places and peoples, something that often gets squeezed out of many undergraduate courses. As may be clear from our descriptions of our own work, we are all human geographers, and this is not a textbook that attempts to engage with the physical geography of the Global South. The natural environments of the South do, however, make their presence felt at various points within the book – in dominant representations of the Global South (Chapter 2), as issues for international governance (Chapter 3), as material contexts of people's lives (Chapter 8) and elsewhere.

In this book, our working definition of the Global South stays with the problematic division put forward by Brandt, but we use this deliberately to highlight the diversity that exists south of the Brandt Line rather than trying to present a singular view of the experiences of its countries, regions and peoples. Within the Brandt Commission's South today are emerging superpowers and failed states, the world's fastest growing economies and the vast majority of the global poor. Whilst being subject to European, Japanese or American colonialism might have been an important part of the history of many Southern countries, it has not been the experience of all (Ethiopia and Thailand, for example, had only the briefest of foreign occupations). The Global South contains massive variations in environmental

BOX 1.1

Glyn's story: arriving in Kolkata

My own understanding of the Global South has been closely shaped by my experiences of living and working in India, and one underlying question guiding my research has been what does 'democracy' mean for people's everyday lives today. Kolkata has a special place within that experience: it was where I learned Bengali before starting my doctoral research, and it was also the first 'proper' city I had ever lived in.

First arriving there in 1992, my mental image of the city was informed by media stereotypes (of Mother Teresa and the 'City of Joy') and the little I knew of its recent history: Kolkata absorbed two tidal waves of crisis immigration (in 1947, and then during Bangladesh's Independence struggle in the early 1970s), it was a declining industrial centre, and had a reputation for radical and often violent politics. Did I find a city engulfed by overcrowding, poverty and political activism? Before even stepping off the train, there was plenty of evidence of all three: a million passengers pass through Howrah station every day, plenty of people sleep rough on its platforms and rival political parties compete to unionize its porters. But these images were quickly overtaken by others: the spontaneous games of street cricket during every strike or public holiday, the city's pride in its annual book fair and its passion for *adda* – a form of free-flowing intellectual and political debate, fuelled by endless cups of sweet tea. This was a city where dollars and Coca-Cola were still traded on the black market (India's economic 'liberalization' was just beginning in the early 1990s),

Plate 1.1 A middle-class neighbourhood in southern Kolkata, India.
Credit: Glyn Williams.

BOX 1.1 (*CONTINUED*)

but also one that possessed a self-assurance and cosmopolitanism all of its own.

In 2008, Kolkata is a city that is changing fast, with its older neighbourhoods (Plate 1.1) being overtaken by frenetic construction. High-rise apartment blocks are emerging from former paddy fields at the city's edge, along with new air-conditioned shopping malls in its southern suburbs (they now *do* sell Coke, alongside far more luxurious Indian and foreign goods). The billboards – no longer hand-painted – advertise up-market property developments and IT career opportunities to its growing middle classes. But alongside these signs of affluence, even within some of the city's richest neighbourhoods, you still find people plying hand-pulled rickshaws and living in squatter settlements. As a friend explained to me: 'Kolkatans, unlike Delhi-ites, have never forgotten that we live in the developing world. The poor live among us: they are a vital part of this city.'

BOX 1.2

Katie's story: researching gender in Oaxaca City, Mexico

I first went to Mexico in 1990 as a Master's student for a three-month fieldwork visit researching women's work in a low-income informal settlement on the edge of Oaxaca City in the south of the country. I had a long-standing interest in gender sparked by my own experiences and reflections on growing up on a farm in North Wales where being male or female strongly influenced the kind of work you did and your future life path. I arrived at university in the mid-1980s, just as feminist geography was beginning to have an impact and chose to write my undergraduate dissertation on gender divisions of labour in North Wales agriculture. I was also greatly involved in the women's groups and feminist organizations which were flourishing at that time.

As a graduate student I wanted to build on my research interests within the context of the 'developing world' as it was largely known then. My proposed research in Mexico would allow me to focus on gender and work, but within a different cultural context. My original perceptions of gender relations in Latin America, and Mexico in particular, were framed largely

BOX 1.2 (*CONTINUED*)

BOX 1.2 (*CONTINUED*)

around constructions of men as *macho* and women as rather vulnerable and passive. While such stereotypical understandings of male and female behaviours and identities certainly existed and still exist in Mexico, the reality was much more complex and challenged my naive assumptions.

During my Master's research and then frequent return research trips to Oaxaca, the diversity of women's experiences was constantly reinforced, as were women's different interpretations of gender differences. Oaxaca is one of the most ethnically diverse states in Mexico; Zapotec women's interpretations of their lives and how their positions of disadvantage could not be disentangled from those of their husbands, fathers and other male relatives, certainly challenged my rather narrow ideas of how to achieve gender equality which came from a particular Western feminist viewpoint. My initial image of passive Mexican women was also completely destroyed through observations of women's involvement in campaigns for a drinking water system in an informal settlement, the educational, health and psychological support provided to women through the 'Rosario Castellanos' Women's Centre and the campaigns against domestic violence run by the 'Mujeres Lilas' Feminist Collective.

Plate 1.2 Altar to the victims of gender-based violence organized by the 'Mujeres Lilas' Feminist Collective, Oaxaca, Mexico.
Credit: Katie Willis.

BOX 1.3

Paula's story: growing up in Durban

I grew up as an Apartheid-era child in Durban, South Africa; a time and place where divisions between fellow citizens were strongly mediated by race, class, language and gender. Being a 'white', middle-class, English-speaking girl meant that my childhood memories of playing on the city's beaches, or eating out in restaurants, are favourable, privileged and very particular. Our state primary school was single sex, all white and highly 'anglo': we were 'made' to watch the wedding of Prince Charles and Diana Spencer. But only as an adult did I register the second generation status of many of my friends, whose parents had moved from the Lebanon, Portugal, Greece, Britain, Malta, France and Germany. South Africa during the 1970s presented a land of opportunity for white immigrants.

Our lives were intertwined with the lives of some black South Africans, structured by unequal working relations. We employed two domestic workers, and encountered janitors at school, gardeners, builders, the milkman, the newspaper deliverer, traders at the Indian food market and cleaners at the theatre. These encounters were never 'normal', but neither were they extreme, they just were. I cannot pinpoint my first awareness that something was not right. I recall as a child feeling grave anxiety about the living conditions of black rural South Africans as we drove through parts of the Kwa Zulu 'homeland' on our way to camping holidays in the Drakensberg Mountains. I used to worry about women being cold in the winter months.

Plate 1.3 Fiona Meth, Paula and Llynne Jones, 1970s, South Africa. Credit: Brian Jones.

BOX 1.3 (*CONTINUED*)

BOX 1.3 (*CONTINUED*)

As a teenager, with a fiercely politically active father, my politicization was rapid and our left-wing, but still largely white University, accelerated my realization of the hideousness of segregation. But I still danced to indie bands on a Friday night, worked at art exhibitions, and stressed over my relationships. Developmental concerns in Durban were all around me, and increasingly a deep part of my consciousness, but so was a life to be lived, music to be enjoyed and cockroaches to be scared of. Since then, my own research on Durban's informal settlements has aimed to get under the skin of a city and to reveal just how poverty and inequality actually play out for different people.

conditions, cultural values and practices, political and economic systems, and ways of life that we can only begin to sketch out here. What *does* link the regions of our definition, however, is that these are the parts of the world that have been commonly described for many decades now as *developing areas*. Our book puts the spotlight on these places both to give them more attention than they normally receive from 'mainstream' social science in the Global North and to challenge standard ways in which they are represented.

Our agenda for rethinking the study of 'developing areas' consists of four main arguments, each of which is intended to challenge assumptions about North–South divisions. The first is that representations of the Global South matter: the way places in the South are written and talked about can have important effects, not just in university classrooms, but in the real world too. The second is that life in the Global South is not separate from the North but intimately linked to it through a range of global processes, and as a result we need to think about what **globalization** might mean from a Southern perspective. The third is that understanding the Global South requires a detailed engagement with its particular histories and geographies: this in turn means a close attention to the **agency** of people living there, and to scholarship on and from its regions. Our final argument is that providing an introduction to the Global South is *not* the same as telling the story of international development. 'Development' may be an important part of the national aspirations of Southern countries, or impact on elements of people's everyday lives, but as Boxes 1.1 to 1.3 show, there is much more to life in the Global South than development alone. Looking at people and places only through categories of development theory, we argue, can blind us to this richness and diversity.

We outline each of these arguments in a little more detail below. Individually, these arguments are not novel in themselves – each has been explored in literature in Geography and elsewhere, including within Development Studies itself – but taken together we hope that they give this text a distinctive approach to teaching on and learning about 'developing areas'. In particular, by placing 'development' in a more peripheral position within our book, we hope that our readers have a more balanced picture of the Global South and its connections (in theoretical and practical terms) with the Global North, rather than simply seeing lines that divide them.

THE POWER OF REPRESENTATION

The first key argument of the book is that the ways in which the Global South is represented – in the media, in academia, in policy documents and elsewhere – is an exercise of **power** (see Concept Box 1.1) that has important, real-world consequences. The idea that images and descriptions of places can help to shape the ways in which people think and act is a long-established theme in Geography (for example, in the field of Critical Geopolitics: Ó Tuathail, 1996), and over the past two decades has also been important for academics writing about the Global South. One way in which **representations** of the Global South can have powerful effects is through the repeated use of restricted forms of description: these can, intentionally or otherwise, 'fix' ideas of a place in our minds, and shape the way we respond to it. Media reporting on Africa in the UK and USA would be one

---

CONCEPT BOX 1.1

Power

Following John Allen (1997), it is possible to identify three aspects of power, two of which are more clearly held by particular individuals or institutions. The first is the ability to command or control the actions of others: examples of this *power over* others would include the control a male household head exercises over other family members within a patriarchal society, or the authority of a national government over its citizens. The second is the ability to control and deploy resources: examples of this *power to* achieve desired ends would include the ability of a multinational company to use its human and financial resources to develop a new product, or to buy out a rival firm. A final aspect of power, associated with the work of Michel Foucault, is more diffuse and sees *power within* the operation of everyday techniques, strategies and practices. The strict timetable and regime of a prison, or the more subtle management practices and performance criteria of a company, have the effect of disciplining their inmates/employees into particular forms of behaviour, often without the need for direct commands. It is this final aspect of power that is drawn upon in work on the power of **representation**.

good example here: TV coverage of Africa is largely dominated by news of famine, warfare or natural disasters, and is often packaged as two-minute clips that give little context or background to the events. The effect on viewers can be that they are blinded to (or unaware of) other images emerging from Africa – whether these are more complex, more positive or just different – and their 'mental map' of the continent as a whole can become subtly constrained by the media's coverage. But the effects can be more serious than simply keeping TV audiences badly informed: Philippa Atkinson (1999:105) has argued that the ways in which the Western media represents conflict in Africa as anarchic or barbaric behaviour also help to shape decision making over aid and international policy.

One powerful way of viewing the Global South since the mid-twentieth century has been provided by the idea of 'development' itself. Early work in development theory often portrayed Southern countries as lacking in important ways (in the structure of their economies, in technological know-how, or in entrepreneurial attitudes), or as lagging behind the West in a series of important processes (industrialization, **modernization**, democratization) – a theme we return to in Chapter 2. Seeing these countries as characterized merely by a series of lacks and lags is, of course, a severe mis-representation, and geographers, anthropologists and others have conducted research that challenges and **deconstructs** the ways in which development ideas, images and writing shape understandings of the South (Box 1.4).

Central to much of this work is the idea that there are important links between having the **power** to represent developing areas, and having some degree of control over them. As Box 1.4 indicates, if USAID can successfully portray Egypt as suffering from a particular development 'problem', that in turn can be used to justify intervention in that country's affairs. Again, we revisit these ideas in Chapter 2 and later in the book, but the central message is that **representations** of the South matter, not least because they can have real impacts on the everyday lives of people living there.

GEOGRAPHIES OF GLOBALIZATION

Our second key argument is that the South is a vital and active part of processes of **globalization**. The use of the term **globalization** has grown massively since the 1990s, and in more simplistic accounts it is often used to present a picture of a 'shrinking world'. Modern technology is helping to reduce the importance of physical distance, with improvements from cheaper air travel through to email effectively bringing us all 'closer' together. At the same time, we are described as living in a 'global village' or as competing in a 'global marketplace': we are part of a world which is increasingly interconnected, with the result that differences between places are apparently being removed. **Globalization** is often presented as an ambivalent process. On the plus side, it means cheaper and more varied products

BOX 1.4

The power of development: America's Egypt

Power of Development (Crush, 1995) was an important collection of essays that questioned the ways in which the development industry (mis)represented the Global South, and Timothy Mitchell's Chapter on Egypt provides a clear illustration of the real-world effects of these representations. Mitchell shows how policy documents of the United States Agency for International Development (USAID) in the 1980s repeatedly portrayed Egypt's agricultural development problems as being about overpopulation. The powerful visual image of the Nile valley – described as a tiny strip of overcrowded agricultural land in an inhospitable desert – and statistics on food insecurity and rising dependency on imported food grains were used to justify intervention in the country's agricultural policy. The policy prescriptions put forward by USAID were to open Egypt's agriculture up to the free market, and invest in technological innovations.

These may seem at first sight to be reasonable propositions, but Mitchell carefully looks at evidence outside the reports to question the way in which USAID had presented a particular vision of Egypt's agricultural crisis, and its proposed solutions. Over the 1980s, Egypt's food-grain imports were increasingly used to feed livestock, rather than people, and the lack of food security among the poor was not simply the result of physical overcrowding, but rather of increasing inequalities of landholding. USAID's policies were themselves contributing to both the growth of export-oriented livestock production and further displacement of small farmers (who were the most space-efficient users of the land). Furthermore, the aid for technological support largely benefited American manufacturers, and made the country as a whole dependent on the USA's continued support. The image of the overcrowded Nile valley, he argues, was a powerful way of 'framing' Egypt's problems as natural, inevitable and internal to the country itself – and of keeping the wider politics of **aid** out of sight.

(Source: Adapted from Mitchell, 1995)

sourced from across the whole world, or a growing sense of 'global community'. On the negative side, the global marketplace can mean job insecurity as industries relocate to areas of cheaper production or collapse in the face of greater competition. Increased connectedness can also bring its own threats, manifest most clearly in the West by the fear of global terror networks (Murray, 2006: 197–201).

Whilst there is some truth in all of these aspects of **globalization**, we would like to raise a few questions about the assumptions that often underlie a 'commonsense' understanding of the term. Many Geography students may already be familiar with a more critical stance on **globalization**, but it is worth thinking through how accurate this picture of a speeded-up, hyper-connected world is, particularly in the contexts of the South. Some particular points that we want to make include the following:

- *Globalization is not a new process.* There is a danger that accounts of globalization present this as a recent phenomenon – whereas the interconnectedness of peoples' lives and places across the world is much older. The spread of Islam (Chapter 5) linked parts of Asia, Africa and Europe together in dense political and economic relationships (Chapter 3) long before European colonial expansion. In the fifteenth century, Spain's Alhambra Palace (Plate 1.4) was located at the northwest border of a Muslim world that extended to present-day Indonesia, a reminder that global interconnections are not all new, or centred on the North. The late nineteenth century also saw 'free' trade (backed by Britain's military strength) driving massive international flows of capital, people and commodities, as European **colonialism** brought distant places into close, if uneven, relationships with each other. Part 2 of this book examines the Global South's position in the contemporary world, and these histories of global interconnection are an important part of this task.
- *Globalization does not mean that the world is becoming uniform.* Ideas of a global village or marketplace may suggest that **globalization** is reducing the differences between places, but this is not necessarily the case. First, the upsurge in the internationalization of the world's economy since the 1970s has been accompanied by ever-increasing inequality: we may be living in one world, but it is a highly uneven one. Second, although places may be increasingly interlinked, this does not mean that differences in culture and **identity** are disappearing. The Global South is made up not just by a set of 'traditional' cultural practices that are about to be steamrollered out of existence by a 'global' culture – international exchange is (and always has been) an important part of remaking local cultures, a theme we return to in Chapter 8.
- *Globalization is not a faceless, or predetermined, process.* A final important part of many commonsense understandings of **globalization** is that it is an unstoppable process driven by abstract forces, such as technological change or 'the global market'. This tends to devalue the **agency** of those shaping global processes, from leaders of the G8 countries or **multinational corporations** down to

ordinary individuals across the globe. When talking about **globalization**, it is important to remember that 'it' is made up of connections between the smaller actions of individuals and groups. As such, the future outcomes of **globalization** are difficult to predict, and there is room for experiences of **globalization** to be more or less equitable, sustainable or empowering. Resisting **globalization** may be a bit like trying to turn back the tide, but reshaping **globalization**'s effects is something that everybody can, to some degree, participate in through their own actions.

ENGAGING WITH THE GLOBAL SOUTH

Our third major argument is that understanding the Global South requires a detailed engagement with its particular histories and geographies. This follows from our criticism of simplistic views of **globalization** – people and places in the South cannot be understood simply by reference to a set of pre-existing 'global processes' – and in turn has important implications for the way we write about and theorize the experiences of people in the Global South. In particular, this means two things for us in this book: emphasizing the **agency** of Southern people and countries, and using context-specific understandings of the South to reflect on wider theoretical debates.

The first of these results of engaging with the South is perhaps the more straightforward to explain. Places in the South are not simply getting 'connected up' to a pre-existing global system, they are actively involved in producing that system. As such, it is vital that our accounts of life in the South pay proper attention to the **agency** of Southern people and countries. This may seem obvious, but dominant accounts of the Global South often fail to take this point in to consideration: people in the South are often represented as merely reacting to – or still worse being the passive recipients/victims of – 'global' changes imposed from 'outside'. Of course, a sense of balance is needed here, and it is important to recognize that many of the relationships that Southern countries and citizens find themselves in are grossly unequal. Whether this is small farmers trying to compete with subsidized US or European imports, or national governments trying to renegotiate their debt repayments with the International Monetary Fund (IMF), their space for action may be severely constrained. But people do make choices and develop strategies, even in difficult circumstances, and a key task of research on the Global South is to recognize and understand both these actions and their wider effects.

Using context-specific understandings of the Global South to reflect on wider theoretical debates involves two different modes of thinking. The first is to consider the particular qualities of places in the Global South – including their histories, cultures, economic structures and political systems – in our explanation of processes happening there. Employing what Doreen Massey (1996) has called a 'progressive sense of place' is important here: we need to see that places do not have

Plate 1.4 The Alhambra Palace, Andalucìa, Spain.
Credit: Margo Huxley.

inherent characteristics which do not change; rather places are being continually reworked through the interactions of economic, social and political processes in particular locations. As an example, it would be impossible to understand the urban geography of Durban (Box 1.3) without first knowing something about the history of colonial settlement in South Africa, and about spatial planning processes during the Apartheid era. This may seem obvious, but because studies of Southern examples have received relatively little attention within 'mainstream' geography in the North, there is a danger of shying away from this careful contextual analysis, and falling back on abstract statistics or (Western-centred) theoretical models. Gearóid Ó Tuathail noted the pressing practical importance of this in the USA, where political geographies of the Middle East (and elsewhere) suffer from 'thin' forms of knowledge:

> [T]he 9/11 attacks have exposed how a culture of thin geographical knowledge characterizes not simply the Presidency of the United States but runs deep in the institutional marrow of the US national security complex. The origins of this go back to the 1960s, when systems analysis and technical intelligence began displacing areas studies and human intelligence within government bureaucracies. The CIA [Central Intelligence Agency], FBI [Federal Bureau of Investigation], NSA [National Security Agency], NIMA [National Imagery and Mapping Agency] and Pentagon, like the National Science Foundation and many geography departments today, are embedded within a culture of technological fundamentalism that seeks technical solutions to challenges that really require cultural, geographical and linguistic learning. 'Thin' and immediate technical knowledge – the ability of satellite cameras and drones to provide real-time video images of obscure Afghan mountain passes – drives out 'thick' long-term geographical knowledge.
>
> (Toal [Ó Tuathail], 2003: 654)

'Thickening' our geographical knowledge of places and cultures of the South (through study, or first-hand research and use of locally produced knowledge) can help to explain *why* things in Kabul are different from Kentucky (or Kinshasa). Beyond this, it can also help us to uncover some of the hidden limitations of supposedly 'universal' social theory that is based around the experiences of the Global North. This second mode of thinking is a less obvious distinction to describe, and a still harder task to perform in practice, but what we mean is that 'mainstream' social science often has some culturally specific assumptions written in to it. To take an example, much economic theory implicitly takes as its starting point the assumption that individuals seek to maximize profit, or other forms of gain. Farmers in Ladakh (Plate 1.5), in the Tibetan plateau, may act quite differently: the household, not the individual, is considered as the basis of society, and farming decisions may pay more attention to environmental conservation (essential in a landscape where water is extremely limited) or maintaining relationships between households (equally important when many agricultural tasks rely on the exchange of labour with neighbours) than they do to maximizing profit. One way of examining Ladakhi farmers' behaviour would be to chart how and why they are 'different' from 'normal' profit-maximizing farmers in the Global North. The more difficult – but ultimately more rewarding – task is to think through the limitations to mainstream economic theory in light of evidence from the South. Understanding farmers' behaviour in Ladakh could, for example, lead to questions of how the idea of the individual become so important within Western economic thought, and what the consequences of this might be. A thorough engagement with the Global South should therefore not only improve understanding of 'distant' places, but also encourage challenges to theoretical models based around the experiences of the North (see Robinson, 2006, for an example of this approach).

CHALLENGING THE POSITION OF DEVELOPMENT

Our final major argument is that we need to challenge current 'common-sense' and academic divisions that equate the study of the Global South with the study

Plate 1.5 Farming in a fragile environment, Ladakh, India.
Credit: © Still Pictures.

of development. As we noted above, Anglo-American Geography has perhaps fallen too heavily in to this trap – academics dealing with the Global South are relatively thin on the ground, and many of them end up teaching courses that introduce students to those parts of the world in terms of development problems and policies. People's actual lives in the Global South do not begin and end with the current concerns of the international development community (whether this is economic growth, technology transfer, or **poverty** alleviation), and that is why we have deliberately left consideration of intentional plans to 'develop' the Global South to the end of this textbook.

As Box 1.4 showed, development **discourses** can oversimplify understandings of Southern people and places, and by doing so can themselves have powerful real-world effects. At its most destructive, development can label societies and areas as 'backward', or whole cultures as impediments to '**modernization**', with the result that they are seen as in need of 'correction' from some outside agency. But before the entire development industry (and Development Studies within academia) is written off as a 'neo-colonial' domination of Southern people and countries, it is important to remember that at its most optimistic, development also contains a sense of possibility, progress and emancipation (Corbridge, 2007). It is in this sense that many Southern leaders and social movements have fought for, rather than against, changes to 'develop' their own societies.

All plans to 'develop' the Global South have the potential to transform societies and landscapes, with gains for some and losses for others. As such, it is important to evaluate the inspiration, content and outcomes of particular visions of development carefully, rather than celebrating or demonizing 'it' as a whole. The aims of development are as varied as human imaginations of a better world can be – from a world without scarcity, or a world without oppression, to a world in ecological harmony – and these different driving ambitions have their own limitations and contradictions. It should therefore be no surprise that the content of development theory and practice is equally varied. In this book, we want to highlight this contested nature of development, and so we don't want to introduce development theory as a singular idea that evolves over time. The story of development could be told in this way, just as some histories of Geography have explained how the discipline has passed through a number of **paradigm shifts** (Johnston and Sidaway, 2004; see Livingstone, 1992, for an alternative account). Instead, we have chosen to focus on three powerful inspirations for change that have been important *throughout* recent history and the forms of development that have accompanied them: wealth creation (and market-led development), rational planning (and the developmental state) and popular participation (and grassroots development initiatives). These three are by no means the only ideas of what 'a better world' is or how it should be created but each has had an important role in shaping parts of the Global South, particularly over the past half-century, and presenting them in this way allows us to highlight the possibilities and problems inherent in each.

Once you have read this book, we hope that you will agree with some of its main arguments: that **representations** of the Global South can have important effects; that the South is an active player in complex processes of **globalization**; that it is important to recognize its diversity; and that 'development' is a contested process. If you also see the study of the Global South not as a narrow specialism concerned with 'development problems', but as something *essential* to the social sciences and the understanding of today's world, then we will have achieved our aims in writing.

USING THIS BOOK

The structure of the book

The structure of the book (see Figure 1.2) reflects these key arguments and key concerns. The first part, Representing the South, looks at the different ways in which the Global South has been viewed both from the North and from within. It looks, in particular, at the problems caused by oversimplified patterns of describing and talking about the Global South that are generated in the North, and aims to show how these can be challenged by alternative understandings that have emerged from Southern writers, academics and institutions.

Part 2, The South in a Global World, looks at the changing ways in which the South has been located within global political (Chapter 3) and economic structures (Chapter 4), and complex social and cultural flows (Chapter 5). As such it develops our argument above about **globalization**, and asks the key question: How has the Global South emerged in its current form? This part of the book looks at large-scale processes and patterns, but we also try to document through case studies some of the diversity of the South. Our arguments about stressing diversity and people's **agency** are developed further in Part 3, Living in the South. The key question here – How are different places and people in the South responding to and reshaping global and local events and processes? – is answered at a more micro-scale. By looking at the ways in which people engage with politics (Chapter 6), make a living (Chapter 7) and experience changing lifestyles and identities (Chapter 8) we aim to highlight their ability to shape 'global' forces and recreate 'local' places. Importantly, given our underlying argument about **globalization**, the difference between these two parts of the book is one of perspective and emphasis, rather than content. We are looking at 'the big picture' and 'the micro-scale' as different starting points from which to investigate what are the *same* sets of processes and connections simultaneously reshaping both global and local aspects of the lives of those in the South.

In keeping with our intention to think about the South in ways that are not limited by development, it is only in Part 4 – Making a Difference – that we turn directly to examine different development strategies. The key questions for this part are: What has inspired different attempts to 'develop' the Global South? and What

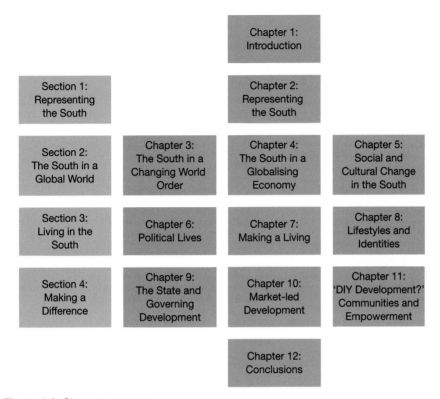

Figure 1.2 Chapter structure.

effects do these interventions have? We look in turn at the developmental state (Chapter 9), market-led development (Chapter 10) and people-centred development (Chapter 11) to see the possibilities and limitations of each. In the conclusions we revisit the four key arguments of the book as a whole. Here, we think about their implications not only for re-imagining the South, but also for the role of the social sciences more widely.

We have structured the book in this way to ensure that our main arguments about the approach to the study of the Global South stand out clearly. As well as being read 'horizontally', however, the different chapters of each part could also be read 'vertically' (Figure 1.2). Broadly speaking, Chapters 3, 6 and 9 develop political themes, Chapters 4, 7 and 10 cover economic debates, and Chapters 5, 8 and 11 look at social and cultural issues. As such, we hope that students (and tutors) using this book will be able to make the links back to debates in political, economic, and social and cultural geography, especially as Anglo-American teaching of each of these sub-fields tends to rely too heavily on examples from Europe and North America.

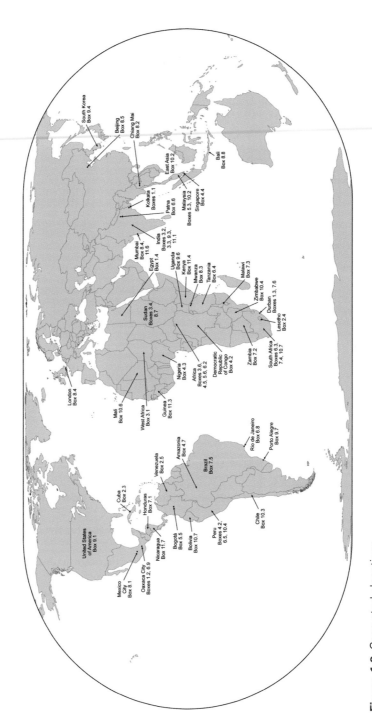

Figure 1.3 Case study locations.

Source: Map data © Maps in Minutes™ (1996).

For the reader in a hurry …

We would, of course, like you to read this book from cover to cover, but we recognize that textbooks rarely work (only) in this way. As well as mapping out our arguments and the book's overall structure, we've aimed to provide a detailed map of different topics, ideas and examples you may want to dip in to. Figure 1.3 shows the detailed case studies we have included within *text boxes*, and where these are located both in the world and in the book. Within each chapter, *key concepts* are explained in separate boxes, so you can access this background information without interrupting the main flow of the text. There is also a more extensive *glossary* with shorter definitions. Glossary terms are in bold throughout the text. Other resources that are included in the book are *suggested readings*, useful *websites* and *review questions/activities*. These are provided at the end of every chapter.

SUGGESTED READINGS

Murray, W. E. (2006) *Geographies of Globalization*, London: Routledge.
     This is a good, introductory textbook to different aspects of globalization

Ghosh, A. (1992) *In an Antique Land: History in the Guise of a Traveller's Tale,* London: Granta.
     This is a rather different introduction to globalization, written by Bengali novelist Amitav Ghosh. It is an account of past and contemporary links across the Global South that challenges the notion that globalization is either recent, or simply based on Northern agendas.

Crush, J. (ed.) (1995) *Power of Development*, London: Routledge.
     This was an important early collection of research showing the real-world effects of representations of the Global South: it raises critical questions about the role of 'development' within the lives of people living there.

Robinson, J. (2006) *Ordinary Cities: Between Modernity and Development*, London: Routledge.
     This explores arguments for rethinking mainstream social science from the perspective of the Global South within the context of urban geography.

PART ONE

Representing the South

Cairo skyline
Source: Steve Connelly.

This part consists of one Chapter which focuses on **representation** as a concept and the diverse ways in which the Global South has been represented, both historically and in the present. The discussions in this Chapter will run through the rest of the book as throughout we are seeking to unsettle 'commonsense' ideas about the Global South and to interrogate the routes through which these ideas are developed and reinforced.

The concept of **representation** often suggests a visual image. While the Chapter will discuss the use of images in, for example, nineteenth-century British art, charity posters and trade promotion literature, the bulk of the Chapter will focus on other forms of **representation**. In particular, we use the notion of **discourse** to encompass the framing of **representations** in words, pictures, sounds or other forms of expression.

Representations of the Global South are important as they both reflect and help constitute relations between North and South. They are not neutral or necessarily harmless; rather they can reinforce negative and offensive opinions about the Global South as a whole, or particular regions or peoples of the Global South. Critically examining **representations** highlights inequalities in **power** which may seem hidden at first as the **representations** are so normalized and seemingly uncontroversial.

The Chapter is divided into three main sections. In the first section we look at how the South has been represented by the North. Given the limited space, we have chosen to focus on four themes which have commonly appeared: the South as exotic and erotic; the close association of the peoples of the South with the natural environment and their construction as 'noble savages'; the South as a place of poverty which requires external assistance; and the South as a source of danger. The themes have not always been applied to all parts of the Global South, and there are clearly differences within the Global North as to who is using such **representations**, but they are widely used in a range of settings in the North.

Having laid out these four imaginings of the South from the North, in the next section we look at how the South has been represented or imagined from the South. Here we focus on three themes: first, interconnections within the Global South, so challenging perspectives which focus on relations between North and South. Second, we consider the South as a place of morality and traditional values, in comparison with the Global North which has been constructed as immoral. Finally, we challenge **representations** of the South as being 'backward' but looking at ideas of **modernity** of different sorts.

The final section of the Chapter looks at the Global South through the lens of 'development'. As we outlined in Chapter 1, an aim of the book is to steer the reader away from focusing on the Global South as a focus for 'development', in recognition of its diversity, but it is also important to remember the influence that development has had on the South. In this third section of Chapter 2, we consider how Northern ideas of development have been implemented in the South; the role of statistics and quantitative indicators on framing the South; and finally alternative forms of development coming out of the South.

By the end of the Chapter the reader should have a clear understanding of the importance of **representations** and their influence. The diversity of images and ideas used to frame the Global South should be recognized and the role of the South in constructing and challenging these **representations** will also be stressed.

2 Representing the South

The focus of this book is on the peoples and places of the Global South. As we outlined in the previous chapter, there is nothing natural about this particular categorization; it is one of many different ways of viewing the world. However, grouping Latin America, the Caribbean, Africa and Asia together and representing them as the 'Third World', 'Developing Areas' or 'Global South' has been and continues to be a common practice. This Chapter examines the ways in which the 'South' has been imagined or represented, and what the implications are for such practices.

In particular, this Chapter will consider the nature of the **representations**, but also who is doing the representing. Ways of thinking about people and places are not neutral, but are, instead, indications of **power** and reflect the cultural, social and political norms within which they are constructed. We therefore want to highlight key types of **representation** (see Concept Box 2.1) of the South and how these are indicative of global **power** relations (see Chapter 1).

First, we focus on how the North has represented the South. Of course, what is really being discussed is how particular groups of people in the North imagine the South, or parts of the South. Thus, it is important to recognize the heterogeneity within the North, as well as the operation of **power** relations within it. As Morag Bell (1994) exemplifies, some of the ways in which the British working class were discussed by the middle and upper classes in the late nineteenth and early twentieth centuries were similar to the language used in relation to populations in Africa. We will also highlight how imaginings of the South have changed over time, although it is also apparent that particular sentiments may remain, albeit in different language.

In the second section, we look at challenges to these Northern **representations**, by considering how the South has been imagined by people in those parts of the world. However, a coherent concept of 'the South' (or an equivalent term) has rarely been recognized or used by the peoples themselves, although there are moments when it has been politically strategic to mobilize an idea of solidarity, as with the **Non-Aligned Movement** (NAM) or the **G77** group of countries (see below and Chapter 3). Instead, this section will focus on the diversity within the South, including the linkages between peoples and places which are often ignored in Northern **representations**. It will also consider the ways in which parts of the South use homogenizing ideas of the North (or more usually 'the West') to frame their own identity (Bonnett, 2004).

Finally, we discuss the ways in which 'development' has been used as a lens through which to frame the South. Using these ideas, peoples and places of the South are viewed as lacking development, a concept which is certainly not neutral (see below and Chapters 9, 10 and 11). Within the **discourse** (see Concept Box 2.2) of development, the South needs assistance from the North in order to develop.

CONCEPT BOX 2.1

Representation

'Representation' refers to the ways in which language, symbols, signs and images stand for objects, people, events, processes or things. Representation is made up of two processes or what Stuart Hall (1997) calls 'systems of representation'; the first involves principles of organization through which objects, people, etc. are conceptualized and represented in our minds. For example, identifying an object as a 'table' relies on the individual understanding the concept of 'table' and being able to apply this concept to the object that they see. These concepts are shared, but in order to share, we need a language to communicate. Hall identifies language as the second system of representation. 'Language' does not refer purely to written or spoken words, but also to images, signs, symbols and sounds for example.

Key to the way in which representation is used in this Chapter and the book as a whole is the understanding that shared meanings about objects and the language used to communicate are not fixed. This is because the meaning is not inherent in the object, event, etc. but is rather constructed in particular social contexts which vary both spatially and temporally. Particular meanings may also become more widespread because of **power** differentials. Thus, the meanings associated with particular objects, people or practices and the language used to communicate about them will reflect and reinforce forms of **power**. This draws on the work of Michel Foucault (see Concept Box 2.2).

(Source: Adapted from Hall, 1997)

Who decides what development is and how it is to be achieved, clearly reflects **power** relations at a number of scales. In this section we consider the **hegemonic** concepts of 'development' used to frame international policy interventions since World War II. However, we also discuss other visions of development and progress which originate from the South.

An examination of how the South has been and continues to be represented, both by those in the North and peoples within the South, is vital to highlight the ways in which **power** is implicated in such processes. It also allows us to reveal networks and linkages, not only between North and South, but also within the South itself. As we discuss later in the book, these global flows of ideas, along with other forms of flow, have been key in shaping the peoples and places of the South as part of what has been termed **globalization**. However, as we also show in this Chapter and throughout the book, the South is not a passive recipient of such ideas; concepts are reformulated in local settings and alternative imaginings are produced.

VIEWING THE SOUTH FROM THE NORTH

In this section we consider four common types of Northern **representation** of the South: exoticism and eroticism; the peoples of the South as 'noble savages'; the South as a place of poverty and in need of help; and the South as a dangerous

---

CONCEPT BOX 2.2

Discourse

The dictionary definition of 'discourse' refers very simply to conversation or speech. In academia the term has increasingly been used in a broader context, drawing on the work of Michel Foucault (1980) who stressed the role of power in knowledge production. Discourse, according to Foucault, is not just what is said or written, but the wider framing of what is meaningful and how particular topics should be approached. The exercise of **power** is clearly relevant in shaping discourse as not everyone or all groups will have the same **agency**. Foucault's work focused largely on sexuality, criminality and mental health and how discourses around these themes were constructed and applied. Within research on 'development', his work has become increasingly important as a way of highlighting how the concept of 'development' has been framed and put into practice by powerful groups.

(Sources: Adapted from Foucault, 1980; Hall, 1997)

place which threatens the North. In all four cases, the construction of the South is based on **Eurocentric** (Concept Box 2.3) assumptions (Amin, 1989; Blaut, 1993). A range of media are considered from written accounts of European exploration, to present-day charity appeals and feature films.

Exoticism and eroticism

As Edward Said argued, European visions of other parts of the world have been based on a process of '**Othering**', where the European experience (or at least parts of the European experience) are placed at the centre and viewed as 'normal', while the lives and cultures of other peoples and other places are abnormal or 'other'. His work on **Orientalism** (see Concept Box 2.4) brought out the ways in which European (particularly British and French) imaginings of the Middle East constructed it as exotic, mysterious and static compared with the dynamism of the West. Thus the West's understandings of itself are partly created through a comparison with the Orient and **Orientalism** is a 'sign of European–Atlantic power over the Orient' (Said, 1978: 6). Said's work dealt with a range of representations, including political discourse and literature; he opens *Orientalism* by discussing a speech in 1910 by the ex-prime minister, Arthur James Balfour, where he justified British occupation of Egypt by referring to Egyptians' inability to self-govern and, by implication, British skills in governing and bringing order.

The concept of '**Orientalism**' has also been applied to paintings and drawings. For example, the picture *The Courtyard of the Coptic Patriarch's House in Cairo* by John Frederick Lewis, painted in about 1864 (see Plate 2.1) shows a vibrant scene incorporating richly robed men wearing turbans, veiled women engaged in domestic tasks and a range of birds and animals including camels. This image is a clear contrast to the dress, architecture and domestic animals in nineteenth-century Britain, stressing the exotic and mysterious aspects of life in the East.

Representations of other peoples and parts of the world as exotic and different were also part of colonial accounts of encounters between Europeans and indigenous

---

CONCEPT BOX 2.3

Eurocentrism

A term used to describe the belief that the European experience is the norm against which non-European experiences are evaluated. Europe is constructed as the 'core' and the rest of the world as the 'periphery' with Europeans viewed as the 'makers of history' (Blaut, 1993:1). Thus, progress and **modernity** develop autonomously in Europe and then diffuse to other parts of the world.

CONCEPT BOX 2.4

Orientalism

A term which encompasses the ways in which the 'West' views the 'East', in particular the countries and peoples of the Middle East. It is based on the work of the Palestinian-American, Edward Said, in particular his 1978 book *Orientalism*, which had the subtitle *Western Conceptions of the Orient*. Orientalist perceptions frame the Orient as traditional and exotic in comparison with a modern and progressive West.

---

populations. Common **tropes** were used to describe what Europeans perceived as societies lacking civilization, particularly due to types of clothing, housing and the absence of Christian beliefs (Nederveen Pieterse, 1992). Archibald Dalzel (1793) in his collection of memoirs about Dahomy in West Africa, discusses 'savage nations' and includes many descriptions of barbarism, human sacrifice and cannibalism, including sketches (see Plate 2.2). In colonial accounts there was sometimes an appreciation of particular cultural practices and ceremonial activities, especially if they were interpreted as exotic and different (see Box 2.1). Of course, indigenous populations also viewed this encounter as a meeting of different peoples, but few of their accounts have survived. A notable exception includes the recollections of Titu Cusi Yupanqui (2005 [1570]) of the Spanish conquest of the Inca empire.

European men's colonial travel accounts also drew on particular **discourses** around **gender** and sexuality, not least through presenting lands which were 'discovered' or 'conquered' as female and virgin (Driver, 2001; Pratt, 1992). While the territories were 'virgin', however, the inhabitants, particularly the women, were often presented as sexually voracious. Anne McClintock (1995) uses the term 'porno-tropics' to describe the way in which Europeans viewed the inhabitants of Africa and the Americas, with special attention paid to women's sexual appetites. She quotes William Smith from his book *A New Voyage to Guinea* where he describes the actions of women; 'if they meet with a Man they immediately strip his lower Parts and throw themselves upon him' (1745: 221–2 in McClintock, 1995: 23). Such behaviour was clearly viewed as highly unusual and inappropriate from the perspective of European middle-class society of the time.

It was not just through writing and paintings that the exotic and erotic world of the Global South was presented to Europeans in the eighteenth and nineteenth centuries. The Great Exhibition in London in 1851 saw the bringing together of people and objects from Britain and British colonies and protectorates to celebrate technological progress in the impressive setting of the purpose-built Crystal Palace in Hyde Park. Notable features of the exhibition included displays representing life for native populations in colonial territories. This involved real

Plate 2.1 The Courtyard of the Coptic Patriarch's House in Cairo, John Frederick Lewis.
Credit: © Tate.

people dressed in 'traditional' clothing recreating everyday life. Such representations were placed alongside the latest technological developments of the industrial revolution suggesting the supremacy and progress of the West in contrast to the traditional South (McClintock, 1995). The 1851 Exhibition was the first of a series of World's Fairs which continue to this day, albeit with very different displays (Expomuseum, 2008).

Forms of **globalization** have allowed greater opportunities for travel for some and a growing awareness of different cultures. However, ideas of the exotic still abound in ways in which the South is represented. Magazines such as *National*

Geographic (Lutz and Collins, 1993) and travel programmes have tended to highlight areas of difference and present images of 'traditional culture'. In an article on travel to Java, Indonesia, the image of the Buddhist temple of Borobudur is used to tempt the reader (see Plate 2.4). The text describes Java as 'one of the world's most serenely restful islands'. While the article does mention Jakarta, the Indonesian capital with a population of nine million, which is also located on Java, the focus of the words and images is of lush green landscapes, remote locations and traditional cultural practices. Of course, communities and governments in the South have also used such images and ideas as part of tourism development programmes (see Chapter 8).

First nations and noble savages

If **Orientalism** has been a mode of presenting the peoples of the Global South as exotic and different from the Global North, an important sub-theme within this mode has been the **representation** of their relationship to nature. In the late seventeenth and eighteenth centuries, European travellers made their first sightings of what they called 'island Edens' (Grove, 1995) in the South Pacific and elsewhere.

Plate 2.2 Victims for sacrifice, Dahomy.
Credit: Dalzel (1793). © The Royal Geographical Society.

BOX 2.1

Accounts of European exploration

A number of European accounts of conquest and exploration were produced and found audiences in Europe who were keen to find out more about other lands and peoples. While these accounts are clearly very partial tellings of the experiences of European expansion into the Global South, they do provide today's readers with insights into forms of **representation** of the South by the North used at the time.

Bernal Díaz del Castillo travelled with Hernan Cortés to what is now Mexico in search of gold and silver in the early part of the sixteenth century. His account of the conquest of New Spain, as the Spanish later termed it, includes tales of the violence and savagery of the indigenous populations, including discussions of human sacrifice. However, he also expresses wonder at both the riches and skills of the Aztecs. On arriving at the Valley of Mexico in 1519, Díaz del Castillo and the other soldiers were in awe at the cities which had been constructed within the lake which covered the valley floor for about 442 square miles, and the causeways and canoes which linked the settlements.

The Spaniards met Montezuma, leader of the Aztecs, in Tenochtitlán on the site of what is now Mexico City. Díaz del Castillo describes in great detail the elaborate clothing and jewels worn by Montezuma, but also the rituals surrounding his meals and the vast array of food which was provided:

> Four very beautiful cleanly women brought water for his hands in a sort of deep basin which they called *xicales* [gourds], and they held others like plates to catch the water, and they brought him towels. And two women brought his tortilla bread, and as soon as he began to eat they placed before him a sort of wooden screen painted over with gold, so that no one should watch him eating. Then the four women stood aside and four great chieftains who were old men came and stood beside them, and with these Montezuma now and then conversed, and asked them questions
>
> (Díaz del Castillo, 1956 [1632]: 210)

A similar fascination with rituals around leadership is apparent in David Livingstone's account of his African travels. In one Chapter he describes his party being welcomed by Shinté, chief of the Balonda peoples in what is now western Angola. The meeting is represented visually in a sketch in the book (see Plate 2.3) as well as in Livingstone's description:

BOX 2.1 (CONTINUED)

We were honoured with a grand reception by Shinté about eleven o'clock. ... The kotla, or place of audience, was about a hundred yards square, and two graceful species of banian stood near one end; under one of these sat Shinté, on a sort of throne covered with a leopard's skin. ... A party of musicians, consisting of three drummers and four performers on the piano, went round the kotla several times, regaling us with their music. The drums are neatly carved from the trunk of a tree, and have a small hole in the side covered with a bit of spider's web; the ends are covered with the skin of an antelope pegged on it; and when they wish to tighten it they hold it to the fire to make it contract: the instruments are beaten with the hands.

(Livingstone, 1857: 292–3)

In both Livingstone's and Díaz del Castillo's accounts, detailed descriptions are provided of formal rituals, clothing, housing and the social interactions both between the local people and the Europeans, and between locals. Such details focus on stressing the exotic and different nature of these practices compared with European expectations of the time.

(Sources: Adapted from Díaz del Castillo, 1956 [1632]; Livingstone, 1857)

Plate 2.3 Reception of David Livingstone's mission by Shinté.
Credit: Livingstone (1857). © The Royal Geographical Society.

Some European commentators at this time saw indigenous peoples of these places as 'noble savages', possessing traits of simplicity, honesty, and above all closeness to nature that they themselves had lost. The 'positive **Orientalism**' of these views was often overshadowed by far more negative aspects of European contact which were often disastrous in human and ecological terms for many indigenous groups: deliberate genocide was practised in parts of North America and Australia, the latter wiping out the entire aboriginal population of Tasmania. Nevertheless, the idea of the 'noble savage' influenced the ways indigenous peoples were treated within some colonial contexts. For example, in British India, there was great anthropological interest in indigenous or 'tribal' groups dwelling in forest areas, and these groups were subject to different forms of rule from mainstream Indian society. This involved elements of both severe repression – such as the designation of entire ethnic groups as 'criminal tribes' – and well-meaning, if patronizing, attempts to protect allegedly primitive groups in their contact with more 'advanced' Indian cultures.

Plate 2.4 Java as an exotic and serene travel destination.
Credit: *The Sunday Times Travel Magazine* and Martin Westlake.

As a mode of representing indigenous peoples, the romantic idea of the 'noble savage' still has its echoes today, with common constructions of indigenous people as 'ecosystem people' (Gadgil and Guha, 1995), at one with nature and protecting authentic cultures and environmental knowledge untouched by the modern world. The anthropologist Candace Slater notes that this is a common **trope** in the treatment of Amazonian people within the Western media, and one that not only obscures the complexities of their everyday lives, but also makes indigenous people appear simultaneously both heroic and less than human. In debates over how the Yanomami people of Brazil's Western Amazonia are to be protected, she notes that American newspaper reports hold the group up as model environmental citizens, and yet also treat them as if they were a non-human endangered species: 'The Yanomami ... come across very much like another particularly noteworthy biological entity whose survival happens to require the protection of a very large amount of land' (Slater, 1995: 120). Ultimately, she argues, such images of indigenous people – like those of the eighteenth-century romantics before them – tell us more about Western society and its nostalgic search for an Eden-like past than they do about the cultures and people they claim to represent. But in doing so, these images can both put barriers to understanding the everyday realities of the South, and more importantly, be representations that have powerful effects on how indigenous peoples are treated. Such **tropes** may, however, be mobilized by Southern populations in order to gain support from the North, as with the environmental protest movement that has opposed the massive hydro-electric power projects in India's Narmada valley (Baviskar, 1997; see also Chapter 6 and Box 9.3). Tania Li (2000) discusses similar issues around who is defined as 'indigenous' in the context of Suharto's regime in Indonesia.

Poverty and pathos

As just outlined, the colonial encounter between Europeans and indigenous peoples was usually framed by Europeans as a meeting between a civilized and uncivilized population. While **discourses** of 'civilization' may not be used in the same way today, there are other ways of representing the South which have similar connotations of inferiority, or passivity. These representations are also based on **normative** ideas of how people should live which are grounded on an idea of the South lacking something, whether that is material goods, or particular social or political structures. This has been the basis for ideas of 'Development' which will be discussed later in this Chapter and in Part 4 of the book. However, in this current part the focus is on how the peoples of the South are represented as unable to do anything for themselves to make their lives better; rather the only way improvements can be made is through assistance from the North. A paternalistic relationship is thus implied.

Such representations can be most easily seen in advertisements and campaigns for public assistance for **aid**. By highlighting ways in which particular forms of **power** are played out in such advertisements, we are not arguing that the problems (such as famine or disease) do not exist, rather we are demonstrating the ways in which certain pictures and language can reinforce ideas of Northern **agency** and Southern passivity. They can also present homogenizing images of the South, focusing usually on parts of Sub-Saharan Africa and South Asia, and often (at least in the past), using children as a focus to illicit pity (and therefore money) from Northern donors. The use of mothers and children in news reporting on crises and disasters in the Global South is also widespread (see, for example, Campbell, 2007, on newspaper coverage of the Darfur conflict). Media reports may also focus on 'White heroes' going to rescue the darker-skinned starving and dispossessed, again reinforcing ideas of Northern **agency** and Southern passivity (Harrison and Palmer, 1986).

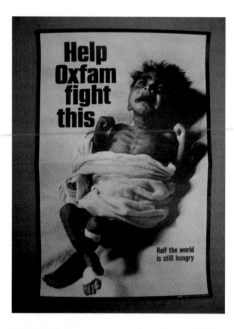

Plate 2.5 Oxfam campaign poster, 1967.
Credit: Oxfam.

Plate 2.6 Oxfam campaign poster, *c.*1998.
Credit: Tricia Spencer/Oxfam.

Development organizations have become increasingly aware of the impacts of the images that they use and there have been significant changes among many (Lidchi, 1999). These changes have been a result of donor fatigue, where the public becomes almost immune to the same images of pathos and feels that their money is not making a difference, but changes have also resulted from shifts within development practice which recognize the **agency**, resources and capacities of people in the South (see Chapter 11). For example, Oxfam has shifted the way in which it advertises, focusing more on how people are helping themselves, rather than presenting them as waiting for external assistance. Around the time of the organization's twenty-fifth anniversary in 1967, Oxfam ran a poster campaign highlighting the continued prevalence of hunger in the world. These posters showed emaciated children and young people (see Plate 2.5) and called on the viewer, most probably in the Global North, to provide assistance. By the 1990s, food insecurity and hunger were still widespread, but Oxfam's publicity material had changed (see Plate 2.6). The woman in the poster is not passively waiting for assistance, but is actively involved in a craft and in the background a field of crops can be seen. The poster's wording highlights the continued problems of food shortages but stresses that 'they are finding new ways to get the most from their land'. The role of Oxfam and public donations is still stressed through the inclusion of 'Thank you' and the Oxfam logo on the poster, but overall the message is one of **agency** on the part of Southern people and communities. This reflects the desire by Oxfam to provide forms of assistance and support which will enable sustainable forms of economic activity, allowing people to help themselves. Such an approach has become increasingly widespread among development agencies (see Chapter 11) and is reflected in their publicity material.

Barbarism and threats

As discussed earlier in the context of colonial encounters, the peoples of the Global South have often been presented as 'savage' or in need of 'civilization'. A related way of imagining the South that continues into the present day is of a place that is dangerous, not only for the intrepid explorers or travellers who venture there, but also by directly threatening lives and lifestyles in the North. Here we review briefly three different ways in which the South has allegedly presented a threat to the North: through 'overpopulation', through the threat of political instability and war, and through the spread of health risks and disease.

The current population of the Global South is both large (around 5.4 billion of a global total of 6.5 billion), and growing faster than that of the North; between 1975 and 2005, annual average population growth was 0.7 per cent in countries classified as high income by the World Bank, compared with 2.3 per cent in the low-income countries (UNDP, 2007). Within the Global South there are, however, significant differences in population growth rates, with many countries of Latin America and East Asia experiencing annual growth rates of less than 2 per cent, while much of Africa, the Middle East and South Asia had annual growth rates of over 2 per cent per annum (see Figure 2.1). Aside from the difficulties of accurate measurement, these facts are relatively uncontroversial. What is far more contentious is the use of statistics such as these to suggest that the South is 'overpopulated' and thus is endangering either the North, or the planet as a whole. This threat can be voiced in two ways, as an ecological/environmental threat to the planet which has roots in Malthusian ways of thinking (Box 2.2), or as a threat of mass Southern immigration that might engulf Northern countries.

Neo-Malthusianism presents current global population growth as leading to a resource crisis in a simplistic, but all-too-convincing way: humans consume precious resources of a finite world, global population is increasing, therefore something must be done to tackle the Global South's 'population problem'. Demographic patterns across the South are, however, highly diverse, and the reasons for population growth (and decline) (see Chapter 8) are far more complex than Malthus's crude resource constraint interpretation. Nevertheless, such ideas still combine with some dominant images of the South as populated by 'teeming millions' of people (see Plate 2.7) to normalize particular political agendas. If 'overpopulation' is accepted as being the problem, a range of 'solutions' – from tightened international immigration controls through to invasive policies to control Southern women's fertility (Hartmann, 1995) – can be made to appear legitimate as a result.

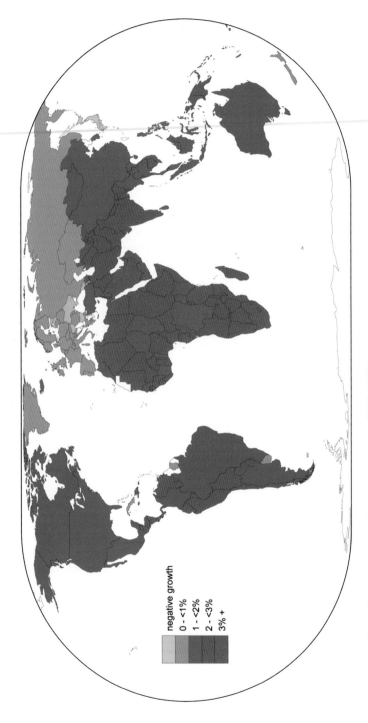

Figure 2.1 Annual average population growth rates, 1975–2005.

Source: Based on data from UNDP (2007). Map data © Maps in Minutes™ (1996).

negative growth
0 - <1%
1 - <2%
2 - <3%
3% +

BOX 2.2

Malthus and (neo-) Malthusianism

The Reverend Thomas Robert Malthus (1766–1834) argued in *An Essay on the Principle of Population* (1798) that humanity faced a crisis of overpopulation. If unchecked by famine, disease or war, he saw human populations as naturally increasing geometrically (1, 2, 4, 8, …), whereas food production could at best increase arithmetically (1, 2, 3, 4, …). If people – and particularly the poor – did not exercise 'moral restraint' and have children later in life, growing human misery and starvation were the inevitable consequence. He argued that the English Poor Laws exacerbated this problem, providing the poor with a degree of social security that undermined their responsibility for their own fertility decisions.

For a century and half after Malthus published his work, technological change seemed to be allowing agricultural production to meet the resource needs of a rapidly increasing global population. By the end of the 1960s, however, the idea of ecological or environmental constraints to population growth were again being discussed seriously. Paul Ehrlich in *The Population Bomb* (1968) argued that 'the race to feed all of humanity is over', and echoing Malthus, argued that international **aid** would ultimately only contribute to larger famines. Elements of Malthusian thinking were also present in the influential report, *The Limits to Growth* (Meadows, *et al.*, 1972), which predicted that population growth would ultimately lead to 'overshoot and collapse': rapid declines in standards of living and life expectancy were inevitable for all. Looking back from the vantage point of the twenty-first century, some of this neo-Malthusian work may at first sight seemed to have got the detail wrong (*Limits* predicted that known world supplies of oil would run out in the 1990s), but the big picture is right. After all, human-induced climate change is now accepted as fact by most people, and perhaps provides evidence that current global population levels have disastrously outstripped our resource base.

Neo-Malthusianism is, however, questionable on various grounds, not least for its simplistic links between population increase and resource use. A focus on population growth, and particularly on the fertility of the poor, detracts attention from the inequality of resource usage between countries and individuals (Table 2.1). Arguably, it is the unsustainable lifestyles of most

BOX 2.2 (*CONTINUED*)

BOX 2.2 (*CONTINUED*)

people in the North and of the growing middle classes in the South that are far more problematic.

(Sources: Adapted from McCormick, 1989; Robbins, 2004)

Table 2.1 Average consumption of resources per capita per annum, India and the USA

Resource	India	USA
Carbon dioxide emissions (tonnes)	1.2	20.6
Energy (kg oil equivalent)	477	7,956
Meat (kg)	4	122
Paper (kg)	4	293
Water (m³)	588	1,894

Source: Adapted from Robbins, 2004: 8; UNDP, 2007: table 24.
Note: as national averages, these figures mask massive differences in consumption *within* both countries.

Plate 2.7 Crowded train, Tongi Train Station, Bangladesh.
Credit: © Still Pictures.

The South is also often represented as a place of conflict and political instability, which both causes problems for the Southern populations, but also potentially threatens the Global North. Northern media reporting of the South tends to be very limited, but among the themes which tend to receive coverage, are those of civil war and political strife (Cleasby, 1995; DFID, 2000). Garth Myers (2001) in his content analysis of ten introductory English-language human geography textbooks, also finds that conflict is a common **representation** of Africa. He goes on to highlight that authors often use 'tribalism' as a key explanatory factor, without recognizing the complexities of African societies and the role of **colonialism** in the creation of these supposedly traditional 'tribal' groups (see also Campbell, 2007 on simplistic media **representations** of the Darfur conflict and Box 3.4).

The so-called 'War on Terror' has involved **representations** of particular parts of the Global South, most notably countries in the Middle East or those elsewhere with a significant Muslim population, such as Indonesia (Figure 5.3), as containing threats to global security, and more specifically to the peoples of the Global North (Gregory, 2004). Matthew Sparke (2007) examines the rhetoric of key US politicians, such as President Bush's State of the Union address in 2002 and Secretary of State Colin Powell's address to the United Nations in the same year, to bring out the manner in which particular geographical imaginations become embedded; '[t]he repetitive conjuring of fear created ... a form of geopolitical spatial fix' (2007: 342). Thus, even when it became clear that Saddam Hussein did not have weapons of mass destruction or links with al-Qaeda, the majority of the American public continued to believe that Iraq posed an immediate threat to the United States as this **representation** of a particular space had been 'fixed' in their minds. Such fears have been reflected in Hollywood films in the post-Cold War period. Whereas the Soviet Union had often been represented as the enemy, new threats have been represented, most notably from Middle Eastern states (Dodds, 2005).

Finally, the South has been represented as a threat to Northern populations as a source of disease. Lucy Jarosz (1992) highlights how **discourses** of the 'dark continent' which were used in relation to Africa in the nineteenth century have also been found in language used to discuss the origin and spread of HIV/AIDS. Paul Farmer (2006) draws out a similar process in his discussion of North American views of Haiti's role in the spread of the virus. In this case, the hegemonic view was that Haitians were the source of the disease, possibly due to 'voodoo practices' and had passed it on to North Americans who visited the island or when Haitians migrated to the USA. Farmer shows that the spread of HIV/AIDS was actually very different, with Haitians catching HIV from North American visitors. His aim of revealing such a process is not to reassign blame, but rather to bring out the ways in which prevailing understandings of the South can hide what is actually going on, so preventing appropriate support.

This section has used four main themes in the **representation** of the Global South to discuss how ideas of Northern superiority have been constructed and reinforced over time. While particular **discourses** may be used originally to refer to particular peoples or places, these have often been more widely adopted to address the Global South as a whole.

REINTERPRETING THE SOUTH FROM THE SOUTH

Having dealt with key themes in the forms of **representation** of the Global South adopted in the Global North, the following section will focus on ways in which the Global South has represented itself. As with the Northern **representations**, **discourses** are derived from particular groups within the Global South and usually refer to specific parts of the region. Three themes are covered in this section: interconnections within the Global South, the South as a site of morality and 'traditional values' in contrast to the 'decadent North', and finally, the South as modern.

Histories of interconnection

As outlined in Chapter 1, a key theme of this book is to see how the peoples and places of the Global South have been incorporated into global economic, political and cultural flows, the impacts of this incorporation and how the South has shaped the networks in which it participates. This focus does, however, often lead to a concentration on interconnections between Global North and Global South, ignoring the importance of interconnections *within* the South itself. Such linkages may be as a direct response to processes involving the Global North, as in the **Non-Aligned Movement (NAM)** and the **G77** grouping, while in other situations (such as international **aid**), interconnections have little or nothing to do with the North.

In Chapters 3 and 4, we discuss in detail the formation of global economic and political systems and the role that the Global South has played within these processes. While the Southern countries have usually been placed in a situation of disadvantage in relation to global trade talks or political influence, forms of cooperation and solidarity in the Global South have, at times, had important impacts. In the post-World War II period, many of the newly independent countries of Africa and Asia sought to carve out an autonomous path for themselves, separate from the **capitalist** road of the West and the **communist** route of the Soviet bloc. The setting up of the **NAM** in 1961 provided a focus of Southern cooperation in a **Cold War** world (see Chapter 3 for more details).

In trade terms, the rise of global organizations to regulate the global economy and trade (see Chapter 4) has led to new arenas for Global South cooperation. In the face of the economic might of the countries of the Global North, Southern countries have set up organizations such as the **G77** (see Chapter 3). In the run up to the World Trade Organization (WTO) talks in Cancún, Mexico in 2003, the G-21 grouping was set up to challenge what were seen as attempts by the European Union (EU) and the USA to protect their domestic agriculture while pushing for agricultural liberalization in the Global South. The G-21 members were largely Latin American, but also included China, Egypt, India, Pakistan, the Philippines, South Africa and Thailand. Despite differences in the size and nature of their economies, as well as their trade policies, these countries cooperated to lobby against the EU and USA at Cancún (Langhammer, 2005). This resulted in the collapse of the trade talks, an outcome similar to that of the WTO trade talks in July 2008 (see Chapter 4).

Economically, the development of regional trading blocs, free trade areas and customs unions have become increasingly important throughout the globe as processes of economic **globalization** have developed. The two most important regional blocs in terms of economic size are the EU and the North American Free Trade Area (NAFTA) which consists of Canada, Mexico and the USA. There are many other regional organizations which seek to support their members in achieving economic growth and development, many of which are in the Global South. For example, the Southern African Customs Union (SACU) consists of Botswana, Lesotho, Namibia, South Africa and Swaziland. Within the Union there is free movement of goods, capital and services, while there is a common external tariff for trade outside the union and funds are redistributed between members (SACU, 2008).

For very small economies, cooperation can make important contributions to national development and can help raise the profile of the region in a way that individual countries cannot. The Caribbean Community (CARICOM) has 15 members (see Figure 2.2), was formed in 1973 and grew out of the previous regional cooperation organizations, the West Indies Federation (1958–62) and the Caribbean Free Trade Association (CARIFTA) (1965–73). As well as promoting regional cooperation, CARICOM also includes a single market so as to facilitate trade in goods and services between the member states. Such connections between Southern countries and the creation of new forms of economic and political space in the Global South challenge the **representations** of this part of the world by the Global North as they come out of South–South relations, rather than North–South ones, so revealing the **agency** of Southern governments.

These challenges to images of the Global South's people as passive can also be seen in forms of international support and **aid** within the South. China's economic role in Africa has increased greatly since the late 1990s (see Boxes 3.6 and 4.5), but even before this, the Chinese government was providing technical and educational assistance to some African governments (Mawdsley, 2007). The Cuban government's focus on international solidarity has been even more pervasive, particularly in the fields of military **aid** and health assistance (see Box 2.3). Governments and peoples in the Global South are also able and willing to provide help at times of disaster, as with Sri Lanka's offer of assistance to the US Government in the wake of Hurricane Katrina in 2004 (Korf, 2007)

In this section we have outlined different forms of interconnection within the Global South to demonstrate how this region can be imagined from the viewpoint

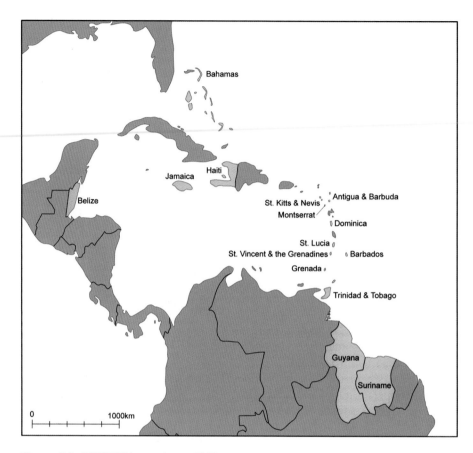

Figure 2.2 CARICOM members, 2008.

of the Global South, rather than being filtered through the **representations** of the Global North. The section has also demonstrated the **agency** of peoples in the Global South; something which is often missing from Northern **representations** of the region.

Morality and traditional values

In comparison with representations of the Global South as 'uncivilized', increasingly it is the North which is being represented as immoral by governments or groups in the South. Such **discourses** are sometimes mobilized by governments as a way of promoting nation-building, but may also reflect a strategy to deal with the tensions arising from the global flows of ideas, commodities and people.

Alistair Bonnett (2004) uses the concept of 'Asian values' to discuss how the 'West' has been constructed in opposition to 'Asia'. In the 1980s, the leaders

BOX 2.3

Cuban health diplomacy

In 2007, over 30,000 Cuban medical personnel, including 19,000 doctors were working in 103 countries in the Global South as part of Cuba's international cooperation programme. In many cases the governments of the receiving countries (largely in Africa and Latin America) provide housing and pay for the air fare and a limited monthly amount for subsistence. Patients do not pay for the medical attention that they receive. While overseas, the Cuban staff continue to receive their salaries from the Cuban government. It is calculated that Cuba provides more medical staff to the Global South than all the G8 countries combined.

Cuba first sent medical staff abroad in the 1960s, when 56 doctors and other medical personnel were sent to Algeria to provide support for the newly independent nation. Since then, medical teams have worked in many parts of the Global South, sometimes as part of disaster response teams (as in 1998 after Hurricane Mitch hit Central America) or through longer-term programmes to poor urban and rural areas. At times assistance has been offered to countries in the Global North. For example, in 2005, Cuba offered to send medical personnel and supplies to the United States to help victims of Hurricane Katrina, but the assistance was rejected by the US Government.

Cuba has also helped build medical capacity in Southern countries through training and the setting up of nine medical schools. The first was in Yemen in 1975. In 1998, the Latin American School of Medicine (ELAM in its Spanish acronym), was set up in Cuba to provide free medical training for students unable to afford fees elsewhere. This includes low-income students from the USA. The training focuses on community medicine and public health, as students are expected to return to their countries of origin and to work with communities that are usually excluded from formal health provision.

Cuba's health diplomacy is an example of South–South cooperation and solidarity. The Cuban government has adopted this policy as part of its ideology of internationalism following the 1969 revolution. However, it has also helped cultivate support from other nations (although medical assistance has been provided to countries whose government ideologies strongly oppose those of the Cuban regime). Such diplomatic support is vital given the US Government's trade blockade against Cuba.

(Sources: Adapted from Feinsilver, 1989; Hamnett, 2007; Huish and Kirk, 2007)

of Singapore (Prime Minister Lee Kuan Yew) and of Malaysia (Prime Minister Mahathir Mohamad) were at the forefront of the mobilization of 'Asian values' in both domestic and international politics. While both had sought to bring rapid economic development to their respective countries (see Box 4.4 on Singapore) by opening up their economies to foreign investment, they were concerned about the social and cultural impacts of Western influences. This was sometimes termed 'Westoxification'. Both Mahathir and Lee contrasted the individualization and liberalism of Western societies with what they saw as the 'Asian' way of doing things, based on Confucian principles of collectivism, respect for authority (both in the family and in wider society) and hard work (Chong, 2004; Perry, *et al.*, 1997; Stivens, 2006).

The governments of other Asian countries, such as China, Indonesia and Vietnam, supported Singapore and Malaysia's mobilization of 'Asian values' as a way of countering Western calls for the implementation of liberal democracies in the region. Chong (2004: 104) quotes a 1994 interview with Lee Kuan Yew when he stated: 'It is not my business to tell people what's wrong with their system. It is my business to tell people not to foist their system indiscriminately on societies in which it will not work.'

For a number of governments in East and South Asia and the Middle East, such claims of different value systems helped create a sense of solidarity in the face of perceived efforts by Northern countries to interfere in the running of Asian countries. However, not all Asian governments were willing to be involved in the use of 'Asian values' in this way. For example, Japan, the Philippines and Thailand did not publicly support Lee's pronouncements, and most Asian human rights organizations also criticized Lee's claims (Chong, 2004).

Despite these divisions within Asia, the way in which Asian values have been mobilized is an example of a **representation** of the Global South from the Global South. Such **tropes** have also been used within specific debates around morality, often in the context of religion. Again, in these **representations** the Global North or West is often portrayed as an immoral and ungodly place where individuals follow their own desires, rather than living according to religious principles (particularly Islamic or Christian morals).

Sexual behaviour is often highlighted as an indication of how the Global South is more moral than the North. Homosexuality, extra-marital sexual relations, childbirth outside marriage and the perceived collapse of the 'traditional family' are all viewed as examples of Northern decadence and immorality (Stivens, 2006). At the Zimbabwean International Book Fair in 1996, Gays and Lesbians of Zimbabwe (GALZ) were to have a small booth to distribute their literature. However, the Government disapproved, particularly as President Robert Mugabe was going to open the fair. The Government's Director of Information, Bornwell Chakaodza, wrote to the executive director of the Book Fair, including the following:

> Whilst acknowledging the dynamic nature of culture, the fact still remains that both Zimbabwean society and government do not accept the public display of homosexual literature and material. The Trustees of the Book Fair should not, therefore, force the values of gays and lesbians onto Zimbabwean culture.
>
> (letter, 24 July 1995, quoted in Dunton and Palmberg 1996: 9, quoted in Murray, 1998: 248)

This letter and later comments by President Mugabe equating homosexual practices with 'sub-animal behaviour', which all Zimbabwean citizens should denounce to the police, highlights how homosexuality is constructed as being external or alien to Zimbabwean culture. This fails to recognize the long-standing existence of same-sex relationships in Africa (Murray and Roscoe, 1998). Following this furore, the GALZ registration was withdrawn.

Representations of parts of the Global South as sharing particular morals or value systems which are contrasted with those of the North have been used by politicians and religious leaders, among others, to create feelings of cohesion around nation-building or group identity. However, as with **representations** of the Global South by the Global North, the images and **discourses** used fail to recognize the contested nature of the **representations** and the diversity within particular parts of the Global South.

Modernity

In contrast to Northern images of the South as poverty-ridden, particular **discourses** of **modernity** (Concept Box 2.5) are presented by countries in the Global South. **Modernity** as a way of living for people in the South will be discussed in Chapter 8, but in this section we highlight how governments, regions and cities in the Global South have used discourses of **modernity**, particularly as a way of attracting investment.

In an era of economic **globalization**, many governments (both North and South) feel that there is no alternative but to 'play the **globalization** game' and make their countries, regions or cities as attractive to international investment as possible (Kelly, 2000). This is done through practices of **place marketing** (Gold, 1994) which has become widespread throughout the world, including attempts by deindustrializing cities in the Global North to attract new investment and residents through urban regeneration strategies. Such **place marketing** includes dimensions which are regarded as 'modern' such as the availability of high quality infrastructure, legal frameworks which will protect foreign investment and a well-trained workforce (Fursich and Robins, 2002). Other factors such low wages, flexible environmental legislation and labour laws (see Chapter 4) may also be highlighted, but will not be framed as 'modern' in the same way as some other factors may be.

The images used to market places in the Global South may also draw on **tropes** which are familiar in Northern imaginations of the South, particularly around nature and pace of life. For example, Jamaica's investment and export promotion agency (Jamaica Trade and Invest) has run a campaign which juxtaposes some of the more common images of Jamaica among people in the Global North with images highlighting economic **modernity**. In the two images shown here (Plate 2.8) the image of the terrace and the clear blue sea is shown alongside a businessman using his laptop. The slogan 'The place you always wanted to visit, is the place you want to do business' reinforces this idea of a desirable location for business because of the scenery and chances to relax, as well as the business opportunities and infrastructure available in Jamaica.

The use of global and regional events, such as the Olympics, the Football World Cup and regional heads of government meetings may also be used by governments to promote their country or city to a wide audience, as well as benefiting from the inflow of people and investment for the event in question. Such images of **modernity** are often associated, however, with attempts to eradicate people and places that do not fit into that image. The displacement of low-income residents as part of the preparations for the 2008 Beijing Olympics is one very high-profile example.

These forms of **place marketing** are usually framed with reference to Northern ideas about **modernity**. It is important to consider them as they counter the frequently presented images of the Global South as 'backward' and economically poor. However, they are usually based on Eurocentric ideas of progress. Other concepts of **modernity** draw upon other sources for inspiration and seek to challenge Northern ideas of how societies and economies should develop.

The Asian values debate outlined earlier is one example of how some countries (in this case Singapore and Malaysia) have sought to achieve economic

Plate 2.8 Promotional material from Jamaica Trade and Invest.
Credit: Jamaica Trade and Invest.

modernization along the lines of the Global North, while preserving what are understood as 'traditional' values around family, respect and authority. There have been similar discussions in the context of what have been termed 'Islamic modernities' (Adelkhah, 2000).

Many countries in the Middle East, most notably the Gulf States of the United Arab Emirates (UAE), have used their oil wealth to fund large infrastructure projects and to develop industrial and service industries to generate economic wealth. However, this form of **modernity** has not usually been associated with a shift towards social liberalism or political democratization; rather, a form of **modernity** which maintains the moral codes of the relevant form of Islam (see Box 5.7) is adopted. For example, in Dubai notable infrastructure projects include the 6-star Burj Al Arab Hotel (see Plate 2.9) and new islands in the shape of palm trees and the countries of the world. These are used to promote Dubai as a destination for investment and also as a location for travel and shopping. However, while the behaviour of foreign tourists and Western expatriate workers is tolerated most of the time, Emiratis are expected to behave (at least in public) in a sober and respect-ful way, including appropriate clothing, particularly for women. For the Dubai Government, **modernity** can be achieved without the loss of religious morality and values.

This discussion of how places in the Global South have been represented as 'modern' contrasts sharply with the **representations** outlined in the section on

---

CONCEPT BOX 2.5

Modernity

At its most basic level, the concept of '**modernity**' refers to the state of being modern or of the time. However, what it is to be modern is not a neutral concept; rather ideas of **modernity** are framed by the exercise of **power**. In the post-World War II period, at a global level, **modernity** has usually been defined in relation to the experiences of the Global North. In economic terms, this has focused on ideas of progress towards industrial, urban and technologically advanced economies (see Box 9.2 on **Modernization** theory). **Modernity** has also been used to refer to particular consumption practices (see Chapter 8) and forms of political participation based on liberal democracy (see Chapter 3).

The **Eurocentric** nature of these constructions of **modernity** have, however, been increasingly challenged, particularly from Islamic and Asian states. These forms of **modernity** seek to decentre the Northern models and expectations to create economically advanced societies with populations living according to appropriate moral codes.

Northern **representations** of the Global South. This demonstrates how ideas about the same parts of the world vary greatly, reflecting **power** differences and the **discourses** which frame the **representations**. Despite positive **representations** of the South, particularly from within the region, the prevailing images of Africa, Asia, Latin America and the Caribbean continue to focus on these parts of the world as lacking something which the North has. In the following section we consider this idea further within the context of 'development'.

Plate 2.9 Burj Al Arab Hotel, Dubai. Credit: Katie Walsh.

DEVELOPMENT AS A MODE OF IMAGINATION

While 'development' is often considered as a way in which the North views the South, we also want to consider alternative views of development from the South. This is not to deny the dominance of particular Northern-centric views of development and how the South is imagined within these, but it is an attempt to stress the possibilities for alternatives, and the contributions that have been made from the South in these imaginings of alternative futures. The development imaginaries discussed in this section are not separate from some of the issues discussed earlier in this chapter. For example, assumptions about Northern superiority play out in many development approaches and racist practices can be found in development practice (Crewe and Fernando, 2006).

Imagining the future: postwar development policies from the North

In the post-World War II period, what Gillian Hart (2001) has termed 'capital D Development' took off as a serious form of international assistance and foreign policy. 'Development' in this sense was a form of intentional intervention to achieve particular goals: in this case 'progress' and **modernity**'. This was a shift from the colonial period where economic, political and social change in the colonized South was geared towards Northern governments and peoples benefiting from the structures of colonial domination. Following World War II, 'development', as a form of conscious intervention and policy making in the independent countries of the Global South, began to gain importance. Key Western international actors, such as the US Government, saw development interventions as a way of both promoting peace and also gaining strategic **power** within the **Cold War** context (see Chapter 3). US President Truman's inaugural speech indicated this new approach:

> For the first time in history humanity possesses the knowledge and the skill to relieve the suffering of these people [the world's poor] ... I believe that we should make available to peace-loving peoples the benefits of our store of technical knowledge in order to help them realize their aspirations for a better life.... What we envisage is a program of development based on the concepts of democratic fair dealing ... Greater production is the key to prosperity and peace. And the key to greater production is a wider and more vigorous application of modern scientific and technical knowledge.
>
> (Truman, 1949, in Escobar, 1995: 3)

Two main premises underpin such sentiments: first, an understanding of desirable economic progress and development based on the experiences of Western nations; and second, a commitment to a liberal democratic, capitalist system. As a result, the form of development assistance provided in the postwar period attempted to replicate the Northern experience of **modernization** (see Box 9.2) through large-scale technical interventions. The role of the state in directing such policies in the Global South was also key (see Chapter 9).

In the 1980s and 1990s, Northern-driven development policies were strongly critiqued by **post-development** theorists (see Concept Box 2.6). These critiques were important in that they stressed the Northern-centric nature of 'development' as it had been implemented in the Global South. There was a strong focus on an analysis of '**discourses** of development' and the **power** of particular forms of **representation** (see Crush, 1995; Box 1.4 on Mitchell's analysis of USAID policy in Egypt and Box 2.4 on Lesotho).

Post-development approaches have received significant attention and the work of theorists such as Arturo Escobar (1995) and James Ferguson (1990c) have been very important in making visible the processes through which peoples and places in the Global South have been represented as requiring 'development' as understood in the North. However, **post-development** has also received significant criticisms itself, not least because the 'development' it criticizes is a one-dimensional **representation** of a particular form of development that has changed greatly since the 1950s and 1960s (Wainwright, 2008). The alternative approaches which **post-development** theorists advocate focus on grassroots, environmentally sustainable and participatory forms of change, policies which have become widespread (albeit in particular forms) in mainstream development policy (see Chapter 11). Finally, **post-development** theorists have often failed to recognize some of the benefits which have accrued from top-down large-scale development projects, particularly in the fields of health (Corbridge, 1998).

This section has provided an introduction to the ways in which the Global

CONCEPT BOX 2.6

Post-development

'**Post-development**' refers to a general set of approaches which seek to decentre and deconstruct the concept of 'development', particularly when applied to the peoples and places of the Global South. **Post-development** theorists, such as Arturo Escobar and Wolfgang Sachs, discuss the **power** inherent in constructions of 'development' and the methods used to introduce 'development' to the Global South. Rather than being neutral, 'development' is a **discourse** and set of practices based on particular ideas about how societies and economies should progress. Escobar, in his 1995 book *Encountering Development: The Making and Unmaking of the Third World*, uses examples of World Bank policies in Colombia to highlight how these policies and programmes were drawn from US ideas about development, rather than those based on local understandings and wishes. **Post-development** theorists advocate grassroots approaches to social, economic and political change which involve significant participation from the communities concerned.

(Sources: Adapted from Escobar, 1995; Rahnema, 1997; Sachs, 1992)

BOX 2.4

The Thaba-Tseka Development Project, Lesotho

James Ferguson (1990c) uses the example of the Thaba-Tseka Development Project in a mountainous area of Central Lesotho to highlight how development interventions are often based on preset categorizations of beneficiary populations. Given such a disjuncture between the framing of the project and how people live 'on the ground' it is not surprising that many projects like this one fail to deliver most of the planned development benefits.

Using a 1975 World Bank Country Report on Lesotho, Ferguson demonstrates how the country was constructed by the report's authors as a 'traditional' country, based on subsistence peasant agriculture, archaic land

BOX 2.4 (*CONTINUED*)

tenure systems and lacking modern infrastructure to promote development. This fitted with the World Bank's views of a 'Less Developed Country'. Given this identification, the World Bank was then involved in funding projects to address this lack of **modernity**, most notably the Thaba-Tseka Development Project which was also funded by the Canadian International Development Agency (CIDA) and the Government of Lesotho. Total funding in the first phase which started in 1975 was about US$15 million.

The main aim of the project was to develop commercial livestock farming in the region. This included training for farmers, a road link with Maseru to help market access and the setting up of a regional centre for warehouses, training facilities and offices. There were also to be village distribution points for inputs such as fertilizers and improved seeds. All these services would help reduce the isolation of the region's population, but the project failed to acknowledge the existing networks of livestock markets, a government livestock improvement centre and the widespread receipt of **remittances** from migrant workers in South Africa's mines. These pre-existing social and economic structures did not fit with the planners' image of the residents and the location, so were ignored.

The lack of engagement with local populations also led to failures in land tenure changes, attempts to limit overgrazing and processes of decentralization. Male farmers saw cattle as a reserve asset not a commodity, so were unwilling to reduce the size of their herds even if this could mean healthier cattle and less pressure on grazing land. Farmers were also unwilling to increase the production of fodder crops because it would reduce land available for food production. They were therefore rejecting the project on rational grounds, not because they did not understand the proposals.

Ferguson concludes that the project largely failed to achieve its objectives because it was based on an unrealistic construction of the people of the region. There was also a lack of understanding of people's existing **livelihoods**. However, it was successful in road development which allowed the expansion of state power through post offices, police posts and health officials (see Chapter 6).

(Source: Adapted from Ferguson, 1990c)

South has been viewed as a site for development; with development interventions being based on Northern ideas of progress and **modernization**. These ideas are discussed in more detail in Part 4 of the book, but they are included in this Chapter as 'development' has been one of the key **tropes** used to define and represent the Global South.

Facts and statistics

'Development', regardless of how it is defined, is usually associated with the use of indices and statistics (Morse, 2004). Such approaches are particularly common in the international development community. A widely used statistic is **Gross National Income** per capita (GNI p.c.) by which the World Bank classifies the world into three income bands: high, middle and low (see Figure 2.3). In the late 1980s, the United Nations Development Programme (UNDP) developed the **Human Development Index (HDI)** (see Concept Box 2.7) to try to reflect a more holistic vision of development beyond the narrowly economistic one used by the World Bank (see Figure 2.4). The assumption is that improvements in these indicators are evidence of 'development'. This fails to recognize the priorities individuals, communities or even whole countries in the Global South may put on other aspects of life, particularly those which are less easily quantified.

Towards the end of the twentieth century, **poverty** alleviation became the key focus of international development efforts (see Chapter 10). **Poverty** in this case is usually measured according to a crude poverty line; most conventionally people living on less than US$1 per day are classified as living in extreme poverty. The **Millennium Development Goals (MDGs)** were adopted as a way of focusing international development attention on **poverty** alleviation. Each of the eight goals (see Table 5.2) has a number of targets which can be measured and countries in the Global South can be assessed annually as to their progress.

The use of the US$1 per day **poverty** line does not recognize that **poverty** exists in the Global North. This does not mean that there are many people living on under a dollar a day in that part of the world, but rather that they are poor in relative terms. People in economically richer countries are often considered poor if they have insufficient income to meet basic shelter, food and clothing requirements for example. The UNDP has developed two indices of human **poverty**; HPI-1 for low-income countries (see Figure 2.5) and HPI-2 for richer countries. These use the same dimensions of human development as the HDI, but they are measured slightly differently to reflect the different nature of **poverty** in richer and poorer countries. For HPI-1, for example, the 'decent standard of living' measure is calculated by looking at the percentage of people who do not have access to safe drinking water and the percentage of underweight children. For HPI-2, the percentage of people living on less than 50 per cent of the median household income for that country is used (UNDP, 2007).

Using these kinds of statistics as a form of **representation** presents the Global South as less developed and thus in need of intervention and assistance from the Global North. The **power** of international organizations, particularly the World Bank in presenting particular development interventions as common sense, means that alternatives are not considered (Mawdsley and Rigg, 2002, 2003).

It is also crucial to note that development statistics are often presented at a national scale (Willis, 2005). Thus whole countries, encompassing millions of

CONCEPT BOX 2.7

Gross National Income and Human Development Index

Gross National Income (GNI) This is the term now used for what used to be called 'Gross National Product' (GNP). It is the measure (usually in US$) of the value of all goods and services claimed by the residents of a particular country. It does not matter where those goods or services were produced. This means that it includes income from abroad, such as profits repatriated by companies from overseas subsidiaries.

Human Development Index (HDI) An index developed by the UNDP in the late 1980s as a way of measuring 'development' which went beyond the purely economic which is the basis of GNI. The HDI identifies three key dimensions of development: a decent standard of living; a long and healthy life; and knowledge. To assess these, the HDI is based on four indicators:

1 Gross Domestic Product (GDP) per capita (p.c.): this measures the value of all goods and services produced within a country divided by the population. Figures are usually provided in US$ and altered to reflect the purchasing power parity (PPP) so as to reflect differences in cost of living.
2 Life expectancy at birth.
3 Adult literacy rate: this is the percentage of people aged 15 and over who can read and write.
4 Gross enrolment rate: this measures the number people who are enrolled in primary, secondary and tertiary education as a percentage of the number who are of the age to be enrolled.

UNDP then uses the measures to calculate the HDI which runs from 0 to 1. The higher the value, the higher the level of human development.

(Source: Adapted from UNDP, 2007; World Bank, 2007b)

people and diverse environments, are collapsed into one statistic. As we argue throughout the book, the diversity of the Global South and the countries within it is something which is often ignored in **representations** of Africa, Asia, Latin America and the Caribbean. The use of national statistics reinforces the image of the Global South as being made up of **nation-state** jigsaw pieces (see Chapter 3) which can be classified as poor and requiring development.

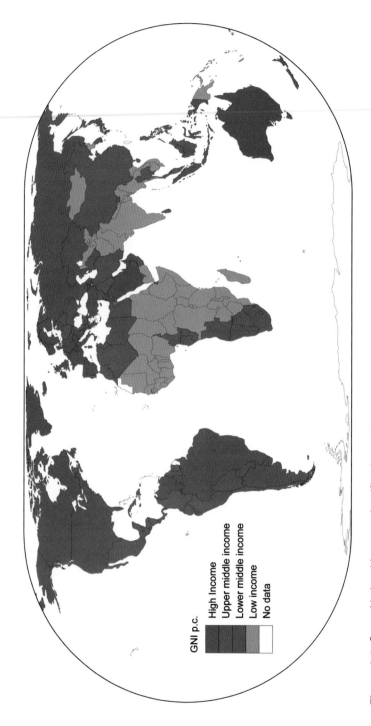

Figure 2.3 Gross National Income classifications, 2007.
Source: Based on data from World Bank (2007b). Map data © Maps in Minutes™ (1996).

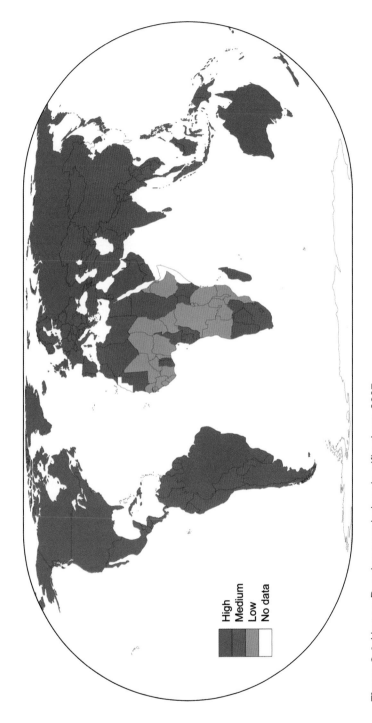

Figure 2.4 Human Development Index classifications, 2007.
Source: Based on data from UNDP (2007). Map data © Maps in Minutes™ (1996).

Imagining the future: alternatives from the South

The previous two sections have focused on the ways in which Northern **representations** of the Global South are as a place that is in need of development intervention. Such representations have often been internalized by peoples in the Global South, such that these Northern visions are viewed as the only option (see Chapter 10 on the **hegemonic** nature of **neoliberalism** as a development approach at the start of the twenty-first century). In this section we consider alternatives to Northern 'mainstream' development from the South. This does not mean that there are no alternatives emerging from the North (see Chapter 11), but as the focus of this Chapter is considering ways in which the Global South has been represented, we want to highlight Southern imaginings of the future.

During struggles for independence and in the early post-independence periods, many liberation movements and independent governments sought to introduce alternative forms of development and progress free from the constraints of colonialism. While for Africa, Asia and the Caribbean, the **Cold War** provided a challenging environment within which to embark on an autonomous development path (see Chapter 3), a number of alternatives were attempted. These were often framed as adapting external, often Northern ideas, to a particular context, as with Julius Nyerere's *Ujamaa* policies in Tanzania which were framed as a form of 'African socialism' (Box 6.4). Kwame Nkrumah, the first president of Ghana at independence in 1957, also embarked on a form of socialist development, arguing that socialist ideas could be appropriately combined with prevailing notions of egalitarianism in Africa. Unlike the focus on rural development in Tanzania under Nyerere, Nkrumah's policies centred on industrialization and attempts to drive forward Ghana's economic development. As the leader of the first independent Black African country, Nkrumah was committed to pan-Africanism and cooperation between African nations and peoples (Zack-Williams, 2006).

It is not just independence that has provided opportunities to introduce alternative forms of development in the Global South. Revolutions, civil wars and elections may also represent moments of change and an opening of new political spaces. While potentially bringing hope, such alternatives may have appalling consequences. For example, in 1975 the Khmer Rouge came to power in Cambodia. Under the leadership of Pol Pot, the Khmer Rouge implemented a **Communist** inspired regime, but one which focused on rural development, similar to some of Mao's policies in China. Between 1975 and 1978 an estimated 1.5 million people died in the repression and the Cambodian economy was left in ruins. In their attempts to construct a new form of society, families were divided and marriages were arranged by the state (Brickell, 2008).

A wide range of other Southern-based development alternatives could be discussed. While these have had differing levels of success (see Box 2.5 on the Bolivarian Revolution in Venezuela), their existence challenges the common assumptions that development can only be initiated by external actors, especially those from the

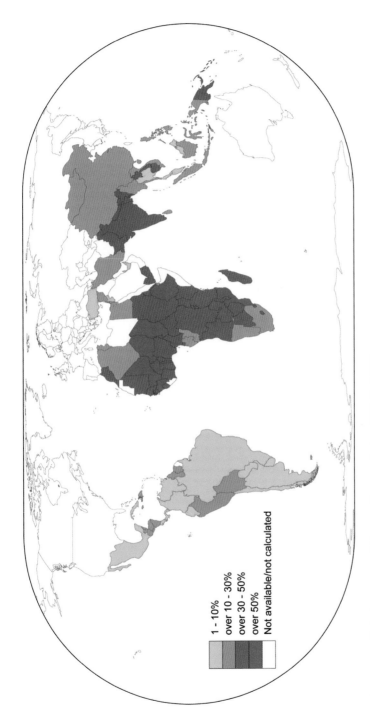

Figure 2.5 Global patterns of Human Poverty Index (HPI-1), 2007.
Source: Based on data from UNDP (2007). Map data © Maps in Minutes™ (1996).

1 – 10%
over 10 – 30%
over 30 – 50%
over 50%
Not available/not calculated

North. One alternative from the Global South that has received particular attention throughout the world is the concept of development as a path to increased happiness. This draws on the concept of 'Gross National Happiness' developed by the King of Bhutan and focuses on the importance of non-material values and sustainable development (Gross International Happiness Project, 2008).

CONCLUSIONS

This Chapter has outlined one of the key themes of the book; that **representations** of the Global South, its people and its places, matter. **Discourses** on the Global South both reflect **power** relations and who has the ability to define and use such words, images and ideas, but these **discourses** also have material effects as they help shape relations between North and South, and within the Global South.

BOX 2.5

Hugo Chávez and the Bolivarian Revolution

In December 1998, Hugo Chávez was elected President of Venezuela and implemented what he has termed a 'Bolivarian Revolution'. This is named after Simón Bolívar who was one of the key leaders in the wars of independence in South America and who advocated a Latin American federal republic.

Venezuela is one of the world's largest oil producers (see Chapter 4), with daily production in 2006 of over 2.8 million barrels a day (EIA, 2008). The wealth generated by the nationalized oil industry has helped Chávez fund large-scale programmes aimed at alleviating the poor conditions of the majority of Venezuela's population. These operate under the heading of *Barrio Adentro* (or inside the neighbourhood) and focus on grassroots-level programmes including primary health care and dental treatment. There are also attempts to promote grassroots democracy and participation and the development of cooperatives.

When he came to power, two-thirds of the population lived below the poverty line and 70 per cent of agricultural land was owned by just 3 per cent of landowners (Buxton, 2005). Chávez's programmes include increased expenditure on health (although at 2.0 per cent of GDP in 2004, Venezuela is still well below Brazil which spent 4.8 per cent of GDP and Mexico which spent 3.0 per cent, UNDP, 2007). The success of the health programmes

BOX 2.5 (*CONTINUED*)

depends largely on the assistance provided by Cuba (see Box 2.3) which sends large numbers of medical personnel to the deprived areas of Venezuela in return for cheap oil. Both UNICEF and the World Health Organization (WHO) have commended the operation of the Venezuelan health projects, particularly in relation to childhood inoculations.

The cooperation with the Cuban government is part of Chávez's attempt to promote greater regional cohesion and solidarity, as suggested by Bolívar. In particular, attempts are being made to develop a regional grouping which could undermine the proposed **neoliberal** Free Trade Area of the Americas which has the support of the USA. Countries which as of July 2008 are in a loose alliance with Venezuela include Argentina, Bolivia, Uruguay and Chile.

Critics of Chávez argue that he rules in an authoritarian manner, promoting changes to the constitution which will give him more **power**. He is also accused of undermining the national economy as he is viewed with distrust by many foreign investors, and land reform policies have often led to land redistribution to people who do not have the skills to farm it effectively. His ability to achieve his goals is also highly dependent on high oil prices and continued support from the Cuban government.

Despite these criticisms, Chávez and the Bolivarian Revolution are viewed by many, especially in Latin America, as providing an example as to how development outside a **neoliberal** framework can be achieved.

(Sources: Adapted from Buxton, 2005; EIA, 2008; *New Internationalist*, 2006; UNDP, 2007)

'Development' has been one of the **discourses** used to frame these relationships, but as we have argued in this chapter, and as we will demonstrate in the rest of the book, it is vital that the Global South and its place in the world is considered from perspectives other than through the lens of development.

In highlighting the role of **power**, we are not denying that some of the **representations** reflect reality, but rather the aim of the Chapter has been to highlight the ways in which such **representations** have been used uncritically with potentially dangerous outcomes. However, we have also sought to stress the role of people in the South in both challenging **hegemonic representations** and developing new forms of **discourse** to describe life in the South and its possible futures. This focus on the **agency** of people and institutions in the South is another theme which runs throughout the book, contrasting with the more common identification of the South as a victim of **globalization**.

Review questions/activities

1   Using newspapers and magazines, compare and contrast the
    representations of people and places in the Global South. What are the
    common tropes used in such publications? What are the potential impacts
    of such representations? How and why do publications differ in the
    representations they deploy?

2   How useful is the Human Development Index in measuring levels of
    development? Devise an alternative composite development measure
    including details of what indicators you would use.

3   Using two or three official government or trade promotion board websites
    from the Global South, discuss the modes of representation mobilized.
    Which audiences are the websites targeting? Who is excluded and included
    from representation on the website?

SUGGESTED READINGS

Blaut, J. M. (1993) *The Colonizer's Model of the World*, London: The Guilford Press.
    An engaging and strongly argued book which challenges Eurocentric
    constructions of progress and modernization.

Power, M. (2003) *Rethinking Development Geographies*, London: Routledge.
    Excellent discussion of the ways in which 'development' has been used as a
    form of imagining the world and the implications for such representations.

Said, E. (1978) *Orientalism: Western Conceptions of the Orient*, Harmondsworth:
    Penguin.
    A key text in the analysis of Northern representations of the Global South.

WEBSITES

www.caricom.org Caribbean Community website

www.expomuseum.com The World's Fair Museum
    Provides information about past and future world's fairs.

www.imaging-famine.org Imaging Famine research project
    Outlines the ways in which famine has been portrayed in the media from the
    nineteenth century to the present.

www.sacu.int South African Customs Union website

www.undp.org United Nations Development Programme homepage
    Follow links for *Human Development Report* and HDR data.

www.un.org/millenniumgoals/ Millennium Development Goals (MDGs)

www.worldbank.org World Bank homepage
    Follow links for *World Development Report*.

PART TWO

The South in a global world

Taulli, Peru
Source: Katie Willis.

In this part, we look at the Global South's position within changing political structures (Chapter 3), economic structures (Chapter 4) and complex processes of social and cultural change (Chapter 5) at the global scale. As a whole, the part addresses the question of *how has the Global South emerged in its current form?* In doing so, we aim to outline an account of some of the macro-level processes that have played a major role in shaping the South as we see it today, but in so doing we want to develop further the ideas about **globalization** that we introduced in Chapter 1.

The first of these was that **globalization** has to be understood as a long-standing historical phenomenon, rather than something that simply emerged in the late twentieth century as a result of technological and other changes. As a result, we begin our accounts of political and economic processes in the Global South almost 500 years ago, to place more recent changes in a wider historical perspective. When looking at civil wars and other **complex political emergencies** in the South (Chapter 3), international divisions of labour and patterns of trade (Chapter 4) or contemporary geographies of health and well-being (Chapter 5) it is important to recognize that the patterns we see today have deep historical roots.

The second argument is that **globalization** is not the same thing as growing global uniformity. Some changes over the modern era – such as a global political system organized on the basis of **nation-states** (Chapter 3), growing interconnection of places through global markets (Chapter 4) or increasing **urbanization** (Chapter 5) – could be argued to be inspired and led by the experience of the Global North. But to recognize the importance of these changes is not the same as claiming that the world is becoming a more homogeneous place; it is not for two reasons. First, these 'Northern-led' global changes are themselves actively producing difference in the contemporary world, and in some instances profound inequality. Without wishing to reproduce **representations** of the Global South as a place of **poverty** criticized in Chapter 2, one important task of this part is to recognize the effects of **globalization** in actively creating imbalances of **power** and resources. Second, other changes emerging from the Global South can counter processes of 'Westernization' – as the discussion of social changes produced through religion and migration/**diaspora** in the South (Chapter 5) clearly shows.

Importantly, this brings us to our third argument about **globalization**: that it is not a monolithic or irresistible process. In this part, we have a difficult balance to strike. We want to outline some important political, economic, social and cultural changes that have been important for many people in the South. But at the same time, we do not want to leave the reader with the impression that the experiences of people and places in the South today are simply the product of a few all-powerful macro-level trends that are being experienced across the Global South as a whole. History and contemporary global power structures may frame the experiences of many in the Global South, but they do not completely remove their **agency** in reshaping processes of **globalization**.

Inevitably, this space for **agency** is highlighted more as we move to the micro-scale and the geographies of people's everyday existence in Part 3, Living in the South. But although this part may be more 'big picture' than that which follows it, Doreen Massey's idea of a progressive sense of place (see Chapter 1) is still important. The focus in this part may be on the macro-level, but the processes of **decolonization** and emerging **global governance**, economic development and market integration, and social and cultural change we describe are important in setting the context in which the character of places in the Global South is being continually reworked. The making of places is, however, always a process in which 'the local' matters alongside 'the global': as the various case studies in this part emphasize, both are important in producing the diversity that exists within the contemporary Global South.

3 The South in a changing world order

INTRODUCTION

This Chapter investigates the central theme of this section, *How has the Global South emerged in its current form?*, by looking at the ways in which the South has been constitutive of a changing world political order. The first part of the Chapter traces the processes that have led to the organization of the Global South in to a system of **nation-states**. Today, countries of the South are part of a global international community, making up over two-thirds of the 192 states taking their places within the United Nations, but in the heyday of European **colonialism** less than a century ago, a world of sovereign Southern states was hardly imaginable. To understand this change, we review some of the major political transformations that the South has undergone over the last 500 years, in particular, looking at processes of colonization, **decolonization** and nation-building. A territorial grid of **nation-states** is often thought of as a 'natural' part of the political order, and is deeply imbedded in today's international structures, but we argue that it does not always or easily match up with the complex cultural and political divisions of the Global South. As a result, many parts of the Global South are still living with the effects of their colonial histories: who or what is included within a country's national **identity** often remains a particularly controversial issue and the legacy of **decolonization** can be a contributing factor in **complex political emergencies** that threaten the very survival of states and their populations today.

The second part of the Chapter looks at the changing position of states of the Global South within the international political system. Although the majority of the South has been free from foreign rule since the 1960s or earlier, its countries have had to position themselves within a world of superpower rivalry and evolving institutions of **global governance**. Especially for the Global South's smaller nations, there can be a gap between formal political independence and genuine self-determination. Whilst some countries of the South have become important regional – or even global – players in their own right, questions remain about whether the official channels of today's systems of **global governance** are sufficient for others to make their voices heard. By looking at the United Nations and international environmental negotiations, we show that expectations of formal equality between sovereign nations are often overridden in practice by differences in power.

FORMING STATES AND NATIONS IN THE GLOBAL SOUTH

Any world political map of today (such as Figure 1.1) appears to be neatly ordered: the surface of the Earth is divided by clearly demarcated boundaries between countries. Of course, this picture hides sites of great *dis*order and conflict, many of which are found in the Global South. Borders and boundaries appear as static and fixed, whereas many have moved over history, and those that exist today can be highly controversial, such as the India–Pakistan border in Kashmir. Stability within these borders can also be illusory, as we will make clear through the example of Sudan's civil wars (see Box 3.4). But although it is a simplification of a more complex reality, this world map represents a view of how political space *should* be ordered: it is a world of internationally recognized states that command **sovereignty** over clearly defined territories. In theory (although here reality often differs greatly from the ideal) it is also a world in which many people have a sense of loyalty, belonging and identification with these political units – they see themselves as Kenyans, Brazilians or Malaysians – such that states' boundaries and national identities match each other neatly.

This way of ordering and viewing the world is hard-wired in to many global institutions and practices today. Sovereign states are represented at the United Nations, which treats them as formal equals (with a few important exceptions – see Box 3.7). Citizens generally accept that their movements across international borders will be governed by passports, visas and checkpoints (Plate 3.1); and links between **identity** and nationality are performed through a range of activities – from Independence Day celebrations to international sporting events – that aim to reproduce a sense of belonging to a nation. Even when conflict arises and a state's **sovereignty** or territory is contested, the fundamental ideals of the system are often defended by all of those involved. For example, whilst the deployment of UK, US and other military personnel in southern Afghanistan in 2006 was justified as re-establishing the Afghan government's rule in these provinces, those opposing this saw it as a foreign violation of national **sovereignty**, and Taliban fighters as legitimate representatives of the Afghan people (for detailed background, see Gregory, 2004). Thus, underlying ideas about the international political order – that it should be composed of states, **sovereignty**, territory and nations of citizens – can be held in common by those holding radically different viewpoints as to how these should be put in to practice. Given its persistence through both peace and war, it is important to remember that the global network of **national states** is a

Plate 3.1 The Mexican–US border at Tijuana.
Credit: David Simon.

relatively recent creation, and the first part of this Chapter traces its origins, and its implications for the South.

Colonialism and the emergence of the modern state

Five hundred years ago, at the beginning of the 'modern' world, the global political map was completely unlike that of today (Figure 3.1). Clearly, states' names and boundaries were different, but more importantly so was the whole idea of what political authority might be. Complex societies with established patterns of rule were scattered across areas of Europe, Africa, Asia and the Americas, but in all cases the relationships between states, territory and citizenship were looser. The areas under the command of political elites were usually indeterminate, with frontier zones rather than fixed borders (Box 3.1). Within these territories, the state's power was by no means absolute: village- or lineage-based forms of authority would often be more important in most people's day-to-day life than that of a distant ruler. At the same time, however, some states in the South were extremely impressive in their spatial extent and control: the Ming emperors ruled an area that

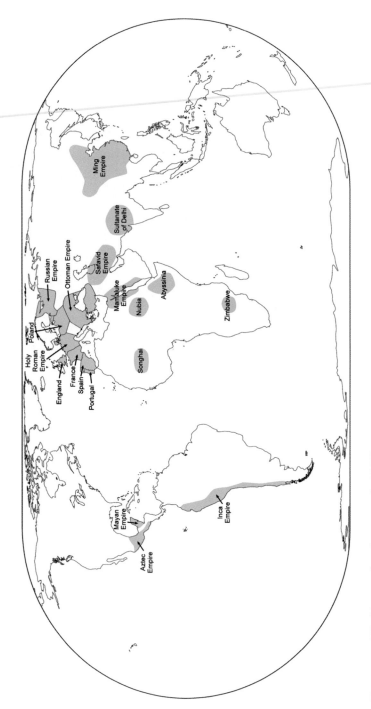

Figure 3.1 World map of major states (c.1500).

Sources: Adapted from McEvedy (1972) and Thomas *et al.* (1994). Map data © Maps in Minutes™ (1996).

BOX 3.1

Early states in West Africa

Alongside the ancient African states of the Upper Nile valley and present-day Ethiopia, West Africa in the middle of the last millennium contained states as complex as those of medieval Europe. Hausaland (northern Nigeria) provides one such example. As a densely settled area with iron-working from the seventh century and irrigated agriculture, it provided the conditions in which complex divisions of labour – and **power** – could emerge. In a period of political and commercial dynamism from the mid-fifteenth century, Hausa rulers built walled cities, traded widely within West Africa and beyond, and adopted Islam. Cavalry forces provided the military power of the state, with the horse-riding heads of great households – the *masu sarauta* – forming a political elite. The maintenance of this cavalry was linked to a trade in slaves, and slaves themselves often became important warriors or office-holding administrators in their own right. The military and economic strength of the Hausa state was impressive, with European merchants in the sixteenth century ranking its capital, Kano, as one of Africa's most important cities. However, its power to intervene in the everyday lives of its peasant citizens was limited. The common people or *talakawa* often practised their own religions, and whilst the rulers could control slaves' labour, they could not directly control that of their peasant citizens. Furthermore with empty lands being available to settle beyond the towns, the limits of the state's territory remained rather elastic: new settlements were founded in the bush during times of plenty, and abandoned when threatened by drought, disease or political instability.

(Sources: Adapted from Iliffe, 1995; Freund, 1998)

encompassed much of China's present-day territory and over 100 million citizens (McEvedy, 1972).

In 1500, the idea of European political dominance over the majority of the South would have seemed fanciful. While Spain and Portugal rapidly established empires in Latin America over the early sixteenth century, destroying indigenous empires such as that of the Aztecs through a combination of military technology and the unintended introduction of smallpox, European traders initially came to Africa and Asia as relatively weak outsiders rather than potential conquerors. If there was a 'global' political system at this time, it was Islamic: the Muslim world extended from present-day Morocco in the northwest (and had included southern Spain until 1492, see Plate 1.4) to Indonesia in the southeast. Its largest political units – such as the Ottoman, Persian and Mughal empires – grew to be globally significant powers in themselves in the early modern era, and trade within this area was very active. Caravans crossed the Sahara to West Africa, and the Indian Ocean linked ports along the East African coast, the Arabian peninsula and South Asia. East of the Muslim empires lay China, which had a highly developed system of administration dating back to the Han Empire (206 BC–AD 221), and importantly trade between China and Europe largely passed through Muslim hands. European engagement with Africa and Asia therefore began from a series of coastal trading posts at the margins of this predominantly Islamic world, where rival European traders struggled to negotiate a foothold with local rulers.

Conflicts and economic change within Europe from the seventeenth century onwards were, however, changing the idea of what states should look like. The Peace of Westphalia (1648), which ended Europe's Thirty Years War, is often credited with being the beginning of a modern political order, as it established the idea of territorial **sovereignty**. This meant that individual states were to be recognized as the ultimate source of authority within their own borders, and to have the right to conduct diplomatic relations with other states beyond. At the same time, Europe began to benefit from increasing economic dynamism, spurred on by growing international trade and innovations in banking (led by the Netherlands), with the result that the wealth and resources at the disposal of their rulers were growing. This wealth enabled new forms of state **power** to emerge that expanded beyond the rather remote forms of rule, such as waging war and holding court, which had been the primary concern of rulers up to this point. Driven in part by the need to raise taxes for wars to defend (or expand) their territory, European states became increasingly interested in transforming the everyday affairs within their own borders, gathering more information on their citizens and aiming to influence or control their economic activities and social norms (in short, they were practising forms of **governmentality**: see concept Box 3.1). By the early eighteenth century, Europe's politics was therefore beginning to look 'modern': states recognized each other through international treaties, there were emerging links between territory, rule and national **identity**, and improving or developing states' populations had become an important part of statecraft alongside diplomacy and warfare.

The growth of European state **power** was, in part, supported by its economic links with the Global South. In the New World, the extraction of precious metals, the slave trade and plantation agriculture were important early mechanisms that transferred wealth to Europe (see Chapter 4), and were accompanied by settler colonization. Spain's American empire was the most extensive, and by the eighteenth century this stretched from Santa Fe in the North to Santiago in the South. The eighteenth century also saw Europe's growing involvement in Asia and Africa, and here again economic and political expansion were closely linked. The Dutch and British East India Companies both started life as commercial organizations, with ambitions to secure direct control of the trade in spices, fine textiles and other luxury goods. As their strength within Asia increased, the Companies (along

CONCEPT BOX 3.1

Governmentality

The French philosopher Michel Foucault used this term to describe the way in which modern states and their citizens interact. It describes the notion that states have a role or duty to actively *govern* their citizens, rather than merely protecting them from outside threats. When a state treats citizens as a *population* that requires its continual intervention, this in turn has important consequences for the form and conduct of the state. Governing effectively now requires the state to know much more about its people – whether about their wealth, education, social habits or beliefs – but also to have the ability to direct or change these aspects of its population. Both parts of governance – knowing and changing a population – in turn usually imply the *expansion* of the state, in terms of its range of functions, and its physical presence, embodied in a growing array of government institutions, buildings and personnel.

with their other European counterparts) began to organize the production of these goods as well as their trade, and this brought them in to the realm of controlling territory and becoming important local political powers in their own right. At this point in history, European expansion was largely driven by economic forces rather than being a preconceived exercise in political control (Box 3.2).

By the late nineteenth century, however, increasing rivalry between European powers was being expressed in attempts to annex territory. The Conference of Berlin (1884–85) was an attempt to 'rationalize' this process within Africa and led to the rapid division of the continent: at the outbreak of World War I, only Abyssinia and Liberia were not under European rule. In Asia, European colonizers also had to contend with the expansionist ambitions of Japan (which gained control of Taiwan (1895), part of Manchuria (1905) and Korea (1910)), and the USA, which claimed various Pacific islands and the Philippines (1898) as well as treating South and Central America as its exclusive 'sphere of influence'. The world political map of 1914 (Figure 3.3) was therefore one where the Northern political domination of the Global South was almost total.

It is impossible to describe the experience of colonial rule in general for the South, as there was so much variation between individual colonized countries, and also between colonizing powers. For example, Britain's primary interest in its colonies was to maintain a political order and administration that would favour its commercial interests, whereas for France, empire was also an explicitly political project of assimilation in which indigenous societies were to be transformed (and 'developed') by taking on French values (as a result, British West African colonies had fewer administrators than their French counterparts, as the British were often

BOX 3.2

The British in India: an 'accidental' empire?

Although India was to become 'The Jewel in the Crown' of its empire, Britain's rule there was never uniform. By the mid-eighteenth century, Britain had used its naval **power** to contain French and Dutch interests and had established itself as the dominant European trading nation in the subcontinent. As the Mughal empire went into decline, the British East India Company increasingly intervened in local politics, using its military strength to support its trading activities. As it played one ruler off against another, it acquired the right to gather taxes over an increasing range of territories. The vast personal fortunes and unscrupulous activities of the Company's agents were, however, leading to public disquiet in Britain and increasing parliamentary regulation of the Company's affairs. In the aftermath of the Indian Rebellion of 1857 (an armed struggle aiming to reassert Mughal rule), the Company was wound up and administration of its Indian possessions passed directly to the British Crown. Even so, 'British India' remained a patchwork of territories, some directly administered by the Governor-General, and others governed by over 500 local rulers. Although the latter had only limited (and sometimes merely ceremonial) **power**, at Independence in 1947 their so-called 'Princely States' accounted for around 40 per cent of India's total land area.

(Sources: Adapted from Wolpert, 2000; Mawdsley, 2002)

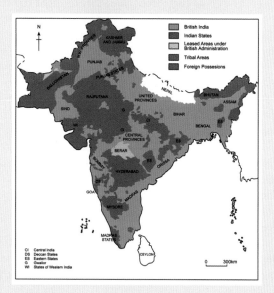

Figure 3.2 Political divisions of British India, 1946.
Source: Adapted from Mawdsley (2002: figure 6.1).

happy to leave aspects of local government affairs in the hands of 'traditional' rulers: Corbridge, 1993). But within all countries, **colonialism** introduced elements of modern statecraft: it redrew (or established anew) firm territorial boundaries, created new political orderings and administrative systems within these borders, and sought to underpin these by a combination of force and persuasion.

Some aspects of state **power** in colonized countries were similar to those we noted above: knowledge-gathering about the population and territory was important, and was conducted through techniques as varied as cartographic surveying, censuses and anthropological studies. This information in turn made the state powerful in controlling and intervening in the lives of its colonial subjects, by gathering taxes efficiently or by deploying its military resources to best effect. These techniques of governance helped relatively small colonial elites to control large colonized populations, for example in British Nigeria in the 1930s there was only one European administrator for every 15,000 inhabitants (Clapham, 1985: 23, cited in Corbridge 1993).

Other aspects were rather different, in particular the attempts to legitimate the relationship between rulers and ruled. In Europe itself, even though democratic rights emerged slowly and unevenly (see Chang, 2002, for details), states sought to justify their actions by claiming to act in the national interest of their populations. For the colonies, external rule was justified instead through ideas of a 'civilizing mission' (see Chapter 2) or a moral duty to improve 'backward' populations, often supported by Christian missionary work (Box 5.6). These claims, themselves often based on overtly racist ideas, were part of the struggle to colonize not only the territory, but also the hearts and minds of people of the South. Frantz Fanon, an anti-colonial scholar writing during Algeria's struggle for independence from France, argued that colonial powers deliberately aimed to instil ideas of European superiority among the people of the South: 'The effect consciously sought by **colonialism** was to drive into the natives' heads the idea that if the settlers were to leave, they would at once fall back into barbarism, degradation and bestiality' (Fanon, 2001 [1961]: 169). Thus while **colonialism** introduced state institutions that were entirely modern, the idea that Southern peoples constituted nations that deserved self-government and their own **sovereignty** was actively denied.

Decolonization and its aftermath in the South

With colonial rule often being experienced in the Global South as a violent and exploitative force, it is no surprise that various forms of **resistance** to this **power** – from tax avoidance to armed uprisings – had taken place throughout this period. Central and South America saw a range of successful movements to gain independence from Spain in the early nineteenth century, even before the 'scramble for Africa' was underway. But for much of the rest of the Global South, **decolonization** came after the end of World War II, beginning with parts of the Middle East (1946), the Philippines (1946) and British India (1947). Fairly swift British and

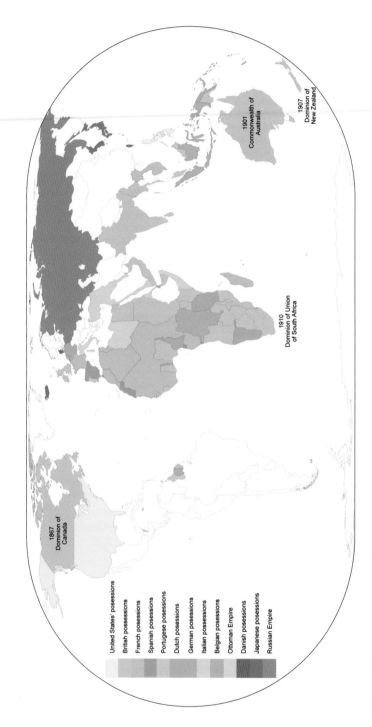

Figure 3.3 The world in 1914.

Source: Adapted from Thomas *et al.* (1994). Map data © Maps in Minutes™ (1996).

United States' posessions
British possessions
French possessions
Spanish possessions
Portugese possessions
Dutch posessions
German posessions
Italian possessions
Belgian posessions
Ottoman Empire
Danish possessions
Japanese posessions
Russian Empire

1867
Dominion of
Canada

1901
Commonwealth of
Australia

1907
Dominion of
New Zealand

1910
Dominion of Union
of South Africa

French withdrawal from Africa and Southeast Asia followed, with the year 1960 seeing a total of 17 countries gaining independence. By the late 1960s, most sizeable colonies were independent: Portugal was fighting losing battles to retain Angola and Mozambique (which both gained independence in 1975); and Zimbabwe did not gain international recognition until majority African rule was established in 1980. With China regaining control of Hong Kong (1997) and Macau (1999), colonial rule in the Global South had all but ended by the close of the twentieth century (Figure 3.4).

For many former colonies, independence was, however, a rather contradictory process: it reasserted local control over the state, but at the same time it often reproduced a framework of imposed colonial institutions and arbitrary colonial borders. One vitally important question for those in charge of the newly independent countries was how could they forge new *nations* around the remnants of colonial *states*. Nation-building, a process that had usually been suppressed or undermined under **colonialism**, was therefore an essential task of the new generation of Southern political leaders if their rule was to have greater legitimacy than the European governments they replaced. Direct involvement in the anti-colonial struggle often gave these leaders a degree of popular support, but along with this came great expectations that their governments would also be able to transform day-to-day living conditions for their people. It is perhaps unsurprising that many Southern leaders attempted to sustain their popularity by making 'modernization' a national crusade, with regimes as varied as Julius Nyerere's Tanzania (Box 6.4); Jawaharlal Nehru's India or Park Chung-hee's South Korea (Box 9.4) all trying to create **developmental states** that would deliver their visions of a better future (see Chapter 9). Whatever the political ideologies of the governments involved, implementing such visions of national **modernization** was not easy. Attempts to deliver rapid change needed firm control of national resources, and required citizens to accept short-term or local sacrifices 'for the greater common good' of development; in practice this meant that modernizing regimes were often under pressure to leave in place more authoritarian elements of the colonial state.

Alongside delivering economic development, governments can also bolster their popular support through appeals to cultural unity. Many states of the Global South have, however, found themselves hampered in this as their colonial-imposed borders encompass a population with a variety of ethnic and linguistic divisions. Symbols of nationhood have to be carefully chosen if they are not to be divisive, and when a government's **power** is threatened the temptation to pander to the **identity** of numerically dominant cultural or ethnic groups is very strong, especially within competitive democratic systems. Considering these constraints, India has been remarkably successful in keeping some of these divisions in check (Box 3.3), but in neighbouring Sri Lanka the Sinhala-Buddhist ruling elite has allowed the social and political marginalization of Tamils – many of whom had been brought to Sri Lanka from India as plantation workers by the British – and this has led to ongoing violence and periodic civil war since 1983.

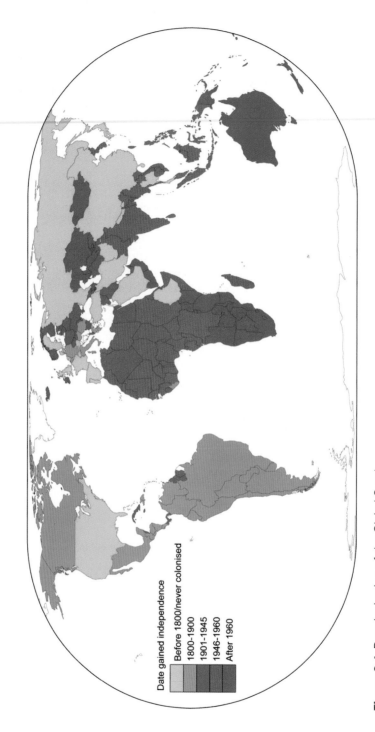

Figure 3.4 Decolonization of the Global South.

Source: Data from Thomas *et al.* (1994). Map data © Maps in Minutes™ (1996).

BOX 3.3

Nation-building in India

In 1947, India gained Independence and was partitioned from Pakistan in a hurried and traumatic process: there was communal violence and massive displacement of refugees, and the British left before the boundaries of the new states had even been publicly declared. India's Constitution has attempted to unify this vast, multi-ethnic and multi-faith country through a combination of *secular politics* (which meant that the government treated religion as a matter for private citizens, and publicly vowed to respect the rights of all religions) and by allowing a degree of regional autonomy within India's *federal system* (where language is important in the demarcation of regional governments, e.g. Maharashtra is a predominantly Marathi-speaking region, and the language is used in schools and government offices there). At the same time, an explicitly *national* **identity** is reinforced through central government and its key institutions (such as the federal parliament in Delhi), and by cultural events that vary from Independence Day ceremonies to India's international cricket fixtures.

At Independence the Congress Party, which had spearheaded the anti-colonial struggle, enjoyed massive political support. Until the 1960s it was largely successful in mediating regional tensions and ideological differences, and was itself a powerful symbol of national unity. Over time, however, its **power** to unify the electorate has declined, and India today has a complex array of local and national political parties mobilizing different caste-, religious- and regional-based loyalties.

It is a significant achievement for such a diverse country that these political differences have been accommodated for over six decades within a system of parliamentary democracy. But the question of who belongs within India remains, in some ways, as divisive as it was in 1947. From the 1990s, there has been a resurgence of explicitly Hindu nationalism, and despite Constitutional guarantees, India's 100 million Muslims have often been made to feel like second-class citizens. Occasional but vicious Hindu–Muslim violence, such as the 2002 riots in Gujarat, and ongoing armed secessionist movements in northeast India are indicative that any search for a unifying 'Indian' **identity** will inevitably leave some groups marginalized and resentful.

(Sources: Adapted from Corbridge and Harriss, 2002; Kohli, 2001)

Warfare and violent challenges to the **power** of the modern state are subjects often left out of development textbooks, perhaps for fear of reinforcing the negative stereotypes of the Global South we noted in Chapter 2, but they are worthy of mention here as political violence profoundly shapes the lives of many millions of people living there (Unwin, 2002). As Figure 3.5 shows, the majority of the battle-related deaths today are located in the Global South, and although international media attention in the first decade of this century has been focused on the 'War on Terror' and its outcomes in the Middle East, there are many other areas where the contested processes of nation-building and state formation have resulted in equally damaging conflict. Arguably, this has been most devastating in Africa, where colonial rule divided the continent rapidly and arbitrarily, and **decolonization** similarly produced a series of 'nations' with little to hold together their culturally diverse populations. This colonial inheritance can become politically volatile when rival factions are willing to exploit ethnic differences to gain **power**, and can be made more so when control of a high-value resource (such as diamonds or oil) is contested or even directly funds armed struggles for the control of territory. Sudan (Box 3.4) is one of a number of states (others include the Democratic Republic of Congo and Liberia) that exemplify these difficulties.

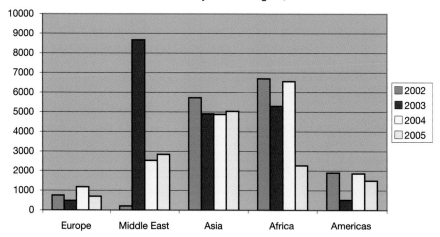

Figure 3.5 Global battle-deaths, 2002–2005.

Source: Data from the Uppsala Conflict Data Program of currently active armed conflicts (UCDP 2008).

Notes: With the exception of the 2003 international invasion of Iraq, all conflicts were categorized as intrastate conflicts rather than direct conflicts between two or more warring governments. Note that all European conflicts recorded in this period were in the former Soviet Union.

BOX 3.4

Sudan's civil wars

Sudan, Africa's largest state, was created in 1956 from what were effectively two separate British colonies, a predominantly Muslim and Arab North, and a black African South, largely with Christian or indigenous African beliefs (Figure 3.6). Arab groups have dominated the Khartoum government, but have faced near-continuous armed **resistance**, leading to over 40 years of civil war. Although the most recent period of North–South conflict (1983–2003) has been ended through a power-sharing agreement, deep cultural and ideological divisions remain between the combatants, the northern Government of Sudan and the southern Sudanese People's Liberation Movement (SPLM). Not only this, but other groups such as African Muslims in the west were marginalized from the peace process (Young, 2005) and this helped to ignite rebellion, followed by repression and a massive humanitarian crisis in Darfur (fighting started in 2003, and is ongoing at the time of writing in June 2008). The internal political dynamics of Sudan cannot be separated from international concerns:

the Sudanese and Ugandan governments have at various points fought a war by proxy by supporting rebel groups within each other's territories (Prunier, 2004), and other neighbouring states have supplied the SPLM with finance and weapons. The Government of Sudan has in part funded the conflict by allowing foreign companies (including Canadian, Malaysian and Chinese) to exploit its oil reserves (Ojaba, *et al.*, 2002).

The human costs of the conflict have been appalling. The Uppsala Conflict Data Program estimates two million deaths from the 1983–2003

Figure 3.6 Civil war in Sudan.
Source: Adapted from Ojaba *et al.* (2002: 671).

BOX 3.4 (*CONTINUED*)

BOX 3.4 (*CONTINUED*)

war, plus four million people internally displaced and another 420,000 international refugees. The first two years of conflict in Darfur claimed over 180,000 civilian lives (Patrick, 2005), and both conflicts have witnessed targeting of civilian populations as a deliberate military strategy (Plate 3.2) – in direct violation of humanitarian law. There has therefore been a pressing need for humanitarian assistance, and Operation Lifeline Sudan (OLS) was a UN initiative to provide humanitarian relief to some 2.6 million civilians caught on both sides of the 1983–2003 conflict. Access had to be negotiated via Government of Sudan and SPLM military leaders, and as Elizabeth Ojaba, *et al.* (2002) report, the OLS's impacts were mixed. Whilst vital food and other **aid** reached many people, and the presence of OLS staff as international observers may have mitigated some aspects of the violence, both warring parties used OLS and its resources to provide breathing space between offensives, or to retain the support of 'their' civilians.

(Sources: Adapted from Ojaba, *et al.*, 2002; Prunier, 2004; Patrick, 2005; Young, 2005)

Plate 3.2 The civilian costs of civil war: a burned-out village, Darfur, Sudan. Credit: © Still Pictures.

Sudan's civil wars show that **complex political emergencies** are not simply the internal product of corrupt regimes or 'bad **governance**': they have intimate links to both colonial histories and wider current geopolitical rivalries, and international humanitarian assistance does not offer any easy solutions. Here, the 'building blocks' that provide political order for much of the Global North – states with clearly demarcated territories, secure control over armed forces and that are seen as legitimate sources of authority by their citizens – are unstable and as such the very viability of Sudan as a political entity is open to question.

The political map of today, with its clear borders and independent nations, is therefore no more 'natural' or 'normal' than the political world of 1500. The South today has modern states, often capable of great intervention in their citizens' lives, but for most the form of the state has been shaped in part by a history of colonial intervention and local **resistance**. All forms of government – North and South – are held in place through a mixture of force and popular support, but for many Southern countries the struggle to establish legitimate rule has occurred in a context where state-building was imposed from the outside, and nation-building both came later and was conducted in reaction to colonial control.

CONCEPT BOX 3.2

Complex (political) emergencies

A term coined by the UN to describe humanitarian crises that are both multi-causal (a breakdown of political, economic and social order – usually involving the complete collapse of normal state functions) and that require multiple forms of response (disaster relief operations, military intervention and diplomatic solutions). They are primarily 'man-made' disasters, with civil war being an important constituent part of most, but can be compounded by 'natural' calamities such as drought or famine conditions. Their victims are largely civilians, with poor and marginalized groups often suffering disproportionately, and as such they present strong grounds for some form of international intervention (whether by government agencies, the UN or the non-governmental organization (NGO) community), However, the very nature of these emergencies makes this intervention difficult: when the state is in such acute crisis, there is a danger that even emergency food **aid** can be used to fuel, rather than defuse, conflict.

BOX 3.5

Superpower politics: from Cold War to American dominance?

Following the collapse of the USSR, the USA is often described as the world's one remaining 'superpower', but how can its relative strength be measured? Traditionally, geopolitical analysis has examined a number of factors that contribute to the relative strength of states: the size of their territories and the wealth of their economies are important alongside the size of their military resources, as all demonstrate their ability to influence events beyond their borders. Table 3.1 shows some of these variables – with oil consumption and electricity production added to give a rough measure of countries' industrial capacity.

The USA has the largest military expenditure – far outstripping that of all the other countries listed – and retains one of the largest stockpiles of nuclear weapons, but in other respects the picture is much more mixed. The combined military and economic strength of the EU is significant, but these countries do not act as a single unit in international affairs, as the serious rifts over the 2003 invasion of Iraq showed. The fact that two of its member states – Britain and France – have permanent seats on the UN's Security Council (see Box 3.7) is more a reflection of their past **power** rather than their current importance. Russia retains a huge nuclear arsenal, but in all other respects has been eclipsed by China. China has perhaps the clearest claim to 'superpower' status today: it has great conventional and nuclear military capability, its economy already influences global market trends, and this is being matched with a growing diplomatic role (Box 3.6). India's massive size, rapidly developing economy and sophisticated military technology mean that it too will be of growing significance in the future. But tables such as these often hide as much as they reveal: various other Southern states, such as South Africa or Indonesia, are important regional powers in their own right, and although North Korea and Iran were labelled by the Bush administration as 'rogue states', the USA is understandably wary of taking military action against either. In short, although America may have 'won' the **Cold War**, its ongoing dominance over events in the twenty-first century is far less certain.

BOX 3.5 (CONTINUED)

Table 3.1 Superpower status

	Global North				Global South		
	USA	Russia	EU	Japan	China	India	Brazil
Land area (1,000 km²)	9,631	17,075	3,976	378	9,597	3,288	8,512
Population (millions), 2006[1]	299	142	457	128	1,312	1,110	189
GDP (US$bn ppp), 2006[1]	13,233	1,656	9,876[4]	4,229	10,153	4,217	1,661
GDP per capita (US$ ppp), 2006[1]	44,260	11,630	31,446[4]	33,150	7,740	3,800	8,800
Electricity production (million mW)[2]	4,062	1000	3,020	1,025	3,256	662	396
Oil consumption (million bbl/day)[2]	20.8	2.9	14.5	5.3	6.9	2.4	2.1
Military expenditure (US$bn), 2007[3]	547	35	>200[a]	44	58	24	15
Nuclear warheads, 2007[3]	5,045	5614	508	0	145	??[b]	0

Sources:[1] World Bank, 2007b;[2] CIA, 2008 (estimates are for various dates 2003–07);[3] Stockholm International Peace Research Institute (SIPRI), 2008;[4] IMF, 2008a.

Notes:

a A composite figure for the EU expenditure is not given within the SIPRI database. The UK (US$59.7bn) and France (US$53.6bn) had the largest military expenditure within the EU in 2007, giving them the second and fourth largest military budgets in the world.

b Estimates for the number of Indian nuclear warheads are not given within the SIPRI database: other sources estimate at least 50 warheads (Norris and Kristensen, 2005).

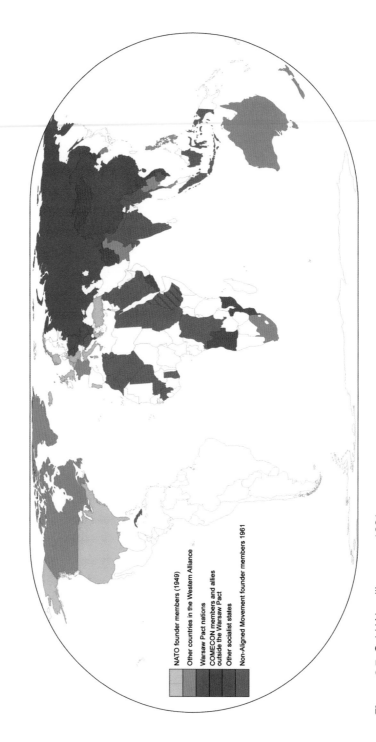

Figure 3.7 Cold War alliances, c.1981.
Source: Map data © Maps in Minutes™ (1996).

NATO founder members (1949)

Other countries in the Western Alliance

Warsaw Pact nations

COMECON members and allies
outside the Warsaw Pact

Other socialist states

Non-Aligned Movement founder members 1961

A CHANGING WORLD ORDER: FROM EMPIRES TO GLOBAL GOVERNANCE?

The break-up of European empires and formal political independence of the Global South has itself contributed to the *impression* of an established political order with which we began this chapter. In practice, as we have argued, stabilizing rule and maintaining order within their own borders has been a pressing task for the leaders of Southern nations. Equally important has been the place of their countries within an evolving, but far from equal, international order. In this section, we look at two related questions: how have Southern nations positioned themselves relative to other (and especially more powerful) countries, and how are they placed within an emerging system of **global governance**?

The South and global superpowers

The first question has traditionally been addressed in international relations and political geography through realist **geopolitics**. This perspective sees the world made up of a set of rival states, the relative strength of which is measured by their economic and military capacity, each pursuing foreign policies (via diplomacy, threats or warfare) to secure their national interests against those of their competitors. In this potentially volatile system, stability is provided by predictable (although not necessarily cordial) relationships between major powers. Militarily weaker states, including much of the Global South, are left to seek alliances with the major powers and/or risk becoming pawns in any conflicts between them.

According to this viewpoint, the world has undergone three major geopolitical orderings since the late nineteenth century (Ó Tuathail, 2002). The first was the imperial system of rival 'Great Powers' that was described in the previous section (Figure 3.3). As these empires broke up after World War II, a second international political order was structured around the rivalry between the USA and the USSR, each superpower being backed by massive nuclear and conventional military capabilities. At the height of the **Cold War**, much of the Global South was drawn in to the superpower rivalry between the USA and the USSR, either through formal alliances (Figure 3.7) or military/financial dependency. For almost half a century, containing the spread of **Communism** was the dominant driver of American foreign policy, with the USA fearing that any Southern state turning to **Communism** could cause a 'domino effect', toppling neighbouring regimes and bringing them under the USSR's influence (Short, 1982). Similarly for the USSR, fear of USA-led aggression shaped its geo-strategic aims: the attempt to secure its position through better access to the Indian Ocean and West Asia's oil fields led to its ill-fated invasion of Afghanistan in 1979.

After the fall of the Berlin Wall in 1989 and the subsequent collapse of the Soviet Union in 1991, there has been no direct rival to the USA's status as the sole surviving superpower (see Box 3.5), and some have argued that this represents a new era of empire (Hardt and Negri, 2000; Harvey, 2005). In this period, the 'peace dividend' hoped for at the end of the **Cold War** has not materialized. The USA's military spending remains high: in 2004 this was around US$400bn, over 20 times its budget for development aid. Armed conflict is concentrated in the South, and has remained a continuous part of the new world (dis)order.

Clearly, this history of superpower succession has been important in shaping the role Southern countries play on the world stage. In some cases, independence struggles in the Global South were themselves supported by global superpowers: for example, in both Angola and Mozambique, Soviet assistance was important in expelling the Portuguese, and the USSR and USA both backed rival forces in the countries' subsequent civil wars. Southern rulers and regimes have frequently been decisively eclipsed by superpower intervention, whether through economic sanctions, diplomatic pressure or outright military force (the USA used a combination of all three to undermine Nicaragua's left-wing Sandinista government throughout the 1980s). But despite the massive military–economic **power** imbalances that are highlighted in these encounters, it is important that Southern states are not treated as 'bit part' actors in someone else's story. Looking critically at realist **geopolitics** accounts of the 'changing world order', three observations suggest that countries of the South have an important role to play.

First, even if realist **geopolitics** 'objective' measures of **power** are accepted at face value, the strength of some Southern nations is becoming more significant. China presently offers a *global* challenge to the USA, but a range of other states – including India, South Africa and Brazil – have a growing regional presence. Second, although such measures are good *general* indicators of a state's **power**, they are rather poor at predicting a state's ability to translate that **power** into effective rule in any particular situation. America's involvement in the Vietnam war (1956–73) and the Soviet Union's invasion of Afghanistan (1979–89) were both pivotal events in **Cold War geopolitics**, and in both cases seemingly weaker Southern regimes emerged victorious.

Third, Southern nations themselves have always recognized **power** imbalances in the geopolitical order, and have made moves to counteract this. Important here are the **G77** and the **Non-Aligned Movement (NAM)**. Emerging from the Bandung Conference (1955) and formalized in 1961, the **Non-Aligned Movement** is an international organization of over 100 states – primarily the ex-colonies of Africa and Asia, but also including several Latin American countries – which aims to ensure the self-determination of its members. India's Jawaharlal Nehru and Indonesia's President Sukarno were among the key founders of the movement, and wished to ensure that newly independent countries were not immediately made subservient to superpowers' political ambitions. The **G77** was formed when 77 developing countries signed a joint declaration at the end of the first session of the United Nation's Conference on Trade and Development (UNCTAD) in 1964. It has subsequently expanded to 134 members (although its original name remains), and has become more a formalized negotiating caucus of Southern nations. It holds annual ministerial meetings and has a permanent institutional structure of liaison offices that enable it to represent its members' interests within core bodies of the UN, the International Monetary Fund (IMF) and World Bank. China has been an additional 'Associate Member' from the outset, and plays an influential role within the group. China's rising economic **power** also means that it is beginning to play a greater independent role on the world political stage (Box 3.6), and for the first time in almost two decades it offers the possibility of a geopolitical counterweight to the USA.

The terrorist attacks of September 11, 2001, and the resulting 'War on Terror' reveal a more fundamental criticism of realist **geopolitics**. Neither al-Qaeda, nor the international alliance to oppose it, were simply aiming to change the borders or government of a particular state, and as a result this 'war' fits rather poorly within conventional definitions of armed conflict. The involvement of the UN, the subsequent global movement against the US-led invasion of Iraq, and indeed the stateless nature of international terrorism itself are all reminders that states are not the only important actors in the current geopolitical system. To gain a richer understanding of today's world order, it is important to look not only at the actions of states themselves, but also at the international systems and ideas that link them together. We therefore turn now to the development of this broader pattern of rule in the postwar era.

The South and global governance

As Chapter 4 will describe in further detail, the **Bretton Woods** Conference (1944) laid the foundations of the key institutions that shape global economic transactions today: the World Bank, the International Monetary Fund (IMF) and the General Agreement on Tariffs and Trade (GATT) which was superseded by the World Trade Organization (WTO) in 1995. The end of World War II also saw the formation of the United Nations 'family' of organizations (Box 3.7), and between them these institutions can be argued to represent part of a emerging system of **global governance** (see Concept Box 3.3) a pattern of rule in which **nation-states**, both large and small, to some degree share **sovereignty** with supra-national bodies. As Box 3.7 shows, these institutions have grown beyond a concern with inter-state security alone and encompass a range of day-to-day activities of government.

In practice, the UN does not act as a final and unchallenged source of authority: individual states can and do ignore its General Assembly or Security Council, and there are many international organizations that fall outside its control (importantly, the world's richest nations have disproportionate **power** within the **Bretton Woods** institutions, and the WTO is completely independent of the UN). So although the UN may be as close as the world gets to a source of 'democratic' international decision making, albeit with serious **power** imbalances in its Security Council membership, **global governance** is better thought of as an evolving system of institutions, agreements and actors of which the UN is only a part. Within this

BOX 3.6

China in Africa: neo-colonialism or South–South cooperation?

China has long had diplomatic links with various African nations: it played an important role in the Bandung Conference, and also supported anti-colonial liberation struggles in Algeria and elsewhere. It has a long history of low-interest **aid**, technical support and infrastructure development in Africa – the most dramatic symbol of which was the Tanzania–Zambia railway, built by 50,000 Chinese labourers. Under Mao's leadership (1949–76), the focus of these efforts was largely political – with **aid** intended to spread the ideals of Chinese socialism, and also to gain the People's Republic of China international diplomatic recognition.

This history is often given barely a footnote in accounts of international development: by contrast China's current economic and political engagement with Africa is a subject of great policy and diplomatic concern in the West. Trade between China and African nations is growing exponentially (from US$3bn in 1995 to US$55bn in 2006), and African raw materials – including iron, platinum and oil – are helping to sustain China's rapid economic growth. China's bargaining to access these resources has led to it being portrayed as a supporter of corrupt regimes (over 60 per cent of Sudan's oil production goes to China), and its refusal to consider the human rights records of its trading partners has led to accusations that it is undermining a post-Washington consensus of making international **aid** conditional on countries meeting 'good **governance**' criteria (see Chapter 9).

For many Africans, however, trade with China seems to offer alternative development opportunities. China promises continued South–South support within international institutions such as the UN and WTO, respect for the **sovereignty** of African countries and non-interference in their internal affairs, low- or no-interest loans for much needed infrastructure projects, and some technological transfer within joint industrial projects. It would be naive to believe that China's increasing interests in Africa are altruistic, and the ability of these links to deliver Africa sustained industrial growth in exchange for its mineral and other resources is as yet unproven. Given the history of **colonialism**, it is however equally unrealistic for Western diplomats to expect that their concerns about Chinese 'exploitation' of the continent's resources will be treated as anything other than hypocrisy by African nations themselves.

(Sources: Adapted from Tull, 2006; Mawdsley, 2007; Sautman and Hairong, 2007)

BOX 3.7

The structure of the United Nations

In 1944, delegates from China, the UK, USA and USSR met in Washington DC and drew up what became the draft United Nations Charter. The UN formally came in to being on 24 October 1945 with 51 member states. Since then its membership has grown to 192 states, Montenegro being the most recent addition in June 2006.The UN currently has five principal organs:

- *The General Assembly* All member states of the United Nations have a seat (and one vote) in the General Assembly, and these member states represent the vast majority of the world's countries (Western Sahara and Palestine are significant exceptions). The General Assembly is the UN's main policy making and representative body – and provides a space for multilateral discussion of a wide range of international issues.
- *The Security Council* is comprised of 5 permanent members (China, the USA, Russia, the UK and France) and 10 non-permanent members, each of which is elected from the General Assembly to serve a two-year term. This is the body that deals directly with security issues, and can request UN members to undertake military action within 'peacekeeping' operations.
- *The Secretariat* has about 8,900 staff, and acts as the UN's civil service. The Secretary-General (Ban Ki Moon in June 2008) is the chief administrative officer of the UN, and plays an important diplomatic role in her/his own right.
- *The Economic and Social Council* has members elected by the General Assembly: it is responsible for a range of issues relating to well-being, including development and human rights, and acts to coordinate the work of the UN's specialized agencies in these fields.
- *The International Court of Justice*, located in The Hague, has 15 judges and acts to resolve disputes between states (it does not try individuals).

Beyond these core bodies are 14 more specialized institutions affiliated to the UN (such as the World Health Organization and the International Labour Organization) and 11 UN funds and programmes, some of which (such as the United Nations Development Programme (UNDP), and the United Nations High Commissioner for Refugees (UNHCR)) are important international actors in themselves. This amounts to a formidable array of institutions,

BOX 3.7 (*CONTINUED*)

BOX 3.7 (*CONTINUED*)

but the world's key instruments of international financial governance are not under the UN's control. The World Bank and International Monetary Fund (IMF) are theoretically answerable to the UN, but in practice they act entirely independently of it.

(Sources: Adapted from Roberts, 2002; United Nations, 2004, 2008)

---

system, there are thousands of different international 'regimes', or sets of 'norms, rules, and decision-making procedures that states (and sometimes other powerful actors) have created to govern international life' (Murphy, 2000: 793). These seek to control the conduct of individual governments, businesses and citizens on subjects that vary from environmental pollution to international migration.

Even though **nation-states** are not the sole actors within these regimes, they remain important in their success or failure. International institutions and agreements can be designed by (and in the self-interest of) particular **nation-states**, and states can themselves choose to ignore or contradict the agreements they have created. The USA is particularly notable in this regard. Despite portraying itself as a global champion of human rights, the US Senate (along with India and China) has not yet ratified the treaty that makes it subject to the International Criminal Court (established in 2002 as a permanent court to prosecute war criminals), and has even cut **aid** to countries such as Brazil that support it (Roberts, 2002). The USA has also used the WTO to attack countries resisting the import of genetically modified crops; its claim that there should be 'free trade' in these products sits uncomfortably next to its own vast agricultural subsidies that undermine the economic fortunes of many farmers across the South (Roberts, 2002). Rather than seeing the **power** of states simply in terms of their military capacity (highlighted in Table 3.1), it is also important to think about the other ways in which influence is expressed – and setting the rules and agendas of international regimes is a key way in which this is achieved.

Despite being under-represented within some of the key institutions of **global governance**, countries of the Global South can and do make their voices heard. Organizations such as the **Non-Aligned Movement** and the **Group of 77** are important in allowing them to voice their interests within and beyond the UN's institutions. Particularly for small and poor nations, they can mitigate some of the costs of participation in complex negotiations, and allow them a degree of collective influence they would not be able to achieve individually. The significance of this can be seen in the Global South's participation in environmental governance regimes. Unity in international environmental negotiations is in some ways unexpected – after all the individual interests of the South's oil exporters differ greatly

CONCEPT BOX 3.3

Global governance

This refers to the organizations (and the relationships between them – often based on partnership rather than hierarchy) that are responsible for holding in place rules of conduct at the international level. Govern*ance* itself is a term that is deliberately broader than govern*ment*: as well as states, private companies, NGOs, or even pressure groups and social movements can be important actors in setting and enforcing these rules. *Globalizing* patterns of governance is inherently contentious: it attempts to make one set of rules have universal reach, and thus challenges the **power** and legitimacy of alternative forms of authority. For example, when the World Trade Organization (WTO) votes to eliminate trade barriers in manufactured goods, this undermines the ability of Southern governments to develop their industrial sector using protectionist measures – and thus threatens their **sovereignty** over domestic economic policy.

from those of its small island states most at risk from global warming – but countries of the South have had some success in establishing an agenda of **sustainable development** (Concept Box 3.4) that has shifted the overall terms of environmental debate (Box 3.8).

The devil, as ever, is in the detail of the individual mechanisms designed to implement this agenda. The Global Environmental Facility (GEF) is the key body that provides the South with finance to take action on climate change and other environmental protection measures. Crucially, control of the GEF and its substantial resources (US$3bn in 2006) lies not with the UN's Environment Programme (UNEP) but with the World Bank (Andresen, 2007) – bringing it more firmly under the control of Northern countries and **neoliberal** ideas (see Chapter 4). As a result, debt-for-nature swaps and other mechanisms can still place the South in a position whereby they are being bought off – and ceding some elements of **sovereignty** – through the call for 'global' environmental action.

Important questions therefore remain about the impact of a changing world order on the Global South. At face value, the shift from colonial or superpower domination to emerging patterns of **global governance** is one that should allow the countries of the South greater scope – both individually and collectively – to represent their interests internationally. However, constant battles need to be fought to ensure that the structures and values of **global governance** regimes fairly represent the interests of the Global South, and these are often battles fought between vastly unequal opponents.

CONCEPT BOX 3.4

Sustainable development

Sustainable development has most famously been defined as 'Development that meets the needs of the present without compromising the ability of future generations to meet their own needs' (World Commission on Environment and Development, 1987: 43). This may seem like a laudable goal, but using it as a guide to action is fraught with difficulty. For example, in the absence of perfect knowledge about the environmental consequences of a decision, how can it be known whether it will compromise future generations? Equally, how can 'future generations' be fairly represented within current decision-making processes? Despite such problems, the widespread acceptance of sustainable development as a broad aim has marked an important shift within global environmental debates. It has moved these away from the neo-**Malthusian** anti-growth approaches of the 1970s, and places human needs – particularly of the poorest – as an environmental concern alongside that of conserving nature, both of which have important practical consequences for the Global South (see Adams, 2008).

BOX 3.8

The Global South and international environmental negotiation

Environmental problems such as climate change or the depletion of biodiversity are inherently global in their effects, and their mitigation requires coordinated international action: as such, establishing structures of **global governance** for environmental protection could be in everyone's common interest. The UN has hosted three conferences on the environment – in Stockholm (1972), Rio de Janeiro (1992) and Johannesburg (2002) – that have been landmark events in the development of environmental governance regimes.

Southern governments were initially sceptical of the motives for environmental protection, and almost boycotted the Stockholm Conference *en masse*. They saw it as primarily based around the Global North's own problems of industrial pollution, and feared that global environmental

BOX 3.8 (*CONTINUED*)

legislation would be used to constrain their own developmental ambitions. Indian Prime Minister Indira Gandhi claimed at Stockholm that 'Poverty is the worst form of pollution', and over subsequent decades, the South's pressing concerns around human development have been more explicitly recognized within the international environmental agenda. Naming the subsequent meetings the Conference on Environment *and* Development (Rio) and the World Summit on Sustainable Development (Johannesburg) was not mere semantics, but showed that a broadening of debate beyond technocratic, Northern-focused definitions of environmental problems was occurring.

This shift in agenda has been achieved in part through collective bargaining in which the Group of 77 has played a significant role. Despite different national interests within the **G77**, as a group they have coherently argued for two principles crucial to Southern interests: the right of the South to development, and the idea of differential North–South responsibility in shouldering the burden of environmental adjustments. Both have shaped the **G77**'s core negotiating position in international environment regimes that vary from hazardous waste disposal to protecting biodiversity, and both are enshrined in the UN Framework Convention on Climate Change, a document that has underpinned climate change negotiations since 1992. The institutionalization of **sustainable development** as the key goal of environmental governance regimes is evidence of the near-universal acceptance of the first principle. The second has provided a means by which the **G77** can bargain for a series of North–South transfers and concessions – including additional resources, 'clean' technology and environmental capacity building – that reduce the costs of dealing with environmental change.

(Sources: Adapted from Mee, 2005; Najam, 2005; Williams, 2005)

CONCLUSIONS

By the end of the twentieth century, modern, independent states had been established across the Global South, and the world as a whole had moved beyond open superpower rivalry. However, as we have seen, independence from colonial rule has not always translated into stable or democratic rule for citizens of the Global South, and the evolving system of **global governance** is both partial and unequal. As such, world maps of neatly divided **nation-states** obscure some much messier geographies of **power** that many Southern states confront on a day-to-day basis, as they deal with internal struggles over nationhood and legitimacy, and negotiate highly uneven international **power** relations.

Armed conflict in the Global South – like that in Sudan (Box 3.4) – needs to be understood as something that is the product of particular historical and geographical circumstances, and not as something 'natural' or inherent to 'barbarous' parts of the world (see Chapter 2). Conflict in the Global South is in part at least a product of the current geopolitical system itself, which like any other geopolitical ordering has its own geography of discontent and disruption where its foundational ideas – like **sovereignty** and the **nation-state** – are contested, or their internal inconsistencies are felt. What is significant about the current geopolitical system is that it has been very effective in placing most of its negative effects, in the form of warfare and chronic political instability, within the Global South. On a more positive note, the gains made by the Global South within international environmental negotiations highlight the opportunities and spaces for collective action that can open up within evolving structures of **global governance**. These gains, and the ambivalent consequences of China's rise to **power** (alongside other Southern countries, such as India, South Africa and Brazil), are an important reminder that whatever shape the world political order takes in future, it is going to be one in which at least some parts of the Global South are going to play a far more active role.

 Review questions/activities

1 Use the Uppsala Conflict Data Program (see websites below) to find details of any currently active armed conflicts within a Southern country of your choice. Search for other background information on the source and nature of the conflict(s). To what degree has the aftermath of colonialism been an important contributing factor to the conflict, and what other issues are at stake? How far has the conflict become internationalized, either in terms of direct participation in the conflict, or in peace-keeping efforts? What does the presence (or absence) of the international community tell us about the conflict and its participants?

2 Look at Table 3.1, which lists some statistics demonstrating the economic and political power of China, India and Brazil. To what degree do you think that these statistics capture the current and future importance of these countries? Which other indicators of their power could be used? Which other Southern countries do you think are also important emerging global/regional powers, and why?

3 Look through the online archives of a good quality newspaper to find reports of a recent international summit meeting. What role did Southern

---

Review questions/activities (*continued*)

governments play within the meeting: were they represented directly or indirectly, individually or collectively? Other than governments themselves, which other organizations were presenting their views inside or outside the formal meetings of the summit, and how did they do this? What might this tell us about the ways in which global governance operates, and the role of the South within this?

SUGGESTED READINGS

Painter, J. (1995) *Politics, Geography and Political Geography: A Critical Perspective*, London: Arnold.
This book remains a very intelligent and readable introduction to some of the key terms and debates within this chapter. See particularly Chapter 2 (on State Formation).

Flint, C. and Taylor, P. (2004) *Political Geography: World-Economy, Nation-State and Locality* (fifth edition), Harlow: Pearson.
A good introduction to political geography that explores debates on empire and **geopolitics** clearly and in depth (although with less emphasis on the impacts of political change for the Global South).

WEBSITES

http://www.un.org United Nations website
Gives details of the structure of the UN 'family' of institutions, and their recent activities in international security, development and other fields.

http://www.g77.org/ The Group of 77
Provides details on both the history of the G77, and its activities in negotiating within the UN.

http://www.pcr.uu.se/research/UCDP/index.htm The Uppsala Conflict Data Program
A free resource giving details of armed conflicts across the world.

4 The South in a globalizing economy

One of the key images of the world in the early twenty-first century is that of an increasingly interconnected globe, with goods and services being traded across national borders with greater intensity than ever before. Being involved in this economic **globalization** is viewed by some theorists and politicians as the main way in which individuals and communities across the world can improve their standards of living. Such ideas will be discussed in more detail in Part 4 of the book when we consider ideas of 'development' more explicitly. The present Chapter considers the spatial and temporal dimensions of global economic interconnectedness with a particular focus on the place of the Global South within the global economic system.

   This Chapter starts with an overview of historical processes of economic **globalization**, focusing particularly on the role of **colonialism**. It then considers the institutions of global economic regulation, before considering different sectors of the economy: primary production and resource extraction; manufacturing and industrialization; finance and services. While each of the sectors will be considered in turn, there are themes which cut across the Chapter as a whole. Most importantly, the Chapter will highlight the spatial manifestations of economic **globalization** processes and how different parts of the world shape these patterns. It will stress the heterogeneity of economic activities, change over time and the distinctive roles of different parts of the Global South. Second, as we saw in Chapter 3, political and economic power can be mutually reinforcing. This theme will continue in this Chapter with an examination of the relationship between economic wealth and political influence on a global scale. Finally, it is vital to recognize the cultural aspects of economic processes. Often 'the economy' or 'the economic' is viewed as a neutral concept, but as this Chapter will argue, this is certainly not the case. Economic activities are conceived, practised and understood in particular cultural contexts, and economic processes can, in turn, affect cultural norms (see Crang, 1997, for an overview of debates around the links between the 'economic' and the 'cultural'). While Part 3 of the book focuses on the 'local' experiences of global change in the South, this Chapter will cover some issues around social differentiation, particularly **gender**, and will also deal with environmental dimensions of economic **globalization**.

ECONOMIC EXCHANGE AND TRADE: HISTORICAL TRENDS

'**Globalization**' as a term describing the growing interconnectedness of the world, has a relatively recent history (see Murray, 2006, for a useful overview). However, this does not mean that the processes which it describes are similarly recent (see also Chapter 3). The importance of trade to access resources which were not locally available has a much longer history, starting from exchanges between local communities with different natural resource endowments. These forms of exchange, originally based on payment in kind, rather than the use of money, were gradually replaced with much larger scale, more organized trading systems, with more complex **divisions of labour** between producers and traders.

These large trading systems were found in many parts of the world, developing out of the local economic, environmental, social and political contexts. Thus, trade as an economic practice is not something which is rooted in one particular part of the world and then introduced elsewhere. Islamic sixteenth-century trading systems were discussed in Chapter 3, and the networks of Chinese merchants also constituted an extensive system of exchange throughout Asia around this time.

As nation-states began to emerge and European colonial projects started (see Chapter 3), **mercantile** activities became increasingly important in the economic growth of European powers. Colonies were viewed as sources of primary products such as food stuffs, precious metals and spices, as well as potential markets for European products. Trade and **colonialism** were therefore intertwined, not least through the operation of larger trading companies, in particular Britain's East India Company and the Dutch East India Company. The role of private companies and entrepreneurs in land claims can be seen more recently in relation to the United States' Guano Island Act of 1856 (see Box 4.2 later). The economic activities of these European based companies, often linking in with pre-existing trading networks in the Americas, Asia and Africa, led to growing links between different parts of the world which can be seen as a precursor to today's more intense forms of economic interconnectedness. As with current networks, some parts of the world were linked more strongly into the international trading systems than others, and even within certain territories, particular regions would have been at the heart of trading activities, while others would be more peripheral. Towards the end of the eighteenth century, key shipping routes included those between Western Europe (particularly Amsterdam and London) and South and East Asia via the Cape, and the triangular trade between Western Europe, West Africa and the Americas (see Figure 4.1).

Of course, these trading links and economic connections were not neutral. Traders were seeking to make a profit from their enterprises, so would be trying to buy goods as cheaply as possible. For some theorists, this represents a process of exploitation, particularly when European traders or their local representatives were operating in the Global South. Theorists following a Marxist interpretation, such as Vladimir Ilyich Lenin in his writings on **imperialism** (1973 [1917]), Andre Gunder Frank (1967) in his work on Latin America and Walter Rodney (1972) in relation to Africa (see Box 4.1), saw European economic activities in the Global South as part of the constant need for **capitalism** to seek out new markets and goods. The underside of this process was one of exploitation and the 'under-development' of the Global South. This does not mean that nobody in these regions benefited from the process, for example, urban elites working within or alongside the colonial administration often amassed great wealth. Frank (1967) describes a chain of exploitation from Europe to small-scale peasant farmers in Brazil and Chile in which local traders and estate owners were in positions of great **power** over producers and labourers. For Marxist and neo-Marxist theorists, the incorporation of the Global South into a more global economy represents a very negative path for the vast majority of the residents of these regions. Such ideas have been challenged for a variety of reasons which will be discussed later in this Chapter and also in Chapter 10.

MANAGING THE GLOBAL ECONOMY

As the globe has become more interconnected, albeit in an uneven way, there have been increasing calls for new forms of **governance** institutions as **nation-states** are no longer appropriate for all forms of regulation (see Chapter 3). Within the economic sphere, such institutions date back to 1944 with the **Bretton Woods** conference in New Hampshire, USA. At this meeting representatives of 45 non-Communist governments, mostly from the Global North, met to discuss global peace and economic stability. This was with World War II and the experience of the 1930s Great Depression as a backdrop.

In an attempt to ensure that such widespread conflict and economic instability never happened again, a number of institutions were established in the aftermath of the meeting and became known as the '**Bretton Woods** institutions': the International Monetary Fund (IMF), the World Bank and the General Agreement on Tariffs and Trade (GATT) (see Table 4.1). Membership of these organizations has grown since the 1940s and in 1995 GATT was replaced with the World Trade Organization (WTO). Not all members of the organizations are equal. For example, voting rights in the IMF are based on contributions to IMF funds. In 2008 this meant that the United States had 16.77 per cent of the votes, while Japan, as the

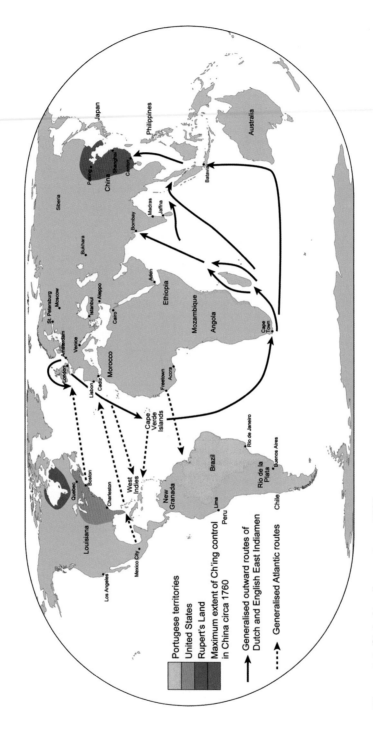

Figure 4.1 Global trading routes *c*.1780.
Source: Adapted from Wagstaff (1994). Map data © Maps in Minutes™ (1996).

BOX 4.1

Imperialism, dependency and underdevelopment

According to classical Marxist theories of development, the key social relations within a society are class based. Karl Marx (1909) argued that societies evolve in a linear fashion from pre-capitalist, through capitalist to the end point of **socialism**. At each stage production is organized in a different way within a broader legal and cultural structure (known as a mode of production). **Capitalism** is based on a separation between the class which owns the means of production, such as land, machinery, mineral resources, and the class which can only earn money for survival by engaging in paid labour. For Marxists, **capitalism** is an unsustainable system as it requires ever-expanding markets to buy the goods which are produced to generate a profit. Vladimir Illych Lenin (1973[1917]) argued that imperialism was a stage within **capitalism** which brought new territories into a system of production and **consumption**, but that eventually the opportunities for new markets would cease and the capitalist system would collapse.

For later theorists also adopting an approach based on the class distinctions used in Marxist theory, the spread of **capitalism** to the Global South not only created inequalities but also led to entrenched positions of disadvantage. For Andre Gunder Frank (1967) working on the South American case following European arrival on the continent in the fifteenth century, systems of exploitation were introduced whereby resources (particularly precious metals, but later agricultural goods) were plundered and taken back to Europe. Systems of agricultural production and trade also involved profits flowing to the core or metropole of Europe. Frank believed that these forms of exploitation caused underdevelopment. For Frank, South America (especially Brazil and Chile which were his focus) was able to develop much more when the continent was less engaged with Europe, for example during World War II.

Walter Rodney (1972) makes a similar argument in relation to Africa. As with Frank's analysis of Latin America, Rodney argues that the development of trade between Europe and Africa in the fifteenth century, the Atlantic slave trade, European exploitation of mineral resources and formal **colonialism** all contributed to a process of underdevelopment for African societies.

(Sources: Adapted from Frank, 1967; Lenin, 1973[1917]; Marx, 1909; Rodney, 1972)

Table 4.1 Bretton Woods institutions

Name	Number of members	Voting system	Role
International Monetary Fund (IMF)	184	Based on financial contributions – greater contributions mean greater share of the vote.	To promote economic growth and global economic stability through surveillance, technical assistance and loans.
The World Bank (originally The International Bank for Reconstruction and Development, IBRD)	184	The five largest shareholders (the USA, Japan, UK, France and Germany) have one representative each on the board of Executive Directors. The remaining 179 members vote for 19 remaining Executive Directors places.	To provide assistance to developing countries. Made up of the International Bank for Reconstruction and Development (IBRD) and the International Development Association (IDA)
General Agreement on Tariffs and Trade (GATT)			To promote free trade of goods and services.
World Trade Organization (WTO)	152 (as of June 2008)	Each member has a representative at the ministerial conference. One member, one vote.	Replaced GATT in 1995. To promote a transparent, rules-based global trading system.

Source: Adapted from IMF, 2008b; World Bank, 2006a; WTO, 2008.

second largest donor had 6.02 per cent of the votes (IMF, 2008b). The WTO has a one member-one vote system, but the unequal nature of global political and economic power means that the Global North has often managed to dominate the WTO agenda. The economic rise of China, India and other Southern nations has, however, led to greater challenges at WTO meetings (see Chapters 2 and 3). In July 2008, negotiations as part of the WTO Doha round of trade talks collapsed due to disagreements between the USA, India and China regarding agricultural subsidies, protection and trade (*The Guardian*, 30 July 2008).

These international financial institutions (IFIs) have been key in shaping the global economy and driving a **globalization** agenda. While there are differences between the organizations, they are all broadly in favour of **neoliberal** policies.

Because of this, they have all received criticism for imposing external ideas on Southern governments and peoples, and also for the negative effects of such policies (see Chapter 10).

'NATURAL RESOURCES' AND PRIMARY COMMODITIES

The Global South is often imagined as a source of agricultural and mineral resources (see Chapter 2). In macroeconomic terms, many countries of the Global South are indeed dependent on the export of primary products for the bulk of their foreign exchange earnings (see Table 4.2). This relates both to their resource endowments as well as to the way in which they have been incorporated into the global economic system as outlined above. Such dependence on primary commodities can have severe effects on the economic stability of a nation; changes in global commodity prices, crop failure or resource depletion can result in extreme economic hardship if countries do not have diversified sources of income. Opportunities for diversification may be limited, making national and regional economies very vulnerable.

The concept of what constitutes a 'resource' changes over time. This reflects changing technologies and cultural norms. A 'resource' is something which is regarded as useful for human activities. This may be to meet new demands, or it may be as a more appropriate or economically viable replacement for an existing substance (see Box 4.2).

While primary production is clearly important for export earnings in many countries in the Global South, reliance on primary production is sometimes viewed as undesirable in the drive for economic development and improved standards of living (see Chapter 10). This is not because of some inherent disadvantage associated with primary products, but because of the way that they have been valued relative to other goods and services in the capitalist economy. For example, the **terms of trade** have increasingly moved in favour of manufactured goods and away from primary products. Commodity prices can also be very variable, meaning that while countries can benefit from high prices (as with the mid-2008 high prices for copper driven by booming demand from China), when global prices fall, significant amounts of foreign exchange reserves are lost (see Figure 4.2). Price fluctuation for many commodities has been increasing since the 1980s because of the policies of multilateral organizations, such as the **World Trade Organization (WTO)**, in promoting **free trade** (see below). Such fluctuations have implications for national economies, but obviously severe effects on small-scale producers.

Exporting primary products also means that the value added to that product, through processing and manufacturing, is not captured by the exporting nation. Primary products are part of global **commodity chains** (see Concept Box 4.1) (Gereffi and Korzeniewicz, 1994) through which different parts of the world are linked together and along which value is added to a product, so that the final good

Table 4.2 Primary products in export earnings, 2005

Country	Primary products as % of total exports of merchandise	Country	Primary products as % of total exports of merchandise
Sudan	99	Sierra Leone	93[c]
Algeria	98[a]	Tonga	93[c]
Nigeria	98[c]	Burkina Faso	92
Yemen	96	Maldives	92
Burundi	94	Turkmenistan	92[c]
Papua New Guinea	94[b]	Ecuador	91
Bahrain	93	Niger	91[b]
Seychelles	93	Zambia	91
Gabon	93[c]	Rwanda	90[b]
Kuwait	93[b]	Saudi Arabia	90

Source: Adapted from UNDP, 2007: table 16.

Notes:
a   One or more component of primary exports missing.
b   2003 figures.
c   Figure from 2000–05.

BOX 4.2

Guano and coltan: the rise and fall of resources

GUANO

Guano, or dried seabird excrement, is an excellent natural fertilizer because of its very high nitrogen and phosphorus content. However, before guano can become a resource, the agricultural need for fertilizer and the potential of guano to meet this need must be recognized. Technology, politics and economics must then come together so that the collection, transport and selling of guano can be organized.

In the early to mid-nineteenth century, Peru was the key source of guano for the expanding agricultural markets in the USA and Europe. While indigenous populations had used guano to improve soil fertility, this

BOX 4.2 (*CONTINUED*)

knowledge was not adopted or understood by the Spanish, so it was only in the early nineteenth century with European scientific discoveries about soil fertility and the increasing demand for fertilizers that the commercial guano market took off. Between 1840 and 1880, guano exports from Peru were valued at more than US$750 million (Gootenberg, 1993: 2). Given the almost complete Peruvian monopoly on existing guano supplies, the United States sought to find new sources. The Guano Islands Act of 1856 allowed for American entrepreneurs to lay claim to possible sources of guano. Other nations of the world were involved in similar activities leading to tensions and conflicts in both the Pacific and the Caribbean. Over time the reserves were depleted and other forms of chemical fertilizer were developed, making guano exploitation unprofitable. Some islands were abandoned, while others, such as Johnston Atoll near Hawaii, became military bases.

COLTAN

Coltan is short for columbite-tantalite and is a metallic ore which is a vital component in capacitors, used in mobile phones and laptop computers. While Australia is the world's largest producer of coltan as of July 2008, the vast majority (over 80 per cent) of the world's known reserves of coltan are in the Democratic Republic of Congo (DRC). Because of the massive increase in the production of mobile phones and laptops, the price of coltan increased greatly in the 1990s, up to a high of about US$600 per kilogramme. This led to a coltan rush in the eastern part of the DRC. Unlike Australia, where production is organized by a large MNC, in the DRC miners often operate individually and illegally. Given the potential economic returns from coltan, it is not surprising that the coltan trade has been implicated in funding for the civil war in the DRC (according to a report from the United Nations Security Council) and has been blamed for significant environmental destruction, including deforestation, water pollution and threats to the survival of local gorilla populations.

(Source: Adapted from Cellular-news, 2006; Gootenberg, 1993;
Redmond, 2001; Skaggs, 1994)

is sold for a much higher price than that paid to the original producer. **Commodity chains** can be incredibly complex, as for example, with electronic goods or cars, but the commodity chain for food stuffs is often much simpler.

Gary Gereffi (1994) identifies two main forms of commodity chain; producer-driven and buyer-driven. In producer-driven global commodity chains **transnational corporations (TNCs)** play a central role in controlling the process. For example, in the case of car production (see Figure 4.3), while research and

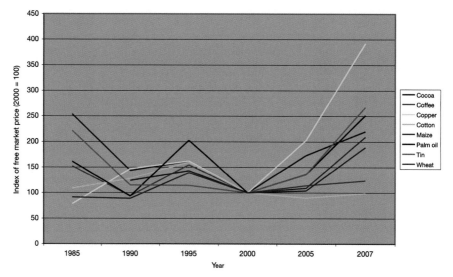

Figure 4.2 Commodity price fluctuations.
Source: Data from UNCTAD (2008b).

CONCEPT BOX 4.1

Commodity chains

Term used to describe how the development, manufacture and distribution of a commodity are linked together. In global commodity chains these processes are spread across international borders, but are strongly integrated. The nature of integration will vary depending on the commodity, the organizational structure of the companies involved and state policies.

(Source: Adapted from Gereffi, 1994)

development, production of components and car assembly will be spread globally, the **TNC** remains in control. For buyer-driven **commodity chains**, such as clothing (see Figure 4.4), it is the buyers, most notably the major retailers and the brand-named companies (such as Nike and Gap) which are directing what is produced. For primary producers, the contribution to such global **commodity chains** will be early in the production process, for example cotton for clothing manufacture and aluminium for car production (see Kaplinsky, 2005: Chapter 5 for a discussion of the changing production networks for textiles, furniture and cars).

For countries in the Global South, trying to trap more of the value added through production can be part of government efforts to generate a particular form of economic development (see later in this chapter). The promotion of '**fair trade**' goods, particularly in relation to primary production, is also a way in which the benefits of global trading can be spread more widely (see Chapter 10).

Oil: economic power in the Global South?

While primary production is sometimes viewed as a problematic basis for economic development, it is important to recognize that not all primary products are the same. At particular historical moments, certain commodities may become increasingly important and in the last half of the twentieth and early twenty-first centuries, oil has become vital in national economies throughout the world. There are significant oil-producing nations in the Global North (most notably the USA,

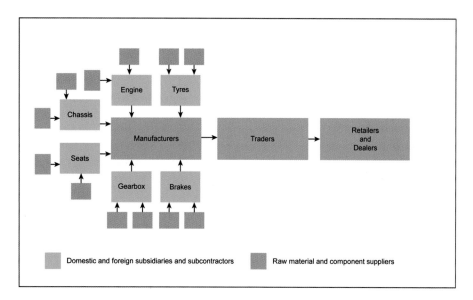

Figure 4.3 Producer-driven commodity chain.
Source: Adapted from Gereffi (1994: 98, figure 5.1.1).

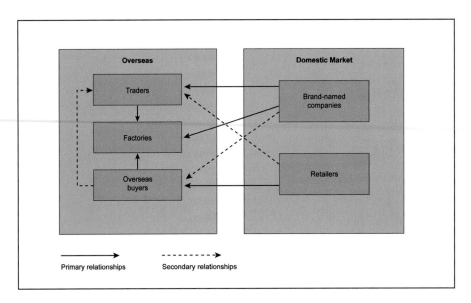

Figure 4.4 Buyer-driven commodity chain.
Source: Adapted from Gereffi (1994: 98, figure 5.1.2).

Russia and Canada), but the majority of oil comes from regions in the Global South. According to figures from the Energy Information Administration (EIA) (2008), about 62 per cent of the oil produced by the top 15 world oil producers in 2006 came from the Global South. Given the global dependence on oil not just for fuel, but also in the petrochemical industries, this geographical pattern represents a contrast to the usual position of the Global South in discussions of the global economy.

Of course, oil is not found everywhere in the Global South (see Figure 4.5), but for those countries with significant reserves, it represents a key economic resource and also, in some cases, a key political resource. The power of oil-producing countries to affect the global economy was keenly felt during the 1970s when in both 1973 and 1979 the members of the **Organization of Petroleum Exporting Countries (OPEC)** increased the price of their oil fourfold. More recently, environmental disasters (such as Hurricane Katrina in the Gulf of Mexico in 2005) and ongoing war and political tensions in the Middle East have resulted in massive increases in oil prices.

As with other primary products, reliance on oil export earnings can be problematic. As a non-renewable resource, the economic advantages which oil revenues may bring must be viewed as temporary and policies should be implemented to promote diversification, to avoid a massive slump when the resource runs out, or when alternatives are found. There are also issues around the way in which oil revenues are spent and how far benefits spread from a small elite to the mass of the

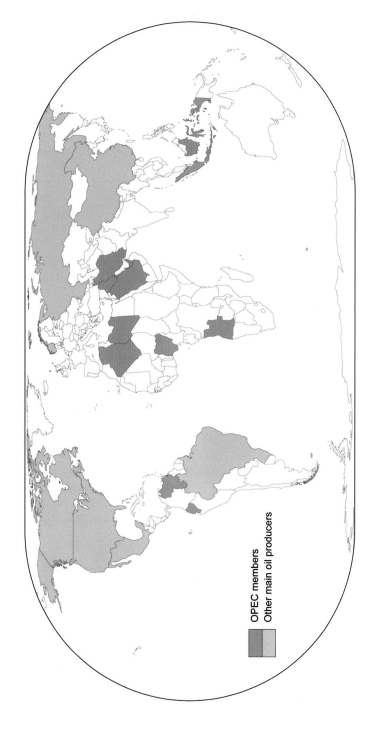

Figure 4.5 OPEC members and major oil producers.

Source: Based on data from EIA (2008); OPEC (2008). Map data © Maps in Minutes™ (1996).

OPEC members

Other main oil producers

population (see Box 2.5 on Venezuela). Oil-exporting nations may appear to be very rich in national income terms, but this wealth may not be distributed equally among the population (see Box 4.3 and Chapter 10).

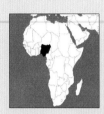

BOX 4.3

Oil revenues in Nigeria

In 2006, Nigeria was the world's eighth largest exporter of oil and the twelfth largest oil producer, with an average daily output of nearly 2.5 million barrels (EIA, 2008). Not only does this make Nigeria an important player in the world's oil markets, it has also had a massive impact on the Nigerian economy. About 80 per cent of government revenue and over 90 per cent of foreign exchange earnings comes from oil (Watts, 2004). Since commercial oil production started in the Niger Delta in 1956, about US$300 billion has flowed into federal government from oil sales and annual oil revenues stand at about US$40 billion (Watts, 2004).

Despite such impressive figures, the majority of Nigeria's population see few benefits. In 2007, Nigeria came 158th out of 177 in the UNDP's Human Development Index rankings (UNDP, 2007). Average life expectancy at birth was 46.5 and adult literacy rates stood at 69.1 per cent. The Gini index, which is a measure of income inequality, was 43.7 in 2007, reflecting the highly unequal distribution of income in the country (UNDP, 2007).

Michael Watts (2004) blames the operation of a form 'petro-capitalism' for this situation and critiques those who see oil (and other products, such as diamonds, iron ore and gold) as inherently leading to what has been termed a 'resource curse' whereby repression and inequalities are widespread. In contrast, Watts argues that it is not oil per se which leads to such situations, but rather the nature of the economic and political structures which evolve around its exploitation.

Watts outlines the 'oil complex' which operates in Nigeria. At the heart of this is the federal government which controls mineral exploitation rights. In addition, the nationalized oil company, the Nigerian National Petroleum Company agrees joint ventures with transnational oil companies and the security apparatus of both the state and companies seeks to ensure that uninterrupted production takes place. The federal government also decides how oil revenues are to be distributed between the states of the country. This mechanism has meant that Rivers State, from which over half the oil production comes, receives very little (less than one-fiftieth) of the value of

---

BOX 4.3 (*CONTINUED*)

the oil it produces. The remaining value has been distributed to other states, the federal government, the private oil companies and the pockets of corrupt individuals who are seeking to benefit individually from Nigeria's oil boom.

Ogoniland in Rivers State has been at the heart of resistance and challenges to the prevailing oil production system. Not only do the Ogoni people see very little benefit from the oil extracted from the land where they live (adult illiteracy rates stand at around 80 per cent), but they also suffer the environmental devastation from oil spillages and gas flaring. In the 1990s, the Movement for the Survival of the Ogoni People (MOSOP) mobilized to challenge the oil companies, particularly Shell, and the Nigerian government. The MOSOP leader Ken Saro-Wiwa and nine other MOSOP members were hanged in 1995 having been accused of murder. Since then MOSOP has collapsed, but other groups, mobilizing around indigenous identities and land claims, have gained prominence. These challenges have created a situation of increasing violence which has done nothing to help improve the daily lives of Nigerians living in the oil-producing areas.

(Sources: Adapted from EIA, 2008; UNDP, 2007; Watts, 2004)

FOREIGN DIRECT INVESTMENT AND THE NEW INTERNATIONAL DIVISION OF LABOUR

As previously outlined, the world has been becoming more interconnected in economic terms for hundreds of years, but the present-day nature of economic **globalization** represents a much more intensive form of exchange, encompassing greater numbers of people and places. The changing nature of manufacturing and what Peter Dicken (2007) has termed a 'global shift' in production processes, is a key part of this process. This has been linked to the rise in **multinational** and **transnational corporations (MNCs** and **TNCs)** and massive increases in the flow of **foreign direct investment (FDI)**. These processes have led to certain parts of the Global South being incorporated into new **commodity chains** and production processes, while other regions and populations are excluded.

In the post-World War II period, but especially from the 1970s onwards, changing technologies meant that the manufacture of many industrial goods could become more 'footloose'. Previous forms of **Fordist production** whereby standardized products were put together on an assembly line, were often being replaced with demand-driven, **'just-in-time' production** (see Concept Box 4.2). Improvements in communication and transport also meant that production did not always have to be based in the Global North; access to cheap, but appropriately

---

CONCEPT BOX 4.2

Fordism and just-in-time production

FORDISM

A term used to describe the organization of the industrial production process which dominated global manufacturing from World War II until the 1970s. Fordism was based on the production of standardized goods with clearly demarcated divisions of labour, both within factories and internationally. The term comes from the assembly-line systems developed by Henry Ford for car manufacture, but can be applied to a range of manufacturing sectors.

JUST-IN-TIME PRODUCTION

Part of what has been termed the 'flexible accumulation' process, compared with the perceived fixity of Fordism. As a result it is also viewed as part of 'post-Fordism' which describes existing forms of global flexible production. Under JIT production, decisions about what is produced, where and when is more flexible and responsive to market demands. JIT involves greater labour flexibility, such that workers are increasingly employed on short-term contracts to meet urgent orders and need to have a range of skills.

---

skilled labour, in the Global South became a key factor in decisions regarding industrial location in some sectors. This has led to a '**New International Division of Labour' (NIDL)**, whereby manufacturing employment is increasingly moving from the Global North to the Global South, leaving areas of high unemployment in the previously heavy industrial areas of the USA and Western Europe (Perrons, 2004). The parts of the Global South which have been able to benefit from such trends are found largely in East and South Asia, Latin America and the Caribbean (Table 4.3). Africa and the Middle East have largely been excluded from these processes. This general global pattern is an outcome of a number of factors, including available infrastructure, access to an educated labour force and political stability.

The shifts in production processes coincided with changes in the approaches many Southern governments took to their industrialization policies, particularly in response to pressure from Northern governments and international financial institutions. The increasing importance of **neoliberal** ideologies (see Concept Box 4.3) meant that more and more governments throughout the world were implementing policies that opened up their economies to foreign investment and made

Table 4.3 Regional patterns of foreign direct investment inflow, 1970–2006

	1970		1990		2006	
	Amount US$m	%	Amount US$m	%	Amount US$m	%
WORLD	**13,418**	**100**	**201,594**	**100**	**1,305,852**	**100**
Developing economies	**3,854**	**28.7**	**35,892**	**17.8**	**379,070**	**29.0**
Africa	1,266	9.4	2,806	1.4	35,544	2.72
L. America & Caribbean	1,599	11.9	9,748	4.8	83,753	6.4
Middle East	147	1.1	456	0.2	59,902	4.6
East Asia	178	1.3	8,791	4.4	125,774	9.6
S. Asia	68	0.5	575	0.3	22,274	1.7
SE Asia	460	3.4	12,821	6.4	51,483	3.9
Oceania	136	1.0	696	0.3	339	0.03
Economies in transition	–	–	**75**	**0.04**	**69,283**	**5.3**
Commonwealth of Independent States	–	–	4	0.002	42,934	3.3
SE Europe	–	–	71	0.04	26,348	2.0
Developed economies	**9,564**	**71.3**	**165,627**	**82.2**	**857,499**	**65.7**
Europe	5,226	38.9	97,044	48.1	566,389	43.3
N. America	3,083	23.0	56,004	27.8	244,435	18.7
Other developed countries	1,255	9.4	12,579	6.2	46,675	3.6

Source: Data from UNCTAD, 2008b: 29.

their country more attractive to **MNCs** and **TNCs**. These policies included reduced tariffs, subsidized premises and infrastructure, for example in **export processing zones (EPZs)** and reduced levels of taxation (Harvey, 2005).

Such policies have been characterized by some as 'a race to the bottom' as it could lead to countries undercutting each other to attract **FDI** resulting in a very poorly paid workforce, environmental destruction and little contribution to the national tax base (Kopinak, 1997). While these market-led policies will be discussed in more detail in Chapter 10, for the purposes of this Chapter it is important to recognize the economic benefits that versions of these policies have created for some parts of the Global South. Rapid economic growth, job creation, infrastructure

---

CONCEPT BOX 4.3

Neoliberalism

A general term used to encompass approaches to economic and political development which stress the role of the market rather than the state as the mechanism for deciding what is to be produced and where, setting prices and wages. It is usually associated with processes of **individualization**, whereby individuals, rather than the state, are responsible for themselves, and collective action around particular forms of **identity**, such as worker, are discouraged.

---

development and improved standards of living have been experienced by some regions and social groups.

In the 1960s and 1970s, the term '**newly industrializing country**' (NIC) was applied to a number of countries in East Asia and Latin America which seemed to be following the pattern of the Global North and moving from largely primary-producing economies to a greater focus on industry and manufacturing. For Hong Kong, Singapore, South Korea and Taiwan, the so-called 'Tiger Economies', this perceived progress reflected an opening up of the economy to **FDI**, particularly in the electronics and clothing industries. In Latin America the **NIC** status of Brazil and Mexico was much more a reflection of industrial development behind high tariff walls, as part of what had been termed '**import-substitution industrialization**' (ISI). Such policies were also blamed for the slowing down of these economies in the 1970s (see Chapters 9 and 10). In reality, the rise of the **NICs** was much more complicated than simply following a market-led or state interventionist pathway (see Box 4.4).

Since the **debt crisis** and the widespread implementation of **Structural Adjustment Policies** (SAPs) in the Global South in the 1980s and 1990s (see below), the vast majority of countries in the Global South have opened up their economies to **FDI**. There are a few exceptions, most notably North Korea, but even countries following a **Communist** political path such as China and Cuba have liberalized their economies to some degree. This means that national economies are increasingly linked together, with **MNCs** and **TNCs** holding significant **power**. The **power** of corporations is sometimes viewed as undermining the **power** of the **nation-state** (Chapter 3). In many cases, particularly in the Global South, it is argued that national governments have little option but to accept **MNC** and **TNC** conditions for fear of losing that investment to another country. In 2003, with its annual sales of US$256.33 billion, Wal-Mart Stores earned more money than all but 19 countries in the world (in terms of **Gross Domestic Product** (GDP)) (Murray, 2006: 131–2; see also Brunn, 2006, for a range of perspectives on Wal-Mart's

BOX 4.4

Singapore's post-independence development

Since gaining its independence in 1965, Singapore has experienced great economic transformation and massive improvements in the standards of living of its population. This has been achieved through a combination of state intervention and guidance, as well as benefits which have accrued from the inflow of foreign investment. In 2007, Singapore was ranked 25th in the UNDP's Human Development Index table. GDP p.c. (PPP) (see Concept Box 2.7) was US$29,663, compared with figures for the USA of US$41,890 and the UK of US$33,238 (UNDP, 2007: tables 1 and 14).

Since its founding in the early nineteenth century, Singapore had grown as a key trading point and port. At independence, the People's Action Party (PAP) government, led by Lee Kuan Yew, embarked on an economic strategy promoting industrialization. This was done through a range of measures, including tax incentives for foreign investment and changing legislation on workers' rights. Global **FDI** was increasing as part of the processes of global shift and Singapore was viewed as safe, well located and with appropriate infrastructure and labour force. In the 1970s, the PAP government gave priority to higher value-added industries such as chemicals and electrical goods manufacture.

While manufacturing industry is still an important part of the Singaporean economy, it is services, particularly financial services, which have become increasingly important. Singapore's port and airport are also highly ranked globally in terms of throughflow, efficiency and quality of service, and tourism and retail are other key sectors. About 70 per cent of employment is now in the service sector (UNDP, 2007: table 21). This shift from low value-added manufacturing to high-level services has been achieved through clear state intervention and direction, but also has arisen during a period of mammoth shifts in the global economy which Singapore has been able to take advantage of. Singapore's location within the rapidly growing Asian region has also contributed to its success. The Singapore government has attempted to develop policies of regionalization since the 1990s, whereby Singapore companies have set up joint ventures in India and China for example. Such policies have had varying success, but demonstrate how a small city-state has tried to gain from the economic opportunities which **globalization** can bring.

(Sources: Adapted from Perry, *et al.,* 1997; Phelps, 2007; UNDP, 2007)

economic, cultural and environmental effects). While **MNCs** and **TNCS** certainly are powerful, the ability of national governments, local communities or consumers to resist or shape their activities is sometimes underestimated (Bickham Mendez, 2002; see also Chapter 8 for a discussion of how the products of **TNCs** are appropriated and consumed in the Global South).

Discussions about **FDI** often assume that the global flow of capital is from the North to the South. Such a pattern is far from the truth. The vast majority of **FDI** comes from the Global North and is invested in the Global North (see Tables 4.3 and 4.4). While the 'global shift' is certainly an important economic change in global production, **TNCs** and **MNCs** still focus their overseas investments in the North. In 2006, just under two-thirds of inward **foreign direct investment** went to Northern countries, while about 84 per cent of outward FDI came from these same countries (UNCTAD, 2008b). **Globalization** in economic terms is therefore, highly spatially uneven.

As the UNCTAD figures demonstrate, the Global South contributes very little in terms of outward **FDI**, but the fact that this exists at all is a challenge to the common **representations** of the Global South purely as recipients of **FDI**. Unsurprisingly, the bulk of **FDI** coming from the Global South originates in a few, largely Asian, countries (see Table 4.4). It is not just Northern **MNCs** which

Table 4.4 Regional patterns of foreign direct investment outflow, 2006

	Amount US$m	%
WORLD	**1,215,789**	**100**
Developing economies	**174,389**	**14.3**
Africa	8,186	0.7
L. America & Caribbean	49,132	4.0
Middle East	14,053	1.2
East Asia	74,099	6.1
S. Asia	9,820	0.8
SE Asia	19,095	1.6
Oceania	5	0.00
Economies in transition	**18,689**	**1.5**
Commonwealth of Independent States	563	0.05
SE Europe	18,126	1.5
Developed economies	**1,022,711**	**84.1**
Europe	668,698	55.0
N. America	261,857	21.5
Other developed countries	92,155	7.6

Source: Data from UNCTAD, 2008b: 29.

Table 4.5 Top 15 non-financial TNCs from developing countries, 2005

Rank[a]	Company	Home economy	Industry	Foreign assets (US$m)
1	Hutchison Whampoa Ltd	Hong Kong, China	Diversified	61,607
2	Petronas-Petroliam Nasional Bhd	Malaysia	Petroleum exploration/ refining/ distribution	26,350
3	Cemex S.A.	Mexico	Non-metallic mineral products	21,793
4	Singtel Ltd	Singapore	Telecommunications	18,000
5	Samsung Electronics Co. Ltd	South Korea	Electronic & electrical equipment	17,481
6	LG Corp	South Korea	Electronic & electrical equipment	16,609
7	Jardine Matheson Holdings Ltd	Hong Kong, China	Diversified	15,770
8	CITIC Group	China	Diversified	14,891
9	Hyundai Motor Company	South Korea	Motor vehicles	13,015
10	Formosa Plastic Group	Taiwan	Chemicals	12,807
11	China Ocean Shipping (Group) Company	China	Transport & storage	10,657
12	Petróleos de Venezuela	Venezuela	Petroleum exploration/ refining/ distribution	8,534
13	Petroleo Brasileiro S.A.-Petrobras	Brazil	Petroleum exploration/ refining/ distribution	8,290
14	CLP Holdings	Hong Kong, China	Electricity, gas & water	6,039
15	Capitaland Limited	Singapore	Real estate	6,017

Source: Data from UNCTAD, 2007: 232.

Note:
a   Rank by foreign assets.

have sought to relocate production to cheaper locations; Southern companies have located factories close to markets or to sources of cheap labour as well (see Table 4.5 and Hart, 2003, on Taiwanese investment in South African factories). In addition, the increasing role of investment from the People's Republic of China in Sub-Saharan Africa represents new forms of South–South transfer linked to resources and **geopolitical** concerns (Box 4.5: also see Box 3.6 on the political dimensions).

BOX 4.5

Chinese investment in Sub-Saharan Africa

As the Chinese economy has boomed, demands for energy and raw materials have experienced a similar growth. In the mid-1980s, China was the second largest exporter of crude oil in Asia, but in 1996 it became a net importer and in 2005 it was the world's second largest importer of crude oil after the USA. It is estimated that China will need to import about 60 per cent of its energy needs by 2020 (Ghazvinian, 2007).

This demand for oil and for raw materials, such as copper and timber, has led to rising Chinese investment in Africa. State-owned oil companies, such as the Chinese National Petroleum Company and the Chinese National Offshore Oil Company, have entered into joint ventures with African governments and companies. China is becoming an increasingly important source of **FDI** for African countries, but the pattern of investment is not even over the continent. For example, China provides over 20 per cent of Sudan's **FDI** (with the associated political controversies about supporting regimes which are implicated in human rights abuses, See Box 3.6) and about 6 per cent of **FDI** in Zambia, particularly within the copper sector. For other African countries, Chinese investment is increasing rapidly, but in terms of share of total foreign investment, countries from the Global North remain much more significant.

Chinese investment in infrastructure projects is widespread throughout the continent, with an estimated 500 projects involving the China Road and Bridge Corporation taking place at any one time. Other areas of significant Chinese investment include the textile industry. However, in some countries domestic textile production has been decimated by cheap imports, particularly from China; thus China's engagement with Africa is not always beneficial in economic terms.

(Sources: Adapted from Carmody and Owusu, 2007; Ghazvinian, 2007; Jenkins and Edwards, 2006; Klare and Volman, 2006; Mawdsley, 2007; Morris, 2006)

Just as improved communications technology has facilitated the relocation of manufacturing activities to certain parts of the Global South, so too have some service activities been moved from the Global North to the Global South. One high profile form of relocation has been the massive rise in the use of call centres in India for customer services (Mirchandani, 2004; Taylor and Bain, 2008). The availability of educated, English-speaking workers, along with the plummeting costs of international telephone calls, means that such relocations are attractive to companies. The sign for English language lessons in Plate 4.1 is for a school which was set up to train call centre workers. For Indian workers, the call centres often seem to represent reasonably well-paid and flexible job opportunities, but this may not be the case, with high staff turnover due to poor working conditions. In some cases, complaints from customers have meant a move back to non-Indian call centres. While the call-centre phenomenon has received significant publicity, other services have been relocated overseas. These include the proof-reading and production of academic journals, market research analysis and data entry and processing.

While these global shifts in manufacturing and services location have obvious spatial implications, incorporating some parts of the world while marginalizing others, there are also social dimensions. For example, the gendered nature of

Plate 4.1 Sign for London School of Speech, Bangalore, India.
Credit: Claire Cowie.

employment is often apparent in the new forms of economic activity available in the Global South (Nagar, *et al.*, 2002). The gendering of work is not a new phenomenon (see Chapter 7), but **MNCs** and **TNCs** have drawn on pre-existing **gender** ideologies and inequalities in the search for new sources of labour. Women are often **socially constructed** as being more reliable and docile than men and more appropriate for assembly-line work where 'nimble fingers' are required. Women have, therefore, dominated factory-floor workforces in **MNC** factories producing electronics, clothing and household goods in **export processing zones** throughout the world (Chant and McIlwaine, 1995).

FINANCE, INVESTMENT AND SPECULATION

While **foreign direct investment** is an important element of global capital flows, it is dwarfed by other forms of international financial transaction. With new communication technologies and the **deregulation** of financial markets, a range of financial products have emerged (see Box 4.6). Deregulation in this sector has been intimately connected with processes of **globalization**.

As with the trade in primary commodities and manufactured goods, the international operation of financial activities has a long history. Banking systems developed to support trading activities in Europe and also to raise money for warfare and state-building activities in the sixteenth century. The period of European **colonialism** also led to the expansion of such activities to the colonies through, for example, the sale of bonds to finance infrastructure projects (Held, *et al.*, 1999: 190–2). The development of an international financial system was greatly assisted in the nineteenth century by the development of the Gold Standard. This was formally established by the main European powers in 1878 and was aimed at stabilizing the world financial system by pegging the value of major currencies to the price of gold, so operating a system of fixed exchange rates. By World War I, the Gold Standard was in use throughout Europe, North and Latin America, Japan and many of the European colonies. World War I led to the collapse of the system, and while it was briefly reintroduced in the late 1920s, the Great Depression of the 1930s, followed by World War II led to its demise (Held, *et al.*, 1999: 192–9). In the postwar period, the **Bretton Woods system** was set up to promote global financial stability (see above).

The pre-war international financial systems were designed to meet the needs of Northern governments and capital and the key sites of activity were in the Global North, with London as the heart of the system in the nineteenth century. Many countries in the Global South were incorporated into these financial systems, but remained marginal to the majority of flows. This general trend continued in the postwar era but, with new technologies and government policies, certain parts of the Global South became involved in global financial transactions in new ways. The **Cold War** and the rising wealth of oil-rich countries provided massive sources of

BOX 4.6

Forms of international capital flow

As financial markets have become more sophisticated and technology has improved, cross-border capital flows have increased and become more diverse. The key flows are:

- *Foreign direct investment* Capital flows where the investor is directly involved in managing the business in the host country. For example, a US car company setting up a factory in Mexico.
- *International bank lending* Loans provided by banks to borrowers overseas.
- *International bonds* Bonds are financial products which guarantee the payment of a set amount at a fixed date, as well as periodic interest payments. International bonds involve the sale of these products to customers overseas.
- *International equities* Company shares issued to foreigners.
- *Derivatives* Encompasses a range of financial instruments which help investors reduce their risks. Since the mid-1980s the derivatives markets have expanded massively. Includes instruments such as futures contracts where agreements are made to pay a set price for a certain commodity (e.g. oil or wheat) in the future.
- *Official development assistance (aid)* Flow of capital between governments for specific development projects or sectoral activities. Includes loans at preferential rates. **Aid** can also include non-capital flows, such as technology, personnel and commodities.
- *Remittances* Transfer of assets from overseas workers to family or community in home country.

(Sources: Adapted from Dicken, 2007; Held, *et al.*, 1999)

funds for banks to lend in the form of **Eurodollars** and '**petrodollars**' (see Concept Box 4.4). Countries in the Global South seeking funds to pay for infrastructure development and industrial expansion were able to borrow money at what were low interest rates. In the 1970s, the Global South received between a quarter and a third of all international bank loans (Held, *et al.*, 1999: 210). This led to a global circulation of capital, often from parts of the South (particularly the Middle East) to other Southern countries with Northern banks as intermediaries. Deregulation within financial services was both necessary for this form of financial **globalization**, but was also a result of greater international connectivity (Dicken, 2007).

CONCEPT BOX 4.4

Eurodollars and petrodollars

EURODOLLARS

The name given to reserves of US dollars outside the USA. Such reserves were originally largely held in Communist countries such as the Soviet Union and China as a way of avoiding American influence over their finances during the **Cold War**. The absence of US control means that Eurodollars could be traded on international currency markets, with the associated possible financial gains (as well as losses). This contributed to greater degregulation of the international financial system (Dicken, 2007).

PETRODOLLARS

Dollar reserves in international banks which came from oil-producing countries, particularly during the 1960s and 1970s when oil price rises generated large profits. Banks used these deposits to loan funds to governments in both the Global North and Global South to fund **modernization** plans.

Rising interest rates, massive increases in oil prices and a slowdown in the global economy in the late 1970s meant that Northern markets for products from the Global South shrank just at the time when repayments on the money borrowed earlier in the decade became much more expensive. This led to many governments defaulting on their debt repayments. In 1992, the Mexican government declared that it would not be able to meet its loan repayments, triggering what has been termed the '**debt crisis**' of the 1980s. Unable to borrow more money from private banks, such governments were forced to turn to the World Bank and International Monetary Fund for assistance. Financial rescue packages were negotiated, but additional funds were conditional on Southern governments implementing **neoliberal** reforms, known as '**Structural Adjustment Programmes/Policies**' (SAPs). These involved a liberalization of the economy with reduced government intervention, opening up of the economy to foreign investment and reduced government spending. Such policies usually resulted in economic stabilization in the short term, but the longer-term implications for economic growth, social inequality and **poverty** are more mixed (see Chapter 10 for more details).

New technologies and reduced government regulation of financial markets in the 1980s and 1990s led to the launch of new financial products (see Box 4.6). The growth in these products, particularly derivatives, was driven by the possibilities for great financial rewards both for the product providers and also for the investors. Speculation on foreign exchange markets and also in what have been termed 'emerging markets' and 'global frontier markets' have meant that particular parts of the Global South have a key part to play in these new patterns of global financial activity. Because of **floating exchange rates**, investors can choose to invest in a particular currency in the hope that the price will improve and they can make a profit. However, such holdings can be liquidated quickly in times of uncertainty. The 'Asian financial crisis' of 1997 was a severe economic downturn in many East Asian countries triggered by the devaluation of the Thai baht. Investors had lost confidence in the currency and withdrew their funds (see Chapter 10). The interconnectedness of the regional economy meant that this lack of confidence spread throughout the region with dire economic consequences (Haggard, 2000).

The stock markets of London, New York, Tokyo, Hong Kong and Frankfurt are well known, and the first three represent the hub of global financial trading. The focus on these key sites hides the fact that stock markets are also found in the vast majority of Southern and transition countries (see Plate 4.2) Those that are more established have been categorized and are classified as 'emerging markets' by commercial financial institutions and are marketed to potential investors. The

Plate 4.2 Mexico City Stock Exchange.
Credit: Katie Willis.

imagery used in such marketing reflects some of the **discourses** about the Global South discussed in Chapter 2, such as the exotic and the unpredictable. However, the countries selected are also viewed as economically and politically appropriate for investments in the context of **neoliberalism** (Sidaway and Prycke, 2000). Standard & Poor's (the most well-known independent provider of international credit ratings and investment advice) identifies both 'Emerging Markets' and 'Global Frontier Markets'. The latter are defined as 'the lesser developed markets even by emerging markets standards' (Standard & Poor's, 2007) (see Figure 4.6). Through emerging markets funds, investors can buy stocks and shares which are traded on the stock exchanges of these particular countries. While such funds can provide excellent sources of capital for companies and governments in the Global South, as with foreign exchange trading, investors can pull out very quickly if confidence levels plummet (Lai, 2006). In June 2006, emerging markets made up 6.38 per cent of global equity trading (Standard & Poor's, 2006).

For some countries in the Global South, the **globalization** of financial services has provided new economic opportunities in the form of **off-shore banking** services. Outside Europe, such services have been concentrated in the Caribbean, particularly the Bahamas and Cayman Islands and the Pacific islands, such as Vanuatu. For some countries with few opportunities for foreign exchange earnings, **off-shore banking** provides an alternative to primary production, as does tourism (see below). Funds are attracted because of the tax-free environment and secrecy laws. While governments of off-shore financial centres offer to relax their fiscal **sovereignty** (tax-raising powers), as is often the case with **export processing zones** for manufacturing, the strength of the state to enforce secrecy laws is an attraction for foreign investors. This complicates the often simplistic view of **globalization** undermining state **sovereignty**, particularly in the South (Hudson, 2000). Increased pressure on off-shore financial centres to be more transparent to prevent the circulation of money for terrorism may, however, undermine this **sovereignty** in the future.

The circulation of money around the global financial system relies not only on organizational structures and technology; the role of individuals' social relations and place-specific knowledge are also key. Jonathan Beaverstock (2002), for example, highlights the ways in which highly skilled financial sector workers in Singapore use their local knowledge and connections to generate business. Thus, although the intensity of capital flows around the world has increased greatly in the postwar period and the global network of financial exchanges is expanding, these flows are embedded in specific contexts.

The importance of cultural context can be demonstrated in relation to some Islamic interpretations of financial exchange. According to *sharia* law, devout Muslims should not be involved in transactions which involve the payment or receipt of interest. Investments in unethical activities such as alcohol production and gambling are also forbidden. This has meant that for some Muslims, financial products provided by Northern financial institutions are unacceptable. In Islamic countries,

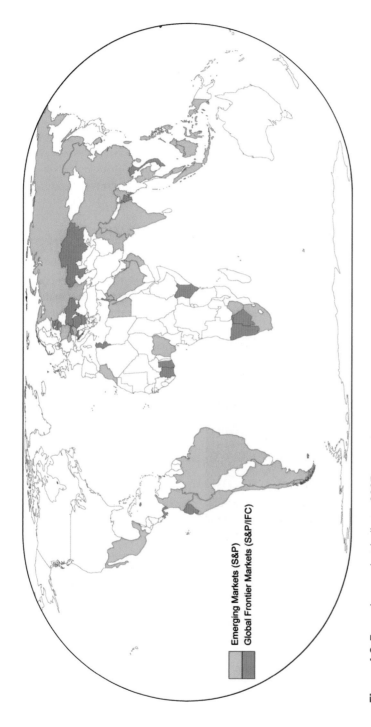

Emerging Markets (S&P)

Global Frontier Markets (S&P/IFC)

Figure 4.6 Emerging market indices, 2007.
Source: Based on data from Standard & Poor's (2007). Map data © Maps in Minutes™ (1996).

banks have developed to provide appropriate products (Khan and Mirakhor, 1990) and with changes in banking regulations these have expanded to meet the needs of some Muslims elsewhere. For example, a bank from Abu Dhabi has been involved in financial services for some of the 1.5 million Muslims in South Africa (*Finweek*, 2006). Seeing commercial opportunities, many Northern financial institutions have now developed their own financial products which meet the requirements of sections of the Muslim community who do not want to use existing services for religious reasons. In the UK, HSBC Bank now has an Islamic banking division called 'Amanah'. The example of Islamic banking demonstrates the way in which economic processes are intertwined with social and cultural ones in particular spaces in both the North and South (Pollard and Samers, 2007).

ENVIRONMENTAL DIMENSIONS OF ECONOMIC GLOBALIZATION

Economic activities have environmental impacts. While these impacts may be limited in some instances, it is important to recognize that agriculture, manufacturing and the trade of goods and services all involve the natural environment to varying degrees. This may be, for example, the use of natural resources as an input, waste products from the production process, habitat destruction to build factories or offices, or pollution as a result of employees' journey to work. Plate 4.3 shows a local resident in Durban, South Africa, with information about measures to deal with the environmental impacts of a petrol spillage. Globalizing tendencies within certain economic processes do not necessarily mean an increase in environmental problems, but this has often been the case.

Environmental protection legislation has become increasingly important in Northern countries since the 1960s. This greater regulation has meant that some production processes have become prohibitively expensive, or are completely illegal. Environmental regulations have also been increasing in many parts of the

Plate 4.3 Resident affected by petrol-contaminated land, Durban, South Africa. Credit: Glyn Williams.

Global South, but required standards are often set at much lower levels and enforcement is frequently ignored by the authorities (see Dwyer, 1995, on Northern Mexico). This means that environmentally damaging production processes are frequently tolerated; for governments seeking to attract **FDI**, flexible environmental legislation may be another incentive they can offer. Of course, it is not just **TNC** or **MNC** production which may be environmentally polluting; local processes such as leather tanning may also contribute greatly to deteriorating environmental quality.

The position of certain Southern countries in global **commodity chains**, may also contribute to environmental destruction in the Global South. As outlined above, many Southern countries rely on agricultural exports for the bulk of their foreign currency earnings. As the demand for certain foodstuffs increases, habitat destruction may rise. Shrimp farming in Thailand (Vandergeest, 2007) and Colombia has destroyed thousands of hectares of mangrove swamps, while vast areas of the Amazon rainforest in both Brazil and Bolivia have been deforested for soybean production (see Box 4.7). In such cases, **globalization** processes have involved the creation of new markets, the provision of financing and transportation technologies to make the movement of the products logistically possible.

From this discussion it is clear that as countries of the Global South become more and more incorporated into the global economy, their natural environments are increasingly threatened. In some cases, the countries themselves are being viewed as 'waste disposal sites' for the world's garbage. As landfill sites in the North become full to capacity and environmental legislation means that certain materials cannot be disposed of, an international trade in waste for recycling or disposal has been growing. In ports such as Alang, India, and Chittagong, Bangladesh, hundreds of ships are broken up every year. This is often done by workers using very basic tools and limited protective equipment, exposing them to toxic substances and also accidents (see Plate 4.4). As the ships are broken up, environmental pollution often ensues, as substances such as arsenic, asbestos and mercury escape into the environment (*The Economist*, 2005).

MIGRATION, REMITTANCES AND PEOPLE TRADING

Globalization has involved greater mobility of goods and capital, but it has also involved the migration of people. Barriers to international mobility tend to be much higher for people than goods or capital, although the global business elite, or '**transnational** capitalist class' (Sklair, 2000) are certainly given greater freedoms than unskilled labourers. Immigration controls and costs have not prevented millions of people migrating to gain employment. In mid-2005, there were an estimated 190.6 million international migrants (UNDESA, 2006). This represents a more than doubling of international migration since 1960 (see Figure 4.7).

Migration can be considered from a range of perspectives that will come

BOX 4.7

Soybean production in the Amazon

World production of soybeans has increased massively from 30 million tonnes in 1965 to 270 million tonnes in 2005. This increase reflects soya use in processed food and, most importantly, as an ingredient in livestock feed. In the early 1980s, 90 per cent of world soybean exports came from the USA. By 2005, Brazil had taken over as the world's main soybean exporter, with significant exports also coming from Argentina and Bolivia.

Increasing soybean demand has led to extensive destruction of the Amazon rainforest in both Bolivia and Brazil. For example, in the period 2004–05, 1.2 million hectares of soya were planted in Brazilian Amazonia. This deforestation has clear environmental impacts in terms of the loss of biodiversity, soil erosion and the release of carbon. The expansion of soybean production is a response to changes in the global economic system. For example, the introduction of **structural adjustment programmes** in the 1980s led to the opening up of both the Bolivian and Brazilian economies to foreign investment and a focus on export earnings. In Bolivia, 70 per cent of soybeans are now grown by foreign individuals or corporations. These corporations are not just those from the Global North; Brazilian firms are also very involved. In Brazil, the key corporations are from the USA (particularly Cargill), France and the Brazilian company, the Amaggi group.

While legislation to protect the Amazon forest exists, there are problems with enforcement. For example, in Bolivia, formal credit is often refused to those wanting to clear forests for soybean production, but money from the coca industry or semi-legal timber activities have been used as investment capital instead. In August 2006, consumer pressure led to the signing of a two-year moratorium on the purchase of soybeans from Amazonia by the five main agrobusinesses and key purchasers of meat from animals fed with soya-based feed from Amazon sources. McDonald's had been particularly targeted in this regard, but Northern supermarket chains and other fast-food companies have also signed the agreement.

(Sources: Adapted from Greenpeace International, 2006; *The Guardian*, 2006; Hecht, 2005; Sauven, 2006)

Plate 4.4 Dangerous work in ship-breaking yard, Chittagong, Bangladesh.
Credit: © Still Pictures.

up throughout the book, but particularly in Chapter 5. However, in relation to the theme of this chapter, the issue of key importance is **remittances**. In 2005, **remittances** totalled US$233 billion, of which US$167 billion went to the Global South (World Bank, 2006b). Given that the total amount of **Official Development Assistance** (ODA) given to the Global South in 2003 amounted to US$65.4 billion (UNDP, 2006: table 19), the importance of **remittances** is clear.

Within the Global South, **remittance** flows are highly concentrated, with India, Mexico and the Philippines constituting the recipients of the largest amount of **remittances** from their emigrants, although accurate figures are difficult to obtain (Table 4.6). In Bangladesh in 2006 the value of **remittances** was equivalent to over 40 per cent of the earnings from the export of goods and services, making it a key component of foreign currency flows into the country. For many countries, **remittances** represent significant contributions to the national economy, as well

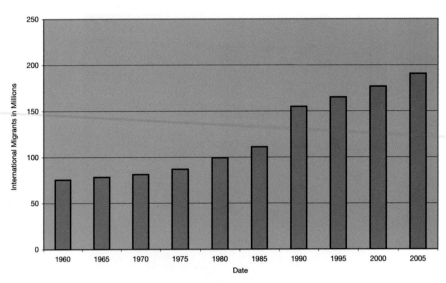

Figure 4.7 Levels of international migration, 1960–2005.
Source: Data from UNDESA (2006).

Plate 4.5 Remittance transfer service,
Dar es Salaam, Tanzania.
Credit: Claire Mercer.

as being of crucial importance to individuals, families and communities. Firms such as Western Union provide facilities to transfer **remittances** throughout the world (see Plate 4.5).

Migration can also create new business opportunities (Salt and Stein, 1997). Moving from one country to another is often a difficult process, regardless of socio-economic status. As migration becomes more common, relocation companies are expanding to meet the needs of expatriate workers, visa-processing services are set up to facilitate the move and people smugglers gain work helping individuals migrate illegally. Other ways in which bodies are commodified for international exchange include marriage brokerage businesses (Constable, 2004) and people trafficking.

Table 4.6 Inflows of remittances, 2006

	Amount US$m	Remittances as % of value of exports of goods & services
India	25,426	12.8
Mexico	25,052	9.4
Philippines	15,239	28.8
China	11,150	1.1
Indonesia	5,722	5.0
Egypt	5,330	14.5
Lebanon	5,202	36.1
Morocco	5,454	25.1
Bangladesh	5,428	42.1
Pakistan	5,121	25.0

Source: Data from UNCTAD, 2008a: 352, table 7.4.

TOURISM

Given limited opportunities for industrial development and wanting to diversify from reliance on primary production, many countries in the Global South have focused on international tourism as a potential source of foreign exchange earnings and job creation. In 2004, international tourism receipts globally amounted to US$622 billion, up from US$273 billion in 1990 (World Tourism Organization, 2006). With increasing air travel, growing disposable income among some sections of the world's population and communications technology (which can highlight the delights of other parts of the world), **globalization** has been key in the massive expansion of international tourism.

Within these flows of international tourists and their expenditure, the Global South is largely positioned as a destination, rather than a source. However, it is worth highlighting that in 2004, African countries made up 2.4 per cent of total outgoing international tourists. In the same year, China was the seventh most important contributor to international tourism expenditure when Chinese tourists overseas spent US$19.1 billion. In per capita terms, this is only US$15 compared with a global average of US$98, but these two examples highlight the diversity of societies within the Global South (World Tourism Organization, 2006). As we will see in Chapter 8, consumption patterns among middle-class residents in the urban areas of the Global South may often match those of their counterparts in the North.

For some countries in the Global South, tourism has become a key part of the economy (see Table 4.7 and Box 4.8). The natural environment, from beaches and reefs to rainforest and wild animals, are all used to entice foreign visitors. The

Table 4.7 Contributions of travel and tourism to GDP and employment, 2008[a]

Contribution to GDP		Contribution to employment	
	%		%
1	Macau	82.5	Antigua & Barbuda	87.1
2	Antigua & Barbuda	76.5	Aruba	81.4
3	Anguilla	69.6	Macau	74.6
4	Aruba	69.3	Anguilla	73.7
5	Maldives	67.0	Seychelles	71.1
6	Seychelles	56.3	Bahamas	63.5
7	Bahamas	50.8	Maldives	57.9
8	Barbados	40.7	US Virgin Islands	46.1
9	St Lucia	40.6	Barbados	45.8
10	Vanuatu	38.8	British Virgin Islands	45.2

Source: Data from WTTC, 2008: 1–3.

Note:
a   Predicted contribution. Includes both direct and indirect economic contributions and
    employment.
    Table shows the top ten countries for each category.

dependence on one source of foreign earnings has similar problems to the primary product dependency described earlier. Political instability, terrorism and natural disasters can all affect tourist flows, leaving a legacy of poor economic performance long after the initial event has passed.

The provision of tourism facilities has also become part of a global industry. The standardization of facilities in many international hotel chains means that food, furniture and uniforms, among other things, will not be sourced locally. Thus, although jobs are directly created, the **multiplier effect** of some forms of tourist development is limited (Hall and Lew, 1998; Telfer and Sharpley, 2008).

CONCLUSIONS

Since the colonial period, economic activities have become increasingly enmeshed in global-level networks. However, such networks are not evenly spread throughout the world; within the Global South, most of Sub-Saharan Africa, Central Asia and Oceania have been particularly excluded, while East Asia and parts of Latin America and the Caribbean have become more integrated. Patterns of economic flows vary depending on the type of economic activity, but in general while there is a net flow of capital from North to South, there remains a net transfer of profits

BOX 4.8

Tourism in small island states

Small Island Developing States (SIDS) are a category of country which has been identified as particularly economically vulnerable. This is because their limited market size, lack of opportunities for economies of scale and often their remoteness means that industrial development and infrastructure investment are costly. Of the 51 SIDS identified by the UN Department of Economic and Social Affairs (UNDESA, 2008), 23 are in Latin America and the Caribbean, 22 in Asia and the Pacific and 7 in Africa.

The climate and natural environment, including beautiful beaches, in many of these SIDS have encouraged governments to focus on tourism as a development strategy. Policies to attract tourist investment include the development of appropriate infrastructure, including airports and tax incentives. As global tourist numbers have grown, SIDS have benefited from the increasing demand for more exotic places and a willingness and ability to travel further for holidays.

In 2008, six of the countries where travel and tourism earnings made the largest contribution to GDP, and seven of those where this sector made the greatest contribution to employment, were Caribbean islands (see Table 4.7). All of these were SIDS. Of the other countries in the top ten for both indicators, only Macau does not appear on the SIDS list. Macau has been a Special Administrative Region (SAR) of China since its return to Chinese rule in 1999. In contrast to regulations in the rest of China, large-scale casinos are allowed in the peninsula of Macau. This has led to the development of resorts and other facilities to attract tourists.

For SIDS, the reliance on tourism, while aiding economic growth, does not greatly reduce economic vulnerability. Some SIDS have been able to develop markets for offshore financing (see above) but for most SIDS, tourism remains the only significant source of foreign currency and formal employment. Changing global economic conditions (such as recession or rises in flight costs), natural disasters (such as the Indian Ocean tsunami in 2004) or changing beach conditions due to sea level rise, could all have a dramatic effect on tourist flows and therefore the economy of SIDS.

(Sources: Adapted from Hampton and Christensen, 2007; SIDSNET, 2008; UNDESA, 2008)

in the other direction. The growing economic importance of Chinese, Indian and other Asian nations, as well as the continued role of oil-producing nations in the South, has meant that economic influence is not purely vested in the governments and companies of the North.

 Review questions/activities

1 How and why have economic globalization processes led to the uneven incorporation of countries in the Global South into the global economy?

2 Compare and contrast the ways in which two countries in the Global South have sought to engage with the global economy. Useful sources to start from are government trade and investment promotion websites and UNCTAD reports.

3 How far would you agree that economic globalization is bad for the environment?

SUGGESTED READINGS

Dicken, Peter (2007) *Global Shift: Mapping the Changing Contours of the World Economy* (fifth edition), London: Sage.
    Very accessible discussion of the spatial dimensions of economic globalization.

Held, D., McGrew, A., Goldblatt, D. and Perraton, J. (1999) *Global Transformations*, Cambridge: Polity Press.
    A clearly written and empirically rich overview of globalization debates.

Murray, Warwick E. (2006) *Geographies of Globalization*, London: Routledge.
    An excellent overview of the debates around globalization and the role of economic processes in globalization.

WEBSITES

www.imf.org International Monetary Fund
    Provides detailed economic statistics on global flows of investment and trade.

www.standardandpoors.com Standard and Poor's
   Good source of information about global investment flows and trends,
   particularly in the context of emerging markets.

www.unctad.org United Nations Conference on Trade and Development
   An excellent source of economic statistics, particularly the *Global Investment
   Report.*

www.world-tourism.org World Tourism Organization
   A very useful source of international tourism statistics and details of global
   tourism trends.

5 Social and cultural change in the South

INTRODUCTION

In the final Chapter of this part, we focus on the diverse ways in which areas of the Global South participate in social and cultural change (see Concept Boxes 5.1 and 5.2). This Chapter makes three key arguments: first, despite their complex links to **globalization**, processes of socio-cultural change in the Global South (and North) are not necessarily homogenous (although at times they can be) and neither are they typified by a singular process of **Westernization** from the centre to the Global South periphery. Second, economic, political and cultural **globalization** does not level out inequalities and differences of opportunity across the world. They are uneven processes which may entrench inequality in different ways (Tomlinson, 1999; Schech and Haggis, 2000). Finally, the tendency to understand social changes through a developmental lens has characterized many conventional analyses of these processes. We argue here for a broader approach which recognizes the role of cultural change in conjunction with social change, so as to explore the Global South as places of socio-cultural activity, rather than simply as spaces defined by developmental failure.

We have chosen to focus on four trends and practices to illustrate our three key arguments: changing patterns of health and lifestyle; migration and **diaspora**; **urbanization** and city living; and religion. Our focus on health and **urbanization** may seem to reinforce negative stereotypes of the Global South that we criticized in Chapters 1 and 2. Indeed, concentrating on HIV/AIDS, malaria and the 'explosion' of cities can generate problematic impressions, but at the same time these problems are very real, and affect millions of people across the world. It is the role of this Chapter to interrogate some of the harsher realities of life in the Global South, while broadening our understanding of how these are produced. These negative trends are also explicit evidence of global inequality: for example, the under 5 mortality rates for children living in Chad was 200 per 1,000 of the population in 2006, compared with rates of 5 per 1,000 in France and

---

CONCEPT BOX 5.1

Social change

Social changes are broad processes of change (or stagnation) which relate to society and people more specifically. Often distinguishable from political, economic or physical change, social changes are trends which shape and are shaped by society, such as education, welfare, crime, health, and **demography**. **Social changes** are more commonly considered within a 'development' framework because they are seen to be important to shaping levels of **poverty** and development. Also, unlike cultural changes (Concept Box 5.2), they are usually measurable or quantifiable in part.

CONCEPT BOX 5.2

Cultural change

Cultural changes relate to the cultural identities of people. These trends link identity with people's ways of living and they give us insight into the dynamics which shape people's lives. Common types of **cultural changes** that are considered are: religion, community or group **identity**, cultural attitudes towards consumption and ideas of nationalism. These cultural trends in particular have been overlooked in earlier Development Studies texts because their connection to questions of **poverty** and development has been poorly understood. They are also more difficult to study because they are often immeasurable. They are however fundamental to the lives of people everywhere and thus shape our understanding of people in the Global South.

---

8 per 1,000 in the USA (WHO Statistics, 2006). Babies and toddlers in rural Chad have a one in five chance of dying compared with negligible chances in Western Europe and the USA, a situation that is unacceptably unequal, and thus we view a critical discussion of these trends as central to the overall politics of our book.

HEALTH AND LIFESTYLE

Conventional analyses of health patterns across the Global South have focused on the spread of infectious diseases and poverty-related illnesses. These approaches risk presenting health matters largely in medical terms, or as a development problem. We want to consider health more broadly, and thus we focus also on cultural issues and **lifestyle** to widen our focus. We use this broader focus initially to consider the distinction between infectious and non-communicable diseases, and then using the examples of HIV/AIDS, maternal health and processes of ageing we illustrate how a broader approach reveals the connections between health and socio-cultural change.

Analyses of health and lifestyle at a global scale illustrate tendencies towards homogenization (the spread of infectious diseases for example) (see Box 5.1 and Table 5.1); the diversity of practice at local and national scales (such as cultural attitudes to unwanted pregnancies); and complex diffusions of trends (such as the spread of smallpox from the Global North to Latin America in the 1500s). Rising political, social and economic inequalities at a global scale suggest that, for many,

Plate 5.1 Living conditions in Makoko, Lagos, Nigeria.
Credit: Muyiwa Agunbiade.

BOX 5.1

Global trends in infectious disease

The following infectious diseases are ranked as those most responsible for death across the world (in order of their severity): lower respiratory infections, HIV/AIDS, diarrhoeal diseases, tuberculosis, malaria and measles. Lower respiratory infections are the leading killer, accounting for 3.9 million deaths annually. Pneumonia is the key disease here and its impacts are most severe in the developing world. HIV/AIDS contributes to about 2.8 million deaths annually, followed by diarrhoeal diseases which account for 1.8 million deaths annually. A key disease within this category is cholera which again is primarily found in the Global South. Tuberculosis (TB) causes nearly 1.6 million deaths per year and its incidence (new cases arising) as well as deaths from TB are largely (but not entirely) concentrated in countries of the Global South. Table 5.1 outlines the global prevalence (number of existing cases) and mortality rates of TB, illustrating the distortion of this trend in the Global South. The reasons why TB is far more dominant in the Global South relate largely to the living conditions of poorer people, and also to the health status of individuals. This is because TB spreads through the air in conditions of overcrowding, and the malnourished and those already infected by other diseases (such as HIV/AIDS) are more susceptible to the illness.

Like TB, malaria is a disease which largely affects people living in the Global South, particularly those living in tropical climates, with Sub-Saharan Africa suffering from a high proportion of malaria-related illness and death. Malaria is also particularly dangerous for young children, particularly the under 5 age group who have not built up their immunity to the disease. Measles again is particularly geographically specific, with the majority of measles-related deaths occurring in Africa and South East Asia. In 2005 an estimated 345,000 people died from measles (most of whom were children). Political and social instability strongly affects measles infection rates. In countries suffering from war or natural disasters (floods, earthquakes, etc.), health services are often interrupted and this reduces the routine immunization of children, leading to disastrous outbreaks (WHO, 2007). To add to the complexity, the connections between these diseases are very strong, for example, people who are already HIV positive stand a far higher chance of contracting TB.

(Source: Adapted from WHO, 2007)

Table 5.1 Global prevalence and mortality rates of TB in 2005

	Prevalence in thousands	Per 100,000 of the population	TB mortality in thousands	Per 100,000 of the population
Africa	3773	511	544	74
The Americas	448	50	49	5.5
Eastern Mediterranean	881	163	112	21
Europe	525	60	66	7.4
South East Asia	4809	290	512	31
Western Pacific	3616	206	295	17
Global	**14052**	**217**	**1577**	**24**

Source: Adapted from the WHO Tuberculosis Fact Sheet, 2007.

health will be further undermined and also that lifestyle choices will be constrained (Last, 1999). Lifestyle practices and health outcomes are also directly shaped by living conditions (often a reflection of inequality), as poor conditions, such as that seen in Plate 5.1 in Lagos, can facilitate chronic illnesses such as cholera, malaria and tuberculosis.

Inequality is not the only component of diverse health trends however. Health, and responses to ill health, are also differentiated at a global scale by what Murray Last describes as particular 'medical cultures'. These are, in part, shaped by government legislation and institutions (Last, 1999: 80) (e.g. policy on HIV/AIDS) but they are also influenced by the everyday practices of people, the choices they make and their beliefs about health. The use of 'traditional or alternative therapies' across the Global South (and North) in complex combination with biomedicine is indicative of the exceptional variety of health practices at the global scale which includes traditional healers, homeopaths, charismatic churches, acupuncturists and others.

Infectious diseases persist across the Global South as Box 5.1 illustrates and the discussion of HIV/AIDS points to the complex causes of these trends. There is however also a more recent rise in non-communicable diseases (those which cannot be passed on) and illnesses shaped by lifestyle factors. These factors are primarily socio-cultural practices relating to diet, exercise, the consumption of alcohol, drugs and tobacco and sexual practice. They directly impact on health and are responsible for illnesses such as diabetes, cardiovascular disease, kidney failure and cancer. The World Health Organization (WHO) argues that countries within Africa, for example, are now experiencing a double burden of disease, namely very high levels of infectious disease and rapidly rising non-communicable disease. Historically, illnesses such as heart disease were assumed to be diseases of wealth, that is associated with a wealthy lifestyle and usually with people in the Global North. This assumption is no longer accurate, as lifestyle practices are shared by many wealthy and poor people across the world, in part a function of **globalization**. A transition or shift in epidemiology is thus occurring. Cardiovascular disease is the key concern here, as it is growing rapidly across the Global South and is estimated to be responsible for 30 per cent of all deaths across the world in 2005 (WHO, 2008a). This disease has explicit lifestyle habits associated with its rise, namely tobacco and alcohol consumption, poor diet and the rise of a sedentary lifestyle (WHO Regional Office for Africa, 2005). Obesity (in contrast to under-nutrition, historically the key concern in relation to malnutrition) is also an important contributing factor in the rise of cardiovascular disease. Dietary shifts, urbanization, urban living and reliance on mechanized transportation have led to more sedentary ways of living. The WHO estimates that globally, physical inactivity causes 1.9 million deaths annually (WHO Regional Office for Africa, 2005: 1). Figure 5.1 illustrates the variations in underweight and obesity levels in different countries and it shows how obesity and malnutrition can exist side by side. Obesity statistics are problematic however, because the Body Mass Index (BMI) calculation (weight/height^2) is not sufficiently sensitive to different ethnic body types and is fundamentally a Northern concept (see Deurenberg, 2001). Figure 5.2 provides a global view of male tobacco consumption and illustrates how widespread this lifestyle practice is.

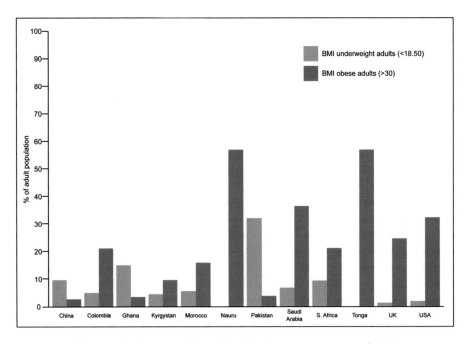

Figure 5.1 Global variations in underweight/obesity as percentage of adult population.
Source: Data from WHO (2008b).

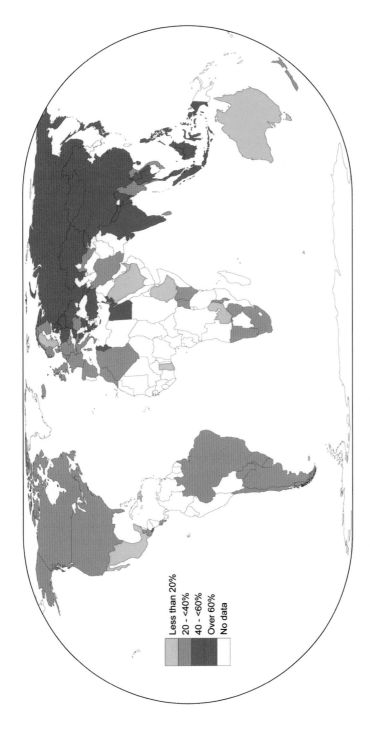

Figure 5.2 Global trends in adult male tobacco consumption.
Source: Data from WHO (2008c). Map data © Maps in Minutes™ (1996).

Analyses of lifestyle practices are very complex, and health outcomes (such as rising diabetes) cannot simply be blamed on the behaviours of individuals. Broader structural processes, namely economic (for example, the ability to afford consumer products), political (such as national alcohol and tobacco taxation rates) and changing physical environments (the rise of malls and the decline of pedestrian pathways) all directly shape people's lifestyle practices. These shifts are closely related to changing global flows in consumption, in terms of how people eat, consume leisure, and their aspirations relating to car ownership and so forth (see Chapter 8).

Ideas about health are shaped by global flows of knowledge and socio-political trends. There is some global agreement about the significance of health as a cornerstone of the concerns about the Global South. The eight UN Millennium Goals (see Chapter 2) are evidence of attempts to homogenize social concerns at a global level – although arriving at these was by no means free of debate (see Freedman, 2003). Three of the eight goals are health specific (see Table 5.2) illustrating the importance of health on the international agenda. The work of the WHO also contributes to the spread of globalized cultural norms about the meaning and achievement of good health. In addition the WHO fosters norms about appropriate and inappropriate lifestyle behaviours (especially around alcohol, tobacco and drug consumption and sexual behaviour).

However, global discourses on health are not equivalent to global socio-cultural practices of health, which remain highly uneven. To examine these differences in more depth, the following sections focus on three issues: HIV/AIDS, maternal health and the implications of an ageing population.

The rise of HIV/AIDS

The UN estimates that in 2007 33.2 million people were living with HIV globally, and regionally the figures vary dramatically: it is evident from Table 5.3 that Sub-Saharan Africa bears the brunt of people living with HIV and dying from AIDS

Table 5.2 The UN Millennium Development Goals

1	Eradicate extreme poverty and hunger
2	Achieve universal primary education
3	Promote gender equality and empower women
4	Reduce child mortality
5	Improve maternal health
6	Combat HIV/AIDS, malaria and other diseases
7	Ensure environmental sustainability
8	Develop a global partnership for development

Source: United Nations, 2008.

Table 5.3 The total number of people living with HIV and dying from AIDS in 2007 by region

Region	Estimated totals	Total deaths due to AIDS
North Africa and Middle East	380,000	25,000
Sub-Saharan Africa	22.5 million	1.6 million
Latin America	1.6 million	58,000
Caribbean	230,000	11,000
Eastern Europe and Central Asia	1.6 million	55,000
East Asia	800,000	32,000
South and South East Asia	4 million	270,000
Oceania	75,000	1,200
Western and Central Europe	760,000	12,000
North America	1.3 million	21,000

Source: Adapted from Joint United Nations Program on HIV/AIDS, 2007.

across the world. In addition, regions of the Global South in general are experiencing the highest prevalence of the disease. Obviously regional statistics entirely obscure intra-national and local differences, of which there are many, and there are also issues about data accuracy.

There are many explanations for the rise in HIV/AIDS across the world. They can be categorized into economic, biological, social, political, cultural and geographical causes. The complexity of trying to understand the AIDS epidemic shows that even though it may seem to be a global trend, in reality, at all scales, it is highly heterogeneous. This relates to all the explanations outlined above which include differing 'medical cultures' (e.g. government attitudes towards contraception) and lifestyle practices discussed earlier. Nandita Solomon notes that biomedical, economic and financial explanations still dominate discussions and she points to the significance of **gender** (see Concept Box 5.3) in understanding HIV/AIDS. Arguing that the general HIV discourse tends to focus on short-term strategies, she explains that 'gender and the related sexual and structural inequalities and inequities between men and women are not normally discussed' (Solomon, 2005: 177) despite their potential long-term benefits. Box 5.2 outlines some of the key components of **gender** relations and how they link to HIV/AIDS.

A focus on gendered inequalities also points to the significance of cultural (and lifestyle factors) in shaping the AIDS epidemic. HIV infection is often linked with alcohol and drug consumption and with the use and practice of sex workers. The adoption of risky sexual practices has concerned policy makers across the world. Catherine Campbell's work in South Africa examines the socio-cultural contexts within which mine workers purchase (and demand unprotected) sex with local sex workers, where both parties are aware of the HIV infection risks. With

---

CONCEPT BOX 5.3

Gender

Gender is used to denote the socially produced differences between men and women, as opposed to the term 'sex' which refers to biological differences. The term '**gender**' indicates that 'real' and perceived differences between men and women are socially derived (such as women are weaker and more emotional, and men are stronger and more rational), i.e. they are traits that are perceived to be appropriated by some men and women through socialization and social interaction. **Gender** is thus both time and place specific and changes across both. The term **gender** allows us to recognize the differences between different men and also between different women, as well as to acknowledge that being gendered, i.e. being a woman or a man, has particular **power** implications.

---

reference to the men, she situates their practices within the broader lifestyles of the mine workers, who reside in cramped although lonely living conditions and risk their lives daily going down the mines, and whose socially constructed sexuality was such that 'going after' women (and demanding skin to skin sexual contact) was fundamental to being a real man. The purchase of sex and alcohol from women was a function of the lifestyles, cultural norms and realities of structural inequality that the men faced. For the female sex workers, their risk taking was a function of their **poverty** and desperation, and their powerlessness to challenge men about condom use. Many of the women had abandoned their children to work in the city and were filled with shame about both this and their working practices (Campbell, 2003). Underscoring these dynamics are unequal **gender** relations, but their explanatory value is strengthened in conjunction with other insights.

The recognition of lifestyle practices has developed alongside globalized debates about how to reduce HIV infection, and widespread campaigns call for varying approaches from abstinence (the ABC approach Abstain, Be faithful and Condomize) to safe sex messages and South Africa's LoveLife advertising brand. Globally, AIDS campaigns adopt particular emphases, often relating to the use or not of contraception. Locally then, campaigns are particular to their national contexts. The support of them by government officials and institutions, and also by everyday people, is an important illustration of the role played by 'medical cultures' in shaping health.

BOX 5.2

Gender and HIV/AIDS

Adopting a gendered analysis of HIV/AIDS reveals that women are more vulnerable than men for various reasons.

1 In the Global South HIV is largely transmitted through heterosexual intercourse and, because of the permeability of the vaginal mucous membranes in comparison to the penis, women are more vulnerable. The number of women infected is higher than that of men.
2 Women have suffered disproportionately from a **discourse** of blame as they are (incorrectly) perceived in different cultures to be the transmitters of the virus. As a result of this **discourse** and the initial focus globally on the homosexual nature of the disease, women's vulnerability has been overlooked.
3 Women are also more vulnerable for physiological and social reasons to sexually transmitted diseases which increase their chances of infection.
4 Socially women have a relatively low status, with reduced access to education, income, health care and legal support. This relative powerlessness shapes their vulnerability to HIV infection as social processes shape sexual relations. In contexts of political and social instability and in relation to coercion these unequal **power** relations are all the more stark.
5 Women (young and old) are disproportionately burdened by caring for HIV patients.

(Source: Adapted from Baylies and Bujra, 2000; Gender and Aids, UNIFEM, 2007)

Maternal health

Although women give birth throughout the world, the socio-cultural realities of pregnancy and childbirth vary dramatically within national and local contexts (see Plate 5.2). We concentrate here on the medical cultures which foster **power** imbalances relating to maternity, but the everyday joy and celebration relating to reproduction for most must not be overlooked. Nonetheless, reproduction is a highly gendered process, not only because women bear children, but also because there are widespread assumptions about being a (proper) woman which relate to the act of reproducing and also the subsequent role of child caring. Cultural assumptions about the appropriate roles and practices of men and women in relation to reproduction are strongly held. Women often have less control over reproductive decision making, shaped further by the status of women nationally. Loss of control affects the numbers of children women bear, the timing between children, who cares for the children and how this impacts on their other tasks (see Concept Box 7.3 on domestic labour). In most countries women are treated as inferior to men, and medical cultures often reinforce this inequality. Maternal health care (a key component of reproduction and an example of a medical culture) reveals a variety in quality and quantity of care (see Table 5.4). Again the region of Sub-Saharan Africa is especially polarized with almost double the rate of maternal deaths compared to the rest of the world.

National **poverty** is argued to be a key explanation for the high rates of maternal mortality, but we illustrate that this relationship is contradictory. Poor countries lack the funds to provide appropriate maternal care, or are forced to make compromises over which health sectors to fund. A comparison of maternal health services across 49 developing countries reveals that four countries scored 'extremely weak' (under 30 per cent of women had access): Yemen, Pakistan, Nepal and Ethiopia. Most African countries aside from Egypt and South Africa scored either 'very weak' or 'weak'. However, gaps between countries with 'moderate' versus 'extremely weak' access to services were not paralleled by an equally wide gap in 'budget adequacy', suggesting little financial dissimilarity between countries (Bulatao and Ross, 2002: 725). Indeed inequalities in maternal health care within countries in Sub-Saharan Africa were more pronounced in countries with better overall (national) maternal health (Magadi, *et al.*, 2003) indicating that national economic performance can be a poor indicator. Factors such as geography, politics, and socio-cultural norms also shape mortality variation.

The urban poor in many Sub-Saharan African cities experience difficulties accessing maternity services. Physical proximity to hospitals does not necessarily provide them with good access, particularly in the case of slum dwellers. The impact of **SAPs** (see Chapters 4 and 10) in Sub-Saharan Africa on health care is overwhelming and the urban poor are particularly affected by cost-recovery programmes (Magadi, *et al.*, 2003). However, living rurally also negatively affects women's access to maternal services (particularly during complicated births). For example, in 49 developing countries, on average only 39 per cent of rural women had access to such services, in comparison with 68 per cent of urban women (Bulatao and Ross, 2002: 722).

In Swaziland, unmarried adolescent pregnancies (accounting for two-thirds of national deliveries) are viewed very negatively by society, and are not offered either family planning services, or maternal and child health services (Mngadi, *et al.*, 2002). The implications of this are higher risks with consequent higher mortality rates. Cultural norms about what is 'acceptable behaviour' explicitly dictate a woman's access to services. Cultural and religious affiliation, as well as class, also shapes women's maternal experiences and their ability to exercise control over the

Table 5.4 Maternal mortality and skilled health care

	Maternal mortality ratios per 100,000 live births in 2000	Proportion of deliveries with skilled personnel 1990 (%)	Proportion of deliveries with skilled personnel 2003 (%)
Sub-Saharan Africa	920	40	41
Southern Asia	540	28	37
Oceania	240	–	–
South East Asia	210	34	64
Western Asia	190	61	62
Latin America and the Caribbean	190	74	86
Northern Africa	130	41	76
Commonwealth of Independent States	68	–	–
Eastern Asia	55	51	82
Developed regions	14	–	–

Source: Adapted from United Nations, 2008.

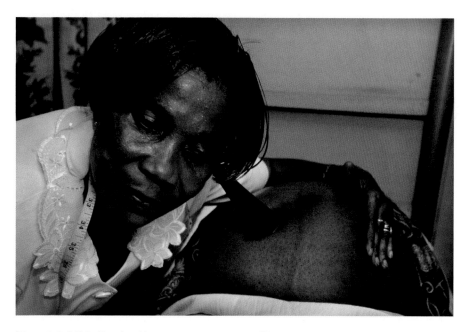

Plate 5.2 Midwife checking pregnant woman, Ghana.
Credit: © Still Pictures.

process of delivery. In Lebanon lower class rural Muslim women feel less able to manage their labour and are largely passive about the over-medicalization of the process (Kabakian-Khasholian, *et al.*, 2000: 104–5, 109).

Health and ageing

Finally, health, lifestyle and demographic changes are closely related. Health is intrinsically tied to social and cultural change, and similarly these changes have health implications. The ageing of populations is a global trend, although there is variation between the rates at which this is occurring between the Global South and North. The Global South is predicted to increase its population of over 65s by 140 per cent between 2006 and 2030 (compared with 51 per cent change in the Global North). Age structure is thus changing very rapidly within the Global South, a process described as the 'compression of ageing'. For example, Sri Lanka is predicted to change its age structure (whereby the proportion of the population in the over 65 category will rise from 7 per cent to 14 per cent) between 2004 and 2027 – a mere 23 years, in comparison to Sweden which took 85 years to change (between 1890 and 1975) (US Department of State, 2007: 7). The causes of ageing are related to declines in fertility, improvements in health and changes to lifestyles. The ageing of populations is evidence of a global demographic transition.

An ageing population presents many challenges for nations, particularly in terms of financial, medical and social resources. The global shift in disease epidemiology is impacting on health care resources. Ratios of workers to pensioners have shrunk, and ageing means that the numbers claiming pensions or other social support grows, without a concomitant growth in resources. This has led to a global rise in **poverty** rates among the elderly (UN Population Division, 2007). Box 5.3 on ageing in Malaysia illustrates the complex interconnections between health and finance.

MIGRATION AND DIASPORA

Many commentators describe the current time in history as 'an age of **migration**'. Although mass migration (see Castles and Miller, 2003) is not new 'what is different about the flows of people in the latter half of the twentieth century is that they are engendered not only by war, dislocation, and poverty, but also by relations of production and **consumption**' (Schech and Haggis, 2000: 59). Migration can be seen to spread **consumption** preferences and social practices across the world. The products of these movements of people and ideas are not homogenous however, and the diversities of **diaspora** (see Concept Box 5.4) populations across the world are testimony to this. Practices of migration are highly eclectic too and they vary by type of migration as well as by cause. Furthermore the experience of migration is uneven and is differentiated by factors such as skills, ethnicity and **gender**.

Migration is also tied to various global flows and processes. Investment patterns and global shifts in production and **consumption** are key and they influence economic migrant decision making.

Causes of migration are often divided into 'push and pull' factors in an effort to identify reasons for relocation with both the place of departure and the place

BOX 5.3

An ageing population in Malaysia

A number of factors have contributed towards the ageing of the Malaysian population: improved health, longer life expectancy, low mortality and declining fertility. By 2020 the aged population (60+) is predicted to constitute 9.5 per cent of the total population, a rise from 6.2 per cent in 2000. The **old age dependency ratio** was 10.5 in 1970 and is predicted to increase to 15.7 in 2020. These trends are differentiated by geography, ethnicity and gender. Rural areas have a higher proportion of elderly residents and of the three official ethnic groups (Bumiputra Malays, Chinese, Indian) it is the Chinese who are ageing at a far greater rate. Finally, older women in Malaysia are at higher risk as they are less educated, less financially independent and have higher health risks.

These trends place a burden on the health care system which is not adequately equipped to respond to this trend. Changing Asian culture also has implications for the ways in which the elderly are cared for (see Chapter 8). Historically the extended family system meant that female family members in particular could be relied upon to offer care of sick or elderly family. However the increase in female labour force participation and also the smaller size of families has meant that families are not able to act as primary care givers. This places a higher burden on (often unaffordable) institutional facilities. Savings schemes and financial planning for an ageing population are also being stretched, and complicated by the reliance of traditional savings systems on formal economy participation. In Malaysia, informal economy employment is very high thus a high proportion of working age people are not formally saving for their future needs. The challenge for Malaysia is to respond to this in adequate time to meet the demands of the future, and to recognize that whilst an ageing population is a positive sign (of improved health, etc.) it is essential to 'put quality into these added years'.

(Source: Ong, 2002)

CONCEPT BOX 5.4

Diaspora

Defining **diaspora** is highly complex and contested. The term **diaspora** refers to the 'scattering of a population' originally relating to the dispersal of Jewish populations in AD 70 but now used more widely to refer to a vast range of population dispersals, including 'victims' of forced resettlement (e.g. African **diaspora** in relation to the slave trade). Movement as labour is another key type of **diaspora** and can be related to imperial **diasporas**, where colonial subjects were encouraged to provide labour in colonies during the nineteenth and twentieth centuries. For example, between 1860 and 1911, 152,184 indentured Indian labourers migrated to South Africa to work on the sugar plantations in the province of Natal, and now form a substantial Indian **diaspora**, numbering 825,000 in 2006 in the region renamed KwaZulu Natal. Trade **diasporas** are a further category of non-state sponsored migrations (unlike imperial **diasporas**). Traders across Africa, Eastern Europe, the Americas and China formed varying **diasporas** pre-, during and post-colonization. Today the migrations of Japanese and Indians for professional and business purposes contribute to new and emerging **diasporas**. A final category of **diaspora** is that of cultural **diasporas** which classify many of the migrations taking place in the modern world (e.g. the movement of migrants of African descent from the Caribbean).

(Source: Adapted from Cohen, 1997)

---

of destination (see Parnwell, 1993; and Chapter 7). Understanding push and pull factors can provide a useful overview but may obscure differences in the significance of factors. Decisions to move between two areas are often based on a multitude of factors and attempts to rationalize the decision-making patterns of migrants can become meaningless. Migration is often not a practice of rational choice (in terms of having positive alternatives); rather it may be a desperate response to intense local, regional or national crises such as sudden economic decline, natural disaster or war and destabilization. The apparent causes and practices of migration also impact on migrant **identity** (see Concept Box 5.5) in both the destination and departure place.

The ways in which migrants' identities are constructed and represented (as deserters, or valuable economic assets, bearers of cultural diversity, scroungers, a source of threat or a human tragedy) explicitly shapes their experience, illustrating quite clearly the links between **representation** and **power** (see Chapter 2). Migration thus can be an important indicator and axis of social polarization, and discrimination against migrants is not unique to the Global North. Extensive violence

---

CONCEPT BOX 5.5

Identity

A focus on cultural **identity** is central to the cultural turn within the social sciences and the humanities. Ideas about **identity** are introduced when we try to define our sense of self, or who we are, usually in contrast or opposition to others. This may be done at the level of the individual (you or I), or it may be at the level of a group (Nigerian migrants in South Africa) or even nation. Identities, previously considered to be fairly stable and distinctive (such as gender, race, nationality and age), are increasingly fragmented, sparking concern over a growing crisis of identity (see Jackson, 2005, for an overview of debates). Ulrich Beck (1992) argues that this fragmentation of identity is linked to changing forms of social and economic organization. These include **globalization** and the growth in the transnational flows of people, knowledge, commodities and media, meaning that more historic ways in which we made sense of ourselves (often around ideas of nation) have now been challenged. This crisis of **identity** plays itself out in relation to how people relate to the world around them, as well as to each other and themselves. For some, the sense that their identities are being marginalized leads to attempts to reassert their 'traditional' sense of self or identity and individuals and groups mobilize around religious, cultural and ethnic identities – witnessed in fundamentalist and other social movements. Others argue that **globalization** has allowed for greater choice over one's identity, opening up new possibilities as individuals express different consumer choices (see Chapter 8) or embrace different subcultures (Bennett, *et al.*, 2005: 174).

---

in South Africa during May 2008 specifically targeted black foreign workers and residents (from countries such as Nigeria, Mozambique and Zimbabwe) in a wave of anti-immigrant hostility related to competition over scarce resources and widespread poverty (see Plate 5.3).

Patterns of migration and the spatiality of migration are mixed. At a global scale migration is a dominant cultural and geographic practice with an estimated 191 million international migrants in 2005, making up 3 per cent of the total world population. Of this total 30–40 million (i.e. 15–20 per cent) are considered to be 'unauthorized' (International Organization for Migration (IOM), 2007) and most nations have strengthened barriers to control immigration more than ever (Dwyer, 1999). Migration patterns are not simplistic, and do not follow an obvious Global South to North pattern, or even a rural to urban pattern within nations. A number of countries from the Global South are key hosts of migrants (see Table 5.5). Remember however that some are also key senders of migrants too, for example, in

Plate 5.3 Collage of South African newspapers depicting xenophobic violence.

2000, India and China sent 20 and 35 million migrants respectively to international destinations.

Practices of remitting money back to home countries is a dominant feature of international (and internal) migration along with the financial institutions that facilitate such exchanges (see Plate 5.4). The practice of remitting is indicative of the force of global **capitalism** and arguably an example of global interdependence (see Chapter 4). Financial flows are only part of the story: international migration also produces **diasporic** communities across the world, many of whom have **transnational** identities (see Concept Box 8.3). Aside from the particularities of these **transnational** identities, analyses of **diaspora** suggest that there are complex interconnections and continuities, as well as breaks, between the cultures of settled migrant communities' host and destination nations. As such, the study of **diaspora**

Table 5.5 The top 10 'host countries' for migration in 2005

Country	Number of receiving international migrations (millions) in 2005
United States	38.4
Russian Federation	12.1
Germany	10.1
Ukraine	6.8
France	6.5
Saudi Arabia	6.4
Canada	6.1
India	5.7
United Kingdom	5.4
Spain	4.8

Source: Adapted from International Organization for Migration, 2008.

Plate 5.4 Money Express remittance agency, Bali, North West Province, Cameroon.
Credit: Claire Mercer.

tells us much about processes of **cultural globalization**, particularly in areas such as art and music (see Dwyer, 1999, for an example of Bhangra music in the UK).

Diasporas not only engage in practices of remittance or artistic fusion, they also can exert a more powerful influence in the spaces they may call 'home'. For example, some engage in political lobbying for their home countries; they are involved in cultural and religious networks and also institutional linkages (Zack-Williams and Mohan, 2002: 205). China has a powerful but also immensely diverse **diaspora** (see Box 5.4): its global reach is extensive, and it clearly illustrates the force and extent of current global migration patterns. The Chinese presence across the world also challenges the simplistic assumptions embedded within ideas of **Westernization** which assume a unidirectional movement of ideas and cultural practices from the West outwards (see Table 5.6 and Plate 5.5).

Although global trends in the significance and presence of **diaspora** have been identified, their impact on cities has been highly uneven. Arguably cities across the world are more diverse now than ever before and this is largely a function of the multicultural and diverse populations that reside within them. Urban spaces however are not simply diverse because of the presence of immigrant communities, rather their diversity is also a function of the increasingly global cultural trend towards city living which encourages diverse cultural practices.

BOX 5.4

Chinese diaspora

Current estimates place the overseas Chinese population at almost 40 million with countries such as Singapore (2,684,900), Indonesia (7,566,200), Malaysia (6,187,400), Thailand (7,053,240) and the USA (3,376,031) hosting substantial numbers of ethnic Chinese. Historically migration grew from the tenth century onwards and Chinese traders began to move in substantial numbers as maritime trade increased. In the 1500s and 1600s many Chinese moved to the Philippines and Japan, trading silks and tools for example. Substantial waves of movement occurred during **colonialism**, such as migration to Singapore where ultimately Chinese immigrants formed the majority population. Elsewhere Chinese migration formed minority communities, often in response to colonial need (as indentured labourers) with the French encouraging Chinese movement to Mauritius.

As **globalization** has proceeded, Chinese migration for professional and business purposes has continued to grow. The Chinese **diaspora** is not

BOX 5.4 (*CONTINUED*)

static but fluid and ever changing and is constantly being shaped and reshaped at a global scale. The global Chinese **diaspora** also carries immense economic **power**, particularly through **remittances** repatriated to China, but also financial **power** invested in host nations. The growth and development of global Chinatowns are illustrative of providing overseas Chinese with an institutional structure through which community and financial needs can be met, although their existence also reinforces beliefs held by host populations about these then being the appropriate places within the city for Chinese immigrants to settle. Chinatowns represent a traditional and quintessentially Chinese space, and are marketed as such by city authorities (Fincher and Iveson, 2008). Chinatowns are spatially significant, illustrating the important connections between place, **identity** and culture.

Although they are marked by particular visual and consumption-related similarities, they are also highly diverse and context rich. Their particularities relate to the local area (such as New York, Kolkata, Liverpool) within which

BOX 5.4 (*CONTINUED*)

Plate 5.5 The Chinese Arch in Liverpool, UK.
Credit: Katie Willis.

BOX 5.4 (*CONTINUED*)

they are placed. Plate 5.5 illustrates the visual significance of Chinese **diaspora** within Liverpool. Less visible however is the growth of Chinese retail investment in suburban parts of the Global North, particularly in the form of suburban malls catering largely to Chinese customers in Toronto for example. These have been less comfortably integrated into the urban fabric as they are seen to challenge commonly held assumptions about where Chinese retailers should locate (Fincher and Iveson, 2008). The geography of **diaspora** is thus constantly changing.

(Sources: Adapted from Cohen, 1997; Fincher and Iveson, 2008; Overseas Compatriot Affairs Commission, 2007)

Table 5.6 Population of overseas Chinese in 2005

Country	Population
Indonesia	7,566,200
Thailand	7,053,240
Malaysia	6,187,400
United States	3,376,031
Singapore	2,684,900
Canada	1,612,173
Peru	1,300,000
Vietnam	1,263,570
Philippines	1,146,250
Myanmar	1,101,314
Russian Federation	998,000
Australia	614,694
Japan	519,561
Kampuchea	343,855
United Kingdom	296,623
France	230,515
India	189,470
Laos	185,765
Brazil	151,649
Netherlands	144,928

Source: Adapted from Overseas Compatriot Affairs Commission, 2007.

URBANIZATION AND CITY LIVING

Urbanization and city living are now dominant features of life in the Global South, although by no means recent (see Figure 3.1 where all the political units identified had their own histories of urbanism). Hania Zlotnik of the UN announced in early 2008 that more than half of all humans across the world will live in cities by the end of that year (United Nations, 2008). Conventional approaches to understanding this trend have concentrated on patterns of **urbanization** and the developmental challenges that this affords. For example, the UN describes global **urbanization** as follows:

> The slums and ghettos, the homeless, the paralysing traffic, the poisoning of our urban air and water, drugs, crime, the alienation of our youth, the resurgence of old diseases, such as tuberculosis, and the spread of new ones, such as AIDS. Every city knows the signs; every city must fight them.
>
> (UNCHS, 1996: xxi–xxii)

This view of cities is problematic, as it prioritizes the developmental consequences of city living without considering other social, cultural and political realities. Jenny Robinson argues persuasively for an approach to understanding all cities which is far broader than that offered through this developmental lens, and instead makes use of theories of urban modernity. In critiquing conventional approaches she suggests that

> understanding cities in the 'Third World' was all about development. Stories about the modernity of poor cities, their diverse cultural practices and complicated political struggles were somehow seen as betraying the need to do something about the terrible circumstances in which many city dwellers lived.
>
> (Robinson, 2006: xi)

In contrast, she argues, we should also be concerned about 'the diversity of cities and the diversity of ways of living in cities' (2006: 171). Here, we look at the links between **urbanization** and city living by providing an overview of **urbanization** trends, and then reflecting on cities of diversity and informality as key illustrations of modern city living across the global South.

The level of **urbanization** across the Global South is substantial, although **urbanization** in Latin America is far higher than in Africa or Asia (see Table 5.7). Variations and diversity within these regions (and often within countries) are significant however: for example, within Latin America and the Caribbean Venezuela had 92.3 per cent urban dwellers in 2005, whereas Belize only 50.2 per cent (UN Habitat, 2007).

Urbanization is not however simply a locally produced and experienced trend.

Table 5.7 Percentages of national population residing in urban areas

	1955		1965		1975		1985	
	% urban	Total pop*	% urban	Total pop*	% urban	Total pop*	% urban	Total pop*
Africa	16.4	250633	21.3	319574	25.7	416446	29.9	554294
Asia	18.2	1550986	21.5	1898591	24.0	2393643	29.0	2896192
Latin America & Caribbean	45.1	192022	53.0	252850	61.1	324834	67.9	404492
North America	67	186882	72	219157	73.8	243417	74.7	269023
Europe	54	575970	60	634811	65.7	676455	69.4	706576
Oceania	64.3	14260	68.8	17788	71.5	21286	70.7	24686

	1995		2005		2010	
	% urban	Total pop*	% urban	Total pop*	% urban	Total pop*
Africa	34.1	726334	37.9	922011	39.9	1032013
Asia	34.4	3451674	39.7	3938020	42.5	4166308
Latin America & Caribbean	73.0	483860	77.5	557979	79.4	593697
North America	77.3	299670	80.7	332245	82.1	348574
Europe	71	728513	71.9	731087	72.6	730478
Oceania	70.5	28995	70.5	33410	70.6	35489

Source: Adapted from UN Habitat, 2007.

Notes:
* Population figures in thousands. Percentage urban calculated based on medium variant. All figures based on most recent data or estimations.

In part it is related to the **population growth rate** and pressures on the viability and carrying capacity of rural regions. It is also interconnected with globalized trends such the spread of **democracy** and **good governance** (see Chapter 9). **Democracy** has represented cities as sites for political freedom and action, a key factor in making cities attractive and modern environments to live in. In reality, however, many spaces within cities across the Global South are sites of repression and intimidation, particularly where residents are judged to be illegal in terms of their citizenship status, housing choices or employment strategies. Finally, the **good governance** agenda has attached much weight to the effective functioning of institutions, many within urban contexts, which has served to situate cities as sites of modern urban management. These more recent trends are underscored by historical associations of **urbanization** and city living with **modernization**. Following World War II it was believed that 'poorer societies had to throw off "tradition-bound" ways to progress. In order to advance the vital processes of economic growth, **urbanization** and industrialization, "modern" attitudes such as competition and individualism were seen as essential' (Murray, 2006: 280). In this postwar era both **decolonization** and shifts from a reliance on agriculture to the manufacturing sector accelerated processes of **urbanization**.

The consequences of rapid **urbanization** for city living are also highly opaque, and the ways in which different cities absorb, celebrate and thwart their residents' actions cannot be easily summarized. The **SAP** policies of the 1980s and 1990s (see Chapters 4 and 10) have undermined the capacity for many cities to manage **urbanization** effectively. The loss of investment in housing, transportation, infra-structure and economic activity has meant that the quality of city living has declined for many residents. But where possible residents have adapted, and home-based enterprises (see Plate 6.5 in Nigeria) have flourished along with other informal economy measures (see Chapter 7). Housing needs have been met through informal housing, now a dominant mode of living across the Global South (see Chapter 8). These choices have environmental implications too, as scarce or vulnerable land is used to provide for rising city populations. Plate 5.6 shows Makoko, a slum built on stilts on the lagoon in Lagos, Nigeria. It is estimated to hold around 25,000 residents and has no sewerage facilities, with all waste entering directly into the lagoon.

An overarching theme then of cities in the Global South is that of informality, which defines for many the shelter within which people live, the nature of employment, the form of political association as well as the cultural practices adopted. This mix suggests that what exists is in a state of ungovernability, but Abdul Maliq Simone argues that given the ongoing survival of African cities, these cities have much to say and inform ideas about urban **governance** in the Global North (2004: 2). Finally, the informality of cities is a central reality of city living, but it is not the only defining feature of city-ness. Cities are also places of mixing, where the spectacular and unpredictable occurs; they are sites of diversity.

The diversity of cities is central to shaping city living. Diversity typifies the

Plate 5.6 Houses at Makoko, on the lagoon, Lagos, Nigeria.
Credit: Muyiwa Agunbiade.

residents of cities, their activities and cultural practices and the spaces of the city. It also points to the inequalities inherent in these diversities. Cities have appeal, the urban vibrancy and vitality of cities offers its residents lives infused with innovation, difference, sociability, anonymity, rapid change and variety. Simone suggests that cities function because of the small everyday collaboratory practices that the residents engage in, and it is the diversity of cities (i.e. their mix of residents) that provides the possibilities for this collaboration. This we would argue is symbolic of modern urban life. Box 5.5 explores some of these ideas in the city of Bogotá, Colombia. It illustrates the important interconnections between politics, place, **urbanization** and culture, and specifically the ways in which city living encourages collaboration.

As the example from Bogotá illustrates, city living creates new spaces for collaboration, and political activity. Religion can be important for both. Cities, through their sheer density facilitate a rise in popular religiosity, and in turn the practice of religion contributes to the diversity and community of city spaces. Religiosity is however not confined to urban contexts, but can play a dominant role in more rural areas too.

BOX 5.5

City life, culture and urbanization in Bogotá

Bogotá is Colombia's capital city and it is a city characterized by diversity. It has historically experienced high population growth rates from 676,000 in 1950 to 7,596,000 in 2005, a result of rural to urban migration fuelled by both high levels of violence and general impoverishment (UNHSP, 2003: 205). This violence has been a function of civil wars in Colombia and refugees have contributed to the densification of Bogotá's poorer areas of housing. Bogotá's wealthier population has expanded to the north, northeast and northwest of the city (MegaCity TaskForce of the IGU, 2008). One of the key consequences of **urbanization** in Bogotá is the proliferation and persistence of poor settlements offering residents inadequate services and facilities. These poor residences fall into three types, namely unplanned **urbanization** on the city margins, squatter settlements and inner city deterioration zones (UNHSP, 2003: 205).

Poor living conditions are however spaces for collaboration and interaction and urban life is intensely diverse. Aside from the political culture this creates, this reality has also spawned a vibrant and diverse cultural sector within the city. The cultural sector is made up of over 8,000 registered organizations performing activities as varied as visual arts, film, dance and archaeology. In addition to serving as expressions of cultural **identity** and practice, they are also politically and socially significant. Cultural practices, organizations and performances work to 'build community' and also facilitate active citizenship within the city (Appe, 2007a). An example of the interconnections between cultural activity and politics is provided by the Chiminigagua Cultural Foundation located in Bosa to the south of the city. Susan Appe (2007b) describes how this foundation has actively engaged with government to further the needs and aims of cultural organizations with a belief that cultural community work will contribute towards the development of the whole nation. This form of networking is unusual, particularly in Colombia, but it is proving beneficial and state policy is very slowly shifting in favour of cultural organizations. This illustration of Bogotá reveals the contradictory forces shaping the city, but specifically the interconnections between trends of **urbanization** and the formation of a diverse culturally

BOX 5.5 (*CONTINUED*)

BOX 5.5 (CONTINUED)

vibrant public. City living (usually in very poor conditions) forms a density across which collaborative networks can take hold.

(Sources: Adapted from Appe, 2007a, 2007b; Gilbert, 1997; MegaCity TaskForce of the IGU, 2008; UNHSP, 2003)

Plate 5.7 Aerial view of Bogotá, Colombia.
Credit: © Still Pictures.

RELIGION

Religion has an extensive history across the Global North and South (see Chapter 3) which long predates **modernity, colonialism** or **globalization**. For example, the rise of Christianity in the countries of Africa is often assumed to be inextricably linked with processes of **colonialism** and the expansion of European empires in the late seventeenth to twentieth centuries. However, it is far longer than this (going back almost 2,000 years in places such as Ethiopia) (see Box 5.6) and is still growing. Indeed it 'has been estimated that Africa will contain more Christians than any other continent' (Ward, 1999: 193) in years to come, although the forms of Christianity within Africa are often different and distinct as ongoing tensions

BOX 5.6

Christianity in Africa

Christianity grew in the north of Africa along the Mediterranean coast in the first 600 years of the Christian era, although shortly thereafter Islam presented a significant sustained challenge which has continued to this day. In the nineteenth century British Protestant Evangelicalism, increasingly hostile to the slave trade, pioneered the growth of missions to Sub-Saharan Africa. Clashes between different powers within the different churches resulted in a history of varied fortunes for the missionary effort (Ward, 1999: 200–8).

Christianity was entrenched with European **colonialism** in the late 1880s, although relations between missionaries and colonial authorities were complicated and the two 'projects' should not be viewed as singular. However, the inculcation of Christian values and practices formed an essential part of the package of **colonialism** and supported beliefs in attempts to 'civilize' and 'enlighten' African peoples. Education was fundamental to this and it attempted to shape socio-cultural relations in ways that were not always regarded as positive. Tensions over the nature of education (vocational versus intellectual) and the conflicts between Christian belief and African customs encouraged a growth in independent African churches.

The label 'African Independent Churches' covers a multitude of different Christian faiths and practices across the continent. Three broad groups can be identified today, namely: (1) Ethiopian or African churches which grew as a political response to European missionary practices, although their organization and interpretation closely relates to the mission churches from which they arose (churches often name themselves Lutheran, Methodist, etc.); (2) prophet healing or spiritual churches which emphasize the power of the spirit and have adapted to popular African worldviews (examples include the Zion Church in Southern Africa and the African Apostolic Church in Central Africa); (3) new Pentecostal churches which have grown since the 1980s and emphasize the power of the spirit. These can be small house churches or substantial organizations and attract younger, wealthier and better educated members. They oppose 'traditional' African values such as polygamy and ancestral cults and have been described as the 'fastest growing expression of Christianity in Africa' (examples include the Deeper Life Church in Nigeria and the Grace Bible Church in South Africa).

(Source: Adapted from Anderson, 1997)

between conservative African bishops and more liberal bishops in the Global North reveals.

The rise and spread of Islam also illustrates the global dimension of religion and the ways in which religious ideas have travelled over time (see Box 5.7). The spread of Islam, historically and currently, directly challenges simplistic assumptions that dominant global flows of culture spread from the Global North outwards. There is now much recognition of the strength and spread of Islam, and the religion is evident in around 50 countries across the world (although by no means dominant in all). Figure 5.3 provides an overview of countries with Muslim populations (greater than 4 million). The global reach of Islam is not a recent phenomenon and, as described in Chapter 3, Islamic trading routes and their influence on cultural and economic change played a key role in the shaping of the modern world. The spread of Islam has had an important influence on the style of architecture historically (see Plate 1.4, the Alhambra), but also on the shape and function of cities today. Mosques are an important structuring element within many cities, and they form a cultural and political presence (see Plate 5.9 in Egypt) as well as providing a site of community for city residents.

Although a number of 'world religions' can be identified, namely, Christianity, Islam, Confucianism; Hinduism; Judaism and Buddhism (Murray, 2006: 228), the spread, transition and interpretation of religion locally and regionally is highly complex. Thus although many religious ideas are global in nature, they are often expressed in particular ways at the local scale. For example, in southern Mexico, pre-European beliefs and practices associated with death have mixed with Christian rituals to produce the celebrations associated with the Day of the Dead. At this time, spirits of departed ancestors are believed to return to their homes. Families decorate altars with flowers and fruit and the departed's favourite foods, they also decorate family graves and the production of sand paintings is common (see Plate 5.8) (Norget, 2006).

The significance of the local or regional spread and interpretation of religion relates to the impact of religion on socio-cultural practices (related to **identity** formation) and one key impact is on **gender** relations. Making sense of unequal and fluid **gender** relations (often as a function of religious belief) provides insights into processes and practices such as the spread of HIV/AIDS and sexuality. However gendered interpretations of religion must be sensitive to the particularities of the meanings of religion at the local level. Research on elite Islamist women by Haleh Afshar challenges Western feminist ideas about the inherent oppressive nature of Islam for women (see Afshar, 1999; and Chapter 8).

The spread and transition of religions at a global scale have also been closely shaped by other processes, and at times flows of religious ideas and practices have been interrupted or transformed because of them. Key impacts on religion have been ideas of **modernity** and more recently **globalization**.

During the eighteenth century, ideas of Enlightenment took hold in the Global North that stressed the value of reason rather than emotion as a key source

BOX 5.7

An overview of Islam

Islam literally interpreted is an Arabic term meaning 'surrender' whereby a believer (known as a Muslim) accepts surrender to the will of Allah (which is the Arabic term for God). Literal translation aside, Islam is a highly complex concept to define. It refers both to the religious faith of a person or group, but also to a political ideology. Both religious faith and political ideology are diverse in themselves, and we can argue that there is not one Islam, but many. Today believers fall primarily into two denominations, namely the Sunnis and the Shi'ites. These have theological and legal differences in their traditions. The Sunnis dominate as they constitute 84 per cent of Muslim populations, compared with 14 per cent Shi'ites. However, despite this evidence of division within Islam at a global scale, it is considered to be one of the least sectarian of all modern religions.

Islam dates from the seventh century AD when it was promulgated by the Prophet Muhammed in Arabia. Muhammed lived from around AD 570 to AD 632. Islam spread from Arabia outwards through conversion and conquest. Its influence was felt across the world and between the eighth and the thirteenth centuries Islam was at the height of its power and influence. Culturally Islam was highly developed and centres of Islam boasted highly skilled engineers, philosophers and scientists. Islamic art and architecture had a widespread influence seen in buildings such as the Alhambra (see Plate 1.4) in southern Spain – an area which was under Muslim rule in the 1300s.

Today Islam is the dominant religion in about 30 to 40 countries across the world. Although it is calculated to be the second largest religion in the world (after Christianity), it is considered the fastest growing. Figure 5.3 illustrates how Islam is particularly dominant in northern African countries, the Middle East and South Asia in countries such as India, Pakistan, Bangladesh, Iran, Egypt and Nigeria. However, its influence is strongly felt in the North, such as the USA with 4.2 million Muslims and France with 4.1 million Muslims in 2007. Elsewhere, Indonesia is a key site of Islam and it has the highest number of Muslims in the world, namely 185 million.

(Sources: Adapted from Grillo, 2004; Encyclopaedia Britannica, 2007 and 2005; Palmer, 2002)

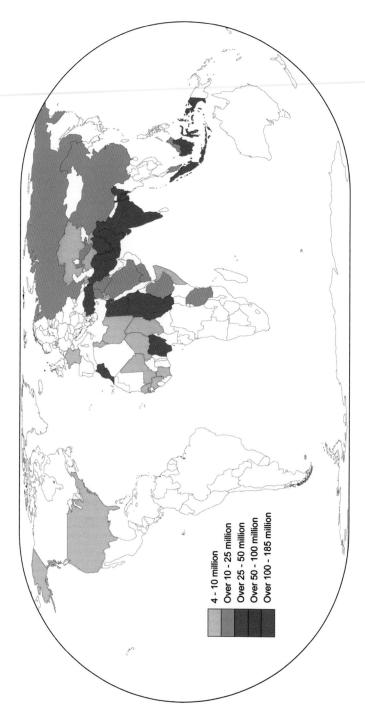

Figure 5.3 Countries with Muslim populations greater than 4 million in 2007.
Source: Data from Encyclopaedia Britannica (2005). Map data © Maps in Minutes™ (1996).

4 - 10 million
Over 10 - 25 million
Over 25 - 50 million
Over 50 - 100 million
Over 100 - 185 million

of authority. This directly challenged the foundations of religion through its questioning of religious authority. Following this, during the course of the twentieth century the rise of **modernization**, **urbanization** and industrialization were widely believed to support modern scientific rationality as the basis for organizing society, and to undermine the significance and value of religion (Haynes, 1999: 224). Indeed the economic models of **capitalism** and **socialism** both contributed to a growth in **secularism** as a result of their parallel but distinct tendencies to value the 'economic' above all else (Raghuram, 1999: 236). However, as Jeff Haynes argues, growing feelings of loss, alienation and relative deprivation (all a function of '**modernization**') (1999: 223) have actually served to make people search for other identities and embrace group formation. This trend ties in with the discussion of the crisis of **identity** (Concept Box 5.5).

Globalization has had an uneven impact across the world, and for some has led to growing inequality. This in turn has caused a rise in religion as a form of protest. This 'revival of religion has been portrayed as expression of socio-cultural particularism against the universalising tendencies of globalisation' (Raghuram, 1999: 238). As a result of this, a dramatic rise in religion is evident, described as a rise

Plate 5.8 Frida Kahlo sand picture and altar for Day of the Dead, Oaxaca City, Mexico.
Credit: Katie Willis.

Plate 5.9 The Fathima Bintu Husain Mosque in Cairo, Egypt.
Credit: Steve Connelly.

in 'popular religiosity' (Haynes, 1999: 223). Popular religiosity is significant because it influences cultural and political practices in different places, and also it shapes people's **identity** formations, that is who they are. Religion also influences people's **consumption** practices (e.g. what they eat or wear) as well as their lifestyles (see also Chapter 8).

Global flows of (often critical) ideas about religion are also evident. Here particular religions (Islam, for example) are represented by specific cultural and political groups (often from within the Global North) in deeply problematic ways. The media and also influential leaders within the Global North play a significant role in shaping many people's views about Islam. Islam is often caricatured by extremism, fundamentalism and irrationality, and is regarded by many in the West as the new enemy in a post-communist world (Akbar and Donnan, 1994: 9). The post 9/11 political climate has meant that analyses of Islam often concentrate on these fundamentalist interpretations of Islam as a political and cultural force. This focus is limited and overlooks the historical complexity of Islam at a global scale, as well as the specifics of Islamic religion and politics at the local level. More importantly, singular representations of Islam as fundamentalist overlook the daily lived realities of Muslims across the world, which for most are about ways of living inscribed by a strong sense of peace-loving humanity, or even indifference (Ruthven, 1997). Representing Islam as fundamentalist points to the **power** of **representation**, as stereotypical images of fundamentalism are used to justify changes to immigration policy and criminal justice systems (in some Western countries). Anti-Islamic sentiments are not restricted to the West, for example, India's BJP party successfully campaigned in 1997 using anti-Muslim issues (Halliday, 2002: 23).

CONCLUSIONS

This Chapter has presented a broad overview of some key socio-cultural trends and practices of the Global South, focusing specifically on health and lifestyles; migration and **diaspora**; **urbanization** and city living; and religion. We have used these trends and practices to argue that **globalization** has not led to the homogenization of social and cultural practice; rather, it has more commonly contributed to the entrenchment of inequalities. Where the spread of global cultural practices are evident, we have argued that at a local level these have often been interpreted and practised quite differently. Avoiding a purely developmental focus on social processes and adopting a view which appreciates the role of culture, we have illustrated that social and cultural practices are intimately connected and directly shape one another. Finally, we have used a historical approach where appropriate, to challenge ideas that trends and practices are fixed in time, or are simply a function of more recent **globalization** events. Many of these ideas are explored further in Chapter 8, where the everyday and local experiences of these broader socio-cultural trends are considered.

 Review questions/activities

1 What does a focus on lifestyles offer and whose responsibility is it to challenge and change lifestyle choices?

2 How do processes of economic migration contribute to the diversity of cities?

SUGGESTED READINGS

Skelton, T. and Allen, T. (eds) (1999) *Culture and Global Change*, London: Routledge. This is an excellent edited collection which covers a diverse range of issues relating to globalized culture. The chapters are relatively short and present critical analytical material in a clear manner.

Schech, S. and Haggis, J. (2000) *Culture and Development: A Critical Introduction*, Oxford: Blackwell.
This text provides a strong overview of concepts and debates relating to culture and development, and is somewhat unique in this respect. The chapters are very readable and they cover many key theorists and debates.

WEBSITES

www.un.org/esa/population/ United Nations Population Division
This particular area of the UN is part of the Department of Economic and Social Affairs (UNDESA) and it can be found on the general website of the UN. It provides annual updates of global population projections, levels of urbanization, etc.

www.unaids.org UNAIDS webpage
See the annual UNAIDS / WHO AIDS Epidemic Updates which provide fairly detailed country by country analysis.

PART THREE

Living in the South

Dress shop keeper in Yemen
Source: Nikky Wilson.

In the following three chapters we change our spatial focus to consider the 'local' scale and people's lives in the Global South. Following the structure of the previous part, the chapters run in the following order: Political Lives (Chapter 6), Making a Living (Chapter 7) and Ways of Living (Chapter 8).

There are five main themes running through these chapters which echo the key arguments in the book as a whole.

First, the chapters consider experiences of **globalization**. As the chapters in Part 2 brought out, **globalization** is a spatially uneven process, and different parts of the world and different groups of people will be caught up in global flows or excluded from global flows. In Part 3 we consider in much more detail how individuals and local communities are affected by global economic, political, social and cultural processes. These context-specific interrogations of **globalization** are vital in understanding how **globalization** is constructed and experienced. However, for most of the people involved, they do not interpret their daily lives through the lens of **globalization**, they are much more interested in the mundane issues of life relating to work, dealing with local state bureaucracies and their social networks of families, neighbours and friends.

Second, the chapters highlight how these local experiences construct global flows. The 'global' only exists through practices at the local level. This may be the **consumption** of particular foods or clothing (Chapter 8), working in a **TNC** factory (Chapter 7) or using global campaign networks to protest against an environmentally destructive project in the local area (Chapter 6).

Third, local processes and experiences are not only spatially differentiated, but they also change over time. How people 'see the state' varies with changing national government regimes (Chapter 6) and labour force opportunities also experience change over time as resources are depleted or the global economy changes, meaning production is shifted elsewhere (Chapter 7). History may also be used as a way of justifying present-day practices. Claims of 'authenticity' or 'tradition' can be mobilized by governments to support political institutions (Chapter 6), the gendering of certain jobs is explained with reference to how things 'have always been done' (Chapter 7) and 'tradition' can be used to enforce cultural practices or to promote economic development, as with the tourist industry (Chapter 8).

Fourth, the three chapters highlight how people in the Global South have sought to challenge or resist processes of **globalization**. This may be intentional in the case of political protests (Chapter 6), workers' demonstrations against transnational capital or changes in civil service pay and conditions (Chapter 7) or a Muslim woman's decision to wear a veil (Chapter 8). In many cases, **resistance** may be less overt or conscious, involving a reworking of supposedly 'Northern' ways of doing things, as with the discussion of **consumption** practices in Chapter 8.

Finally, in all three chapters we are keen to highlight the possible limits to individuals' **agency**. An aim of the book is to challenge common assumptions and **representations** of the Global South and its people by demonstrating diversity and the **power** and ability many people have to shape their lives. However, it is important to recognize the potential limits to such **agency** and how wider structures of economic and political inequality have left millions of people with few chances to improve their lives in the direction they would want. The insecurity of many informal sector jobs leaves millions vulnerable to economic changes and exposed to poor working conditions (Chapter 7). Food insecurity and dreadful housing is a daily reality for many people in parts of the Global South (Chapter 8) and repression and exclusion from political processes are also widespread (Chapter 6).

6 Political lives

INTRODUCTION

Chapter 3 examined the evolution of **nation-states** in the Global South, and their relationship to a changing international political order. Here, we change the focus to the local scale, and look at how these macro-level changes interact with people's lived experience of politics and **power**. In doing so, we need to examine both formal and informal politics (Box 6.1) and the overlapping ways in which both influence structures of **power**. The Chapter starts by looking at the different patterns of rule that have been established across the Global South, and asks two main questions. The first section addresses the question: *How do governments establish rule at the grassroots of societies in the Global South?* Governing populations always involves a series of contested relationships, whereby rulers attempt to define their authority over people, and to acquire some measure of legitimacy to justify this authority. In the Global South, the contrast between 'modern' forms of government and 'traditional' authority may often make the institutions of the state and their legitimate spheres of operation particularly open to question. The second section asks the question: *How is the power of government experienced at the grassroots?* What are the ways in which the state makes itself 'visible' to its citizens in the Global South, and how do people respond to it? The experience of rule draws people into a series of relationships that extend over space: as citizens belonging to a national territory, as voters within particular electoral wards, or more mundanely as people trying to access services or benefits provided by the local state. These experiences can in turn, and in conjunction with other economic, social and cultural relationships, play a part in shaping people's sense of **identity** (see Concept Box 5.5).

Those on the receiving end of **power** in the South are not simply shaped by local patterns of rule. Important within the definition of formal and informal politics in Box 6.1 is the fact that people's own actions can challenge the actions of formal government institutions in a number of different ways. The final section of this Chapter asks the question: *How is the power of the state contested by 'grassroots' actions?*

BOX 6.1

Formal Politics and informal politics

Joe Painter (1995) provides a useful definition of the differences between formal and informal politics:

Formal Politics: 'the operation of the constitutional system of government and its publicly defined institutions and procedures' (Painter, 1995: 8).

Informal politics: 'forming alliances, exercising **power**, getting other people to do things, developing influence and protecting and advancing particular goals and interests' (Painter, 1995: 9).

The formal political sphere is what we normally think of as 'Politics' (with a big 'P'): it is concerned with governments, political parties, elections and public policy, or about war, peace and 'foreign affairs'. Informal politics (or politics with a small 'p'), on the other hand, is a far broader concept: it is part of what we all do as part of our normal social interaction all of the time, and as such is an all-encompassing definition. Painter argues that the two concepts are closely intertwined. Formal politics extends into everyone's lives to a much greater degree than we often think: it's not a separate 'political world', but impinges on how and where individuals and groups can exercise **power** in day-to-day activities. Equally, informal politics has important effects on the institutions of formal politics: the interests and **power** struggles of everyday people 'outside the system' help to shape the organizational cultures and internal dynamics of these institutions, alongside those of the politicians and civil servants within them.

This Chapter is primarily concerned with 'formal politics' – the institutions of the state – but following Painter's arguments, it takes a particular interest in the ways in which these institutions shape, and are shaped by, people's exercise of **power** in their everyday lives.

(Source: Adapted from Painter, 1995)

Three different ways of contesting rule are covered here. The first is the variety of strategies people employ to resist the state by ignoring its commands or failing to comply with its regulations: although not usually aimed at revolution or any other explicitly 'political' goal, it can have substantial effects in blunting or reshaping state **power**. Second, there are more conscious attempts to replace particular functions of the state, where groups carve out territories or subjects of operation in which forms of authority other than the state are seen as legitimate, and can (re)instate their own orderings of space. This helps us to remember that state **power** can retreat in the face of alternative forms of grassroots organization, which may themselves spring up in response to wider economic or political processes that have undermined state institutions or capacity. The third way of contesting patterns of rule is through direct attempts to challenge the state in the form of social movements and mobilizations. Such challenges require far more organization and investment of energy on the part of their participants than more 'passive' forms of **resistance**, and potentially expose them to far greater risks of repression and reprisal should they fail. Crucial questions here are therefore when, where and why do people choose to mobilize against the state. As we shall see, questions of social and cultural **identity** are often key to these oppositional movements, as are the spatial strategies used to resist state **power**.

Throughout, the Chapter aims to highlight the *spatial* nature of **power**, both within the institutions of government and in practice within people's everyday experiences. In Chapter 9, we will turn to explicit attempts to 'develop' the South through reforming the rules and practices of government, but this current Chapter aims to show that these international agendas for '**good governance**' are always interacting with local patterns of rule that have complex geographies (and rich histories). Not only that, but people's political lives – alongside their economic and social activities that we examine in Chapters 7 and 8 – play an important part in constantly reshaping these patterns of rule.

ESTABLISHING RULE

In Chapter 3, we noted that in 'pre-modern' states the relationships between governments, their citizens (or subjects) and particular territories were often far looser than the Western-based ideal of a 'modern' **nation-state** with fixed boundaries and borders. Furthermore, the centralized authority of a king or emperor often remained a marginal influence on the population in general. Rulers might wage wars or levy taxes, but did so with the acceptance that their authority was shared with a variety of more localized sources of **power**. For example, Sudipta Kaviraj (1991) has noted that in pre-colonial India, much day-to-day conduct was governed through localized and community-specific customary laws, enforced by a range of caste and village councils. This produced a 'cellular' society in which small communities were effectively self-governing, with a centralized state being only a dim and distant presence. He argues that the model of the state the British brought to India was far different: especially over the nineteenth century, colonial plans to 'improve' or 'modernize' India gained momentum and required the state to intervene to a far greater extent in Indians' everyday lives, in ways that varied from the direction of agricultural production to the outlawing of the practice of *sati* (widow suicide). This resulted in a dense network of state institutions – court rooms, barracks, administrative offices – spreading out from centres of imperial **power** such as Mumbai and Kolkata to a range of provincial towns, enforcing rule over their surrounding populations.

Through this more ambitious engagement with their colonial subjects, the British (and other European powers) were significantly changing the relationship between governments and their populations. In doing so, they did not start with a blank slate, but were instead intervening in areas of people's lives that would previously have been controlled by a variety of indigenous institutions. Colonial regimes were therefore faced by a pressing political question: How could they gain legitimacy for this increased intervention in society, and ensure that a relatively small governing elite gained the support, or at least compliance, of their colonial subjects? One strategy was by consciously projecting an air of **colonialism**'s absolute authority, both in major state occasions (such as the proclamation of Queen Victoria as Empress of India in 1877), but also in far more mundane encounters between representatives of colonial government and their subjects. In this sense, the **performative** aspects of the colonial state's everyday behaviour were a very important part of reinforcing its **power** and control (see Concept Box 6.1). Colonial officers clearly marked themselves out through their dress, speech and deportment as distant from their subjects, and encounters were often 'staged' to highlight this sense of **power** difference. Plate 6.1 shows one such meeting, in the Indian Himalayas in the late 1930s, with four local rulers sitting cross-legged at the feet of the British political officer while they discuss affairs within the region.

Another widely used strategy was to incorporate forms of 'traditional authority' within emerging patterns of governance, rather than to supplant them altogether. This was seen particularly clearly within the colonization of Africa at the turn of the twentieth century, where European rulers recognized existing 'chiefs', or installed new ones where they did not exist, and used these individuals as key

---

CONCEPT BOX 6.1

Performativity

Performativity refers to speech or other action that performs or 'acts out' elements of an individual's **identity**, or her/his relationship to wider social norms. As used within geography, the term is particularly associated with Judith Butler's work on **gender**. Butler (1999) argued that social categories such as **gender** do not have a fixed, prior status: rather than something we 'are', it is something we 'do', and as such performative acts are important in the continual recreation of our own individual identities. Importantly, however, people are not free to perform **gender** or other roles: dominant social norms and the historical contexts in which performative acts are made will place limitations on their interpretation or effect.

figures to link local populations in the countryside to colonial authorities in the cities. Mahmood Mamdani (1996) has argued that the resulting patterns of rule in Africa are best described as 'decentralized despotism' (Box 6.2), with stability at the upper levels of colonial government being bought at the cost of authoritarianism at the grassroots.

This history is important for people's political lives in the South in a number of different ways today. First, modern states across the South remain ambitious in their intended engagement in people's everyday lives in the ways discussed above, even if their capacity to intervene does not fully meet these intentions. For example, the direct state control over the markets and public services people use on a day-to-day basis has been eroded in many countries by the spread of **Structural Adjustment Policies** and other attempts to open Southern societies to market forces (see Chapters 4 and 10), but this rolling back of the state itself remains contentious. Questions of when and where government *should* have a role in society – and how they link the informal politics of everyday life to the Politics of formal state structures – therefore remain live issues across the South. Second, in those areas where modern state structures were introduced through **colonialism**,

Plate 6.1 The British colonial political officer meeting local rulers, Assam, India.
Credit: © The Royal Geographical Society.

BOX 6.2

Mahmood Mamdani: decentralized despotism and the invention of 'traditional authority' in colonial Africa

Mahmood Mamdani, a Ugandan-born scholar, provides a very important account of colonial government in Africa and its aftermath in his book *Citizen and Subject* (1996). He argues that colonial powers used experiences from older parts of their empires (the French in Indochina, the British in India) to address the question of how to establish their rule. In the towns and cities, formal institutions and laws directly mirrored those of Europe, but their associated rights and citizenship were largely restricted to white settlers. The 'proper' place for black Africans was deemed to be the countryside, and here the appropriate form of government was seen to be through 'traditional authorities'. As Mamdani notes, however, chiefs created under **colonialism** had far more absolute **power** than most previously existing indigenous rulers: judicial, legislative and executive **power** over a 'tribe' and area were all located in a single person. At a stroke, various forms of indigenous checks and balances on the chief's **power** were removed, as were other forms of traditional authority such as local tribunals and councils of elders who would have had jurisdiction over various matters from the settling of domestic disputes to the allocation of land.

The British in particular claimed that by recognizing 'traditional authorities', they were practising an enlightened recognition of 'native culture'. Mamdani questions this interpretation, and shows that the pattern of rule that was established was a form of decentralized despotism that directly suited colonial needs. With **power** devolved to the chiefs, colonial authorities were spared the costs and friction of engaging with 'native' affairs; at the same time, chiefs were wholly dependent on the colonial state for their recognition and support, and as such had to answer its demands to collect taxes, or provide labourers for the colonial economy. Not only this, but by establishing chiefs as the authorities over their 'tribes', colonial powers hardened the **identity** of existing ethnic groups (and competition between them), and ensured that collective opposition to Europeans' overall domination was far harder to organize.

(Source: Adapted from Mamdani, 1996)

the transition to independence at the national level often left largely intact the internal structures of colonial forms of rule which had been designed primarily with the intent of controlling dissent. Here, breaking free of this legacy requires a democratic remoulding of local sources of **power**, something that is difficult to achieve in practice. Old habits of 'performing' **power** in ways that emphasize the distinction between the ruler and the ruled die hard, as will be shown below. Undoing institutional structures that have created a role for 'traditional authority' can be still harder, even for governments with a strong national mandate (Box 6.3). Accordingly, questions remain about how people in the South experience government: Is this as an authoritarian or a democratic force, a source of **power** that is arbitrary and partisan in its actions, or one that is bound by predictable rules?

Finally, questions of how to *justify* rule remain contentious: as Box 6.3 shows, appeals to 'traditional' values may hide forms of authority that are discriminatory, and open to popular challenge. At other times, states have argued that their rule is legitimate because they are acting to achieve 'progress' or defend 'the national interest' (see Box 9.3 on dams in India). Such claims rarely (if ever) enjoy universal support, and exposing the particular interests that lie behind these claims is an important part of **resistance** for those who are marginalized or oppressed by government action. But before looking at how **power** is contested, this Chapter will first turn to the different ways in which people encounter the state in the Global South today. It is here that informal politics of everyday social interactions, along with the inheritance of past government actions, are continuously involved in reshaping local patterns of rule.

BOX 6.3

Undoing 'traditional authority' in rural South Africa

Under South Africa's Apartheid system (*c.*1948–94), black Africans were denied voting and citizenship rights and were also supposed to reside in Bantustans, the primarily rural areas beyond the cities and major centres of South Africa's mining and commercial agricultural wealth (Figure 6.1). The Bantustans had their own government structures and separate codes of law, supposedly representing their own customary values. The spatial separation of these 'traditional authorities' and the modern state institutions of the white-majority areas was a direct

BOX 6.3 (*CONTINUED*)

BOX 6.3 (*CONTINUED*)

development of earlier British practices to divide and rule indigenous populations in South Africa and elsewhere.

When Nelson Mandela's African National Congress (ANC) won the country's first multiracial elections in 1994, it had a mandate to introduce a uniform rule of law and install democratic local government in the former Bantustans. In practice, however, this has proved difficult. Under Apartheid, local government and land administration in the Bantustans was undertaken by Tribal Authorities, all-male unelected bodies that enjoyed immense discretionary **power** over the internal affairs of their territories. In the anti-Apartheid struggle, popular civic organizations challenged these 'traditional' sources of **power**, particularly over the allocation of land, where chiefs and headmen had been using their **power** arbitrarily and amassing wealth through bribery. Although many civic activists were elected as councillors in the post-Apartheid system of democratic local government, Apartheid-era laws over land administration still had not been repealed. A resident of rural Transkei explains the effects of both systems of **power** operating in parallel:

> This is the reason why we still use chiefs. Rural councillors run in circles. This makes us a laughing stock and divides us. People will tell you 'Go to your rural councillor, you won't succeed'. You end up going to the chief, even if you did not want to.
>
> (Ntsebeza, 2004: 76)

Rather than fading away, the traditional authorities in the countryside have become politically organized: their threat of widespread violence before the 2000 local government elections stalled legislation that would have transferred key areas of their **power** to the elected councillors. Customary law is inherently undemocratic and patriarchal: as Haripriya Rangan and Mary Gilmartin (2002) note, it does not recognize women's individual access to land, and traditional authorities often ignore women's needs in favour of allocating grazing rights to men for cattle (a key symbol of wealth). Challenging this bias through the courts is almost impossible for the most vulnerable in society, and the achievement of local democracy envisaged by the ANC in 1994 remains a distant dream in many parts of rural South Africa.

(Sources: Adapted from Ntsebeza, 2004, 2003; Rangan and Gilmartin, 2002)

BOX 6.3 (*CONTINUED*)

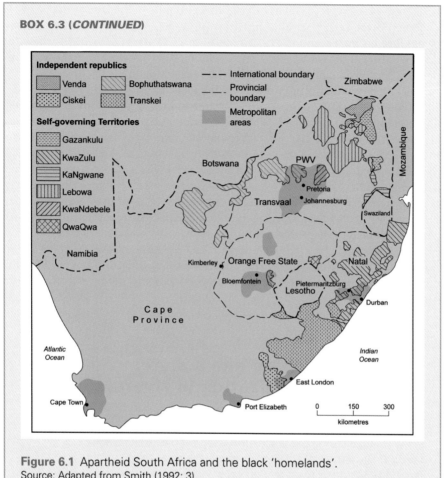

Figure 6.1 Apartheid South Africa and the black 'homelands'.
Source: Adapted from Smith (1992: 3).

EXPERIENCING STATE POWER

Although the macro-level changes to the formal political system that were examined in Chapter 3 are important in the evolution of the Global South, many of the people living there may be less concerned with processes of **decolonization, global governance** or regime change in the abstract than they are with the way in which these developments are experienced within their everyday lives. Occasionally, such macro-level changes do manage to make themselves felt directly and dramatically at the grassroots, either through profound instances of state failure (see Box 3.4 on Sudan's civil wars), or through a powerful ruling coalition that has both the ambition and the coherence to try to force through a particular vision of how society should be altered. James Scott (1998) has argued that examples of the latter – such as Mao Zedong's 'Great Leap Forwards' in China (1958–62), or Julius Nyerere's attempt to produce his own brand of African socialism in Tanzania (Box 6.4) – show the dangers inherent in 'seeing like a state'. The modern state may be an incredibly powerful institution, he argues, but it is one that 'sees' only a fraction of society's complexity, and as a result top-down master plans rarely work in practice.

Chapter 9 returns to Scott's arguments, and their implications for the **developmental state**. In this section, however, it is not state plans for directing dramatic change that are the focus, but rather the far more commonplace activities of routine government: how the state is *seen by* people in the Global South, through their far more mundane contacts with government offices, services and personnel (Corbridge, *et al.*, 2005). Here, what goes on at the lowest levels of the state may, as in Nyerere's Tanzania, be very different from the plans of its central ministries. But whether or not they work as intended, people's everyday encounters with government help to shape relationships between states and their citizens.

One important way in which people meet the state in their everyday lives is through the delivery of key services. The provision of education is a fundamental duty that modern states have taken upon themselves to perform in the Global

BOX 6.4

Seeing like a state: villagization in Nyerere's Tanzania

Independent states in the South have often been at least as ambitious in their plans to transform the lives of their subjects as the colonial regimes they replaced. Julius Nyerere, the first president of independent Tanzania, wanted the country to follow a path of African socialism, linking the best parts of indigenous culture (which he saw as inherently egalitarian) with rapid technological modernization. James Scott's book *Seeing Like a State* (1998) examines Nyerere's attempts to transform Tanzania's countryside from 1967 until the late 1970s, modernizing its traditional agriculture and re-settling large sections of the rural population in planned, nucleated villages where they would practise collective farming.

By moving populations to 'rationally' planned villages, the Tanzanian state disrupted old social networks and arrangements, and attempted to insert new ones in their place. In practical terms, the physical dislocation of farmers from the environments they knew well often meant that their indigenous agricultural knowledge was of little value in the new villages: this made it

BOX 6.4 (CONTINUED)

easier for state extension officers to enforce 'scientific' agricultural practices and cropping systems upon them. The rearrangement of people into planned, nucleated settlements made it easier to provide services – from tractors to schools and clinics – but it also made it far easier for the state to measure, monitor and direct elements of their everyday lives. Village collective farms were given annual work plans and production targets by the state, and all farmers were to comply with agricultural extension officers' advice on the types and acreage of crops to be grown.

Scott argues that the physical rearrangement of the rural landscape was underpinned by a desire to *order* the countryside, something that Nyerere's brand of socialism shared with other development models of the time. But with ordering, however, comes dangerous simplification. The shifting cultivation, scattered homesteads and nomadic pastoralism that Nyerere wished to change may have symbolized chaos and inefficiency to him, but they were themselves a reflection of farmers' finely tuned adaptations to East Africa's changeable physical environment. The state had the **power** to dictate where people should live and what they should grow, but by turning its back on the localized, practical knowledge of its subjects, it could not make its ordering economically successful. Despite replacing voluntary villagization with forced relocation, the expected transformations in production never materialized and by the mid-1980s the entire experiment was abandoned.

(Source: Adapted from Scott, 1998)

North and South alike, and it develops links between the state and its citizens in a number of ways. As buildings, government schools are important markers of the state's physical presence: they spread out across its territory and extend into a range of peripheral locations. The rural primary school can be seen as the modern equivalent of a military outpost, symbolically linking far-flung areas and their communities to the capital and the nation (Wilson, 2001). Teachers hold an important position at the interface of 'the community' and 'the state' that can make them important intermediaries between the two. Particularly in poor or marginal areas, teachers may take on a degree of responsibility for and leadership within their communities that far outstrips their classroom duties, acting as key figures in helping people deal with 'officialdom' (Bhattachrayya, 1999). Schools' daily functioning – in educating children and in engaging parents with various school-sponsored events – is also a powerful form of socialization, as both parents and children learn to conform with the schools' activities and the behaviour expected of them (Plate 6.2). This socialization can also include the deliberate promotion of powerful messages about nationhood and citizenship. In Peru's education system, all three of these roles have become explicitly politicized (Box 6.5).

Through its involvement in highly contested ideas of national **identity**, schooling in Peru is clearly tied into the world of formal Politics. Everyday meetings with the state are also interwoven with the informal politics of day-to-day social relationships, and here the ways in which government representatives and officials behave in their interactions with the public are very important. In some instances, they may perform their public roles deliberately to project an idea of inclusivity, such as the government official sitting cross-legged in a village open meeting in Plate 6.3. In other circumstances, the same official may echo the authoritarian performance of **power** seen in Plate 6.1; when a widow requests help from the local government office, rough forms of address and being repeatedly pushed to the back of the queue may be deliberately used to 'put her in her place' (Corbridge, *et al.*, 2005). Although the state *should* be above such concerns, **gender**, age, wealth and status clearly do affect people's encounters with everyday officialdom. As Akhil Gupta (1995) notes, there are blurred boundaries between government officers' public position and their private **power**, and in negotiating everyday contacts with the state in the South, people often expect that they will have to work hard,

Plate 6.2 Girls in a village school near Laboos, Yemen.
Credit: Nikky Wilson.

BOX 6.5

The 'everyday' state: primary education in Peru

In Peru, the political elite in Lima are distanced from Andean rural society through sharp racial, cultural and economic divides. As such, it is no surprise that in linking the two, the school system has been subject to fierce battles over its curriculum and its control ever since the state first declared that free primary education should be made available in 1870. From the early twentieth century, increased central control of rural schools was used in a fairly heavy-handed attempt to integrate children from indigenous communities within dominant cultural norms via the teaching of 'proper' habits and behaviour, and nationalist messages overlaid with a strong militaristic content. Rural school teachers – themselves often *cholo* (mixed race) and from poor backgrounds – were increasingly critical of these aspects of the curriculum from the 1970s, and were politically active both in the classroom and beyond in challenging the elitist ideas of a good citizen it embodied. However, schools and teachers themselves were frequently attacked as being 'reactionary agents of the state' by the Maoist guerrillas of *Sendero Luminoso* during its armed struggle (1980–92).

At the end of the uprising, the national government used the school system to symbolize the return of the state to areas that Sendero had controlled by initiating a mass rural school-building campaign, and the revitalization of schools' role in the militaristic parades of *Fiestas Patrias* (Peru's Independence Day). Caught between the Maoists, a **neoliberal** state and the local community's own expectations, school teachers expressed profound ambivalence about this role:

> Although we are not in agreement with the political system of government of the time, nevertheless we believe in a state, in a nation, in an identity. The teacher has to understand there are ancestral custom in the *pueblos* (villages). ... [o]ne of these is that people are accustomed to make parades for Fiestas Patrias. This is when the mothers and fathers feel they are somebody in the community, when they see their children participate in a civic-patriotic ceremony.

> (quoted in Wilson, 2000: 14)

(*Sources:* Adapted from Wilson, 2000; 2001)

showing deference to officials, operating through political brokers or even making bribes (Table 6.1).

The spread of formal democratic structures in many countries of the South provides people with a different way of encountering the state. Elections are important events in the collective political life of a country, and in some ways reverse 'normal' relationships of **power**: they are moments when politicians and the political system as a whole seek to gain legitimacy by actively engaging with their citizens, rather than the other way round. Universal adult suffrage – everyone having an equal say in choosing who is to rule them through their vote – is a key part of many democratic systems across the South, and elections should be points at which everyone's active role as a citizen is affirmed. The reality is, however, often rather different: formal political equality of all individuals can contrast quite strongly with longer-established local perceptions of who should have a voice and who is a 'proper' citizen. As a result elections are not always experienced on the ground as a process that reaffirms the ideals of citizenship and **democracy** (Box 6.6), but rather can bring to the surface underlying tensions between groups within society, and highlight existing **power** differences.

At first sight, the election process in Bihar, India described in Box 6.6 would appear to contrast strongly with the ways in which **democracy** operates in the North: Scheduled Caste voters were physically threatened, and some gross violations of electoral rules occurred. However, it is important to remember that elections

Plate 6.3 Local government official at a village open meeting, West Bengal, India.
Credit: Glyn Williams.

Table 6.1 The cost of negotiating officialdom, Jharkhand, India

Village Council	Revenue Department (Circle Office)	Forest Department (Range Office)	Forest Department (Superior Office)	Police Department (Police Station)	Other expenses
Headman [political donation]	Clerk [1,000]	Range Forest Officer [4,000]	Clerks [4,000]	Officer-in-Charge [2,000]	Miscellaneous payments [1,500]
	Amin [1,000]	Clerk [500]	Officers [5,000]	Check posts [500]	Daily expenses and transport for *dalaal* [6,000]
	Circuit Inspector [1,000]	Block Officer [1,000]	Others [500]		Transport for verification officers [1,050]
	Circle Officer [2,000]	Forest Guard [1,000]			Contingency [1,000]
		Check Posts [500]			Logging and transport [20,000]

Source: Adapted from Corbridge and Kumar, 2002: 778.

Notes:
This table shows the different visits and cash payments (in Indian Rupees) made to various state officials in Jharkhand State, India, in order for a teacher to gain the legal permission to sell timber from 10 jack fruit trees on his homestead land to the State Trading Office. To negotiate this array of local state offices, the teacher employed a broker or *dalaal*.
The total value of the timber was Rs100,000: of this 20 per cent was spent on logging and transport, around 26 per cent went to the broker as profit and the teacher received a mere 20 per cent. Bribes to officials made up a staggering 34 per cent of the total value of the timber.

within established democracies can also be controversial, such as the 2000 US Presidential elections in Florida, USA. Although there was none of the blatant 'booth capturing' seen in Hajipur, the polls raised questions of whether other more subtle forms of political exclusion had been used to tip the vote in favour of George W. Bush. After the elections, Florida Governor Jeb Bush was asked to appear before a civil rights tribunal to face allegations that there had been a deliberate attempt to discourage African Americans, who were largely Democrat supporters, from voting (Brabant, 2001). Certainly, political systems in the North have their own problems (such as voter apathy – which is certainly not an issue in India).

BOX 6.6

Elections, democracy and disempowerment in Bihar, India

In *Seeing the State* (Corbridge, *et al*., 2005), the authors include an eye-witness account of the 1999 Indian parliamentary elections in Hajipur, Bihar. Political allegiances in Bihar are often sharply divided on caste lines, and here, as elsewhere in India, there are close connections between elected politicians and their wider networks of brokers and 'fixers', who mobilize supporters on polling day. The elections themselves are major public spectacles: colourful party banners and political graffiti appear everywhere (Plate 6.4). Politicians tour the area, addressing mass rallies and visiting the houses of senior figures in the countryside, in the hope that these respected people will swing the voters of entire villages behind their campaigns. For the brokers and fixers, this is a time where past 'favours' that politicians have done for them are called in: they are kept busy ensuring that they can demonstrate a popular 'wave' of support for their candidate, and often pay out considerable sums of their own money in campaign expenses such as for providing transport for supporters to get to rallies, or providing food and lodging for candidates and their entourage. Elections also involve considerable disruption of everyday life: the operation of schools, local government and even medical services is interrupted as staff and offices are seconded for election purposes, and street-vendors, rickshaw-pullers and other workers find themselves forced to provide their services for free to the ever-present fixers.

The discussions in Hajipur's teashops and marketplaces in 1999 showed how India's communication technology revolution of the 1990s has transformed voters' engagement with the elections. The dress, speech and behaviour of the candidates – as seen on TV screens, or observed in person at rallies – were all hotly debated, perhaps more so than their policies, as voters tried to make up their minds about who had true leadership qualities, and

Plate 6.4 Political graffiti in Kolkata, India. Credit: Glyn Williams.

BOX 6.6 (*CONTINUED*)

how a new government in Delhi would affect their daily lives. On polling day itself, however, the limitations of this popular engagement were clearly seen. The strongmen of a local dominant caste captured their neighbourhood polling booth, and let it be publicly known that Dusadhs, members of an 'untouchable' Scheduled Caste and followers of a rival candidate, would be beaten up if they attempted to vote there. The local presiding officers – school teachers deputed to election duty – expressed frustration at their inability to control the malpractice, but with only four police officers to ensure law and order over a wide area containing several booths, what could they do?

(Source: Adapted from Corbridge, *et al.*, 2005)

Both the flaws and the successes of elections are important in reshaping people's sense of identity, and their relationship to the state. For everyone in Hajipur, the election was a moment at which national questions about the nature of government were discussed, and the links between their locality and India as a whole were stressed. For the Scheduled Caste voters in particular, the differences between what *should* happen during elections and the ground realities of middle-caste domination of the locality came sharply in to focus. Their physical exclusion from voting and the state's inability to protect them highlighted the insecure nature of their formal rights. These were painful and formative experiences for those involved, and ones which can prompt people to seek out alternative relationships to the state and other structures of **power**: it is these ways of contesting rule we turn to next.

CONTESTING POWER

For the scheduled caste voters of Hajipur, and for many others in the South who find themselves in positions of relative political powerlessness, direct confrontation with state institutions or local political elites is often not a realistic option. Instead, they must find ways of working with or around those in **power**. At its most simple, this can be through 'passive' forms of protest – subverting rule by publicly complying with the commands of those in **power**, but nullifying these in practice through foot-dragging, evasion or minor acts of sabotage. When sufficiently widespread, these acts of everyday **resistance** can significantly undermine patterns of rule (Scott, 1985). An example would be Tanzanian farmers' individual responses to President Nyerere's plans for rural collectivization described in Box 6.4, where relative neglect of the collective farm in favour of their private plots, engaging in unofficial trade or even flight into the bush had the combined effect of undermining an entire rural development strategy. Attempts to enforce collectivization were abandoned in 1976, and following President Nyerere's retirement in 1985, a major shift towards market-based economic policies was quietly introduced.

As Asef Bayat (1997) has noted, many people living and working in the informal sector of cities in the Global South undertake similar, individualized actions to the Tanzanian farmers in order to survive and maintain a dignified life. As such, street hawkers, squatters and others are often engaged in 'quiet encroachment' on the **power** of the state and dominant groups. By testing out what they can get away with – such as building unregulated housing on areas of vacant land, trading illegally from their homes (see Plate 6.5) or setting up pavement stalls – they chip away at formal regulations and governmental authority. These people often do not see themselves as acting 'politically', and may do everything they can to present themselves as unthreatening to those in power, such as keeping trading 'invisible', or bribing local state officials to keep their houses or businesses out of government records. But, as Bayat notes, when the state aims to crack down on encroachment these simple and everyday practices can transform into more open and collective acts of **resistance**. By attempting to reassert its rule, government can force hawkers to become organized, or squatters to lobby to defend their de facto 'ownership' of housing. More generally, people's everyday strategies to make unofficial claims over land, housing or ways of making a living are part of what Partha Chatterjee (2004) has called 'the politics of the governed' (Box 6.7). In negotiating these claims, institutions and representatives of the state often act outside their formal remit and guidelines, and as a result, patterns of rule are constantly being restructured.

At times, however, this everyday renegotiation of patterns of rule can be replaced by more open challenges to the state. When the state begins to lose control over parts of its territory, parallel or alternative forms of rule can emerge. Where the state's lack of control coincides with a population that sees itself as politically 'different' from the rest of the country this may facilitate open struggles for political autonomy, and these are important background conditions within numerous civil wars and armed movements for secession across the Global South. The state's lack of control may, however, be a much more local and prosaic affair, as Box 6.8 shows. Drugs bosses in Rio de Janeiro generally do not have ambitions either to gain formal political **power** or to rearrange the internal or external borders of the Brazilian state. Instead, they are playing the role of a quasi-government within the *favelas* almost by default: they provide a degree of social support to residents, and crucially offer some form of social order in a context where the police are more feared than respected. Importantly, in instances like this, formal government institutions have been marginalized within local patterns of rule, a situation that is only likely to change through sustained attempts to re-establish a more positive and inclusive relationship between city government and the *favelas*.

When the risks of getting politically organized are low – or there are pressing needs to defend interests collectively – citizens in the Global South can and do

Plate 6.5 Trading from home, Jos, Nigeria.
Credit: Hassan Sani.

participate in a second type of challenge to the state, that of open political protests. When these take the form of social movements, they can be important in that they both oppose existing forms of rule, and provide opportunities to voice alternative political ideas. Social movements may be targeted at opposition to a particular element or policy of the state (Box 6.9), but their protests can quite easily spill over the boundaries of normal contests between parties within the formal political system, making their composition, forms of protest and the resources and networks they draw upon all of potential interest.

The composition of social movements is often far more diverse than that of formal political parties, because engagement with a movement may be issue-specific, and can thus cut across the 'normal' identities or affiliations – such as class, region or ethnicity – that are mobilized in elections (Box 6.5). Thus APPO has drawn together a range of different groups opposing the Oaxaca government's **neoliberal** policies (Box 6.9), and the *Narmada Bachao Andolan* (NBA, translated as 'Save the Narmada Movement') involves both highly educated professionals and a range of farmers and fisherfolk threatened by flooding in its struggle against the building of the Sadar Sarovar dam in India (Box 9.3). These differences may cause tensions and debate within movements, but they can also be productive

BOX 6.7

Partha Chatterjee's *The Politics of the Governed*

Partha Chatterjee's book *The Politics of the Governed* (2004) looks at the relationship between political participation and forms of governance in the Global South. Chatterjee argues that rather than being treated as full rights-bearing citizens, most people in the South are dealt with by the post-colonial state primarily as governed *populations.* By this Chatterjee means that people are viewed by the state as groups of individuals with particular characteristics (such as 'people living below the poverty line', 'unemployed youths' or 'backward castes') which in turn make them targets of government policy, often aimed at 'correcting' or 'developing' aspects of their lives. By doing so, current Southern governments are bringing the colonial state's project of 'improving' society into the present day, but with an important twist. The range and breadth of these intended interventions has by far outgrown the state's capacity to act: there is no chance of all 'below **poverty** line' people benefiting from **poverty**-alleviation schemes, or all 'unemployed youths' gaining from skills training programmes.

These extensive (but unfulfilled) obligations to the state's governed populations produce a particular relationship between it and society more widely. Lacking full citizenship, the vast majority of people in the South make claims on the state's resources by complying with government policies: they present themselves as deserving cases for support according to the state's definitions (for example, as a 'community group' that fits with aims of the latest **poverty**-alleviation policy), rather than being able to assert more generic rights expressed in their own terms. Not only this, but as there are many more people deserving of support than the state can assist, individuals or groups need a degree of political backing to ensure that *their* claims are prioritized over others'.

This need for political support or an 'inside track' to gain access to the state's resources brings in to being a murky field of political brokerage. This is 'the politics of the governed': a pattern of rule whereby a largely marginalized populace aims to use whatever political connections it has at its disposal to voice and pursue its claims on the state. As might be expected, this involves politicians and lower-level state employees bending official rules as much as enforcing them, but Chatterjee does not see this as an entirely negative process. Rather, by fighting for their place within contradictions of the post-colonial state, marginalized groups learn to create new forms of representation, and aim to stretch the everyday practices of the state to their maximum advantage.

(Source: Adapted from Chatterjee, 2004)

BOX 6.8

Replacing the state: drugs, violence and authority in Rio's *favelas*

Rio de Janeiro has had a significant number of shanty settlements or *favelas* for well over a century. Today, their total population is estimated to be in the region of two million people. Due to their informal and semi-legal status, their inhabitants have had difficult relationships with the rest of the city. Despite playing a vital role in reproducing a cheap labour force for industry and the service sector, the *favelas* have often been marginalized by 'mainstream' society and government alike, suffering from poor social service provision and often brutal policing. Rocinha – the *favela* that appears on the front cover of this book – is a typical example: in the mid-1990s its population of 150–200,000 had four elementary schools, no secondary school and only two health centres, one of which was run by one of its residents' associations (Leeds, 1996). From the 1980s, Rio became a major transit point in the international cocaine trade, and sales of the drug within the city itself also began to affect life in the *favelas*. The massive profits available from cocaine transformed the *favelas*' earlier low-key marijuana rackets into a hierarchically organized system, with each *favela's* trade under the control of a boss (*dono*), under whom work various runners, lookouts, traders and *soldados* – security guards armed with increasingly powerful weaponry.

　　The relationship between the traffickers and the wider community is complex. The trade provides lucrative jobs at a time when unemployment is rife, and a 'good' *dono* will recycle some of his profits within the community in various forms of social support. But the most significant and controversial relationship is in the provision of law and order. The *donos* protect the community from external threats (including those of rival drug gangs), resolve internal disputes and offer a form of rough justice enforced through violence. In return, the community is expected to provide the traffickers with a safe space in which to trade, hiding them within the community if the police seek them out. In recent decades, residents have largely accepted this bargain as less bad than state neglect, but it has severe consequences for ordinary people's ability to express themselves politically. Caught between the police violence and that of the drug traffickers, the local residents' associations – the

BOX 6.8 (*CONTINUED*)

BOX 6.8 (*CONTINUED*)

only democratic institutions able to represent the *favelas* to the municipal government – have often been unable to voice their needs freely. It is estimated that around 100 community leaders were assassinated between 1992 and 2001 for failing to comply with traffickers' wishes.

Plate 6.6 Policing the favelas of Rio de Janeiro, Brazil.
Credit: © Still Pictures.

(Sources: Adapted from Leeds, 1996; Dowdney, 2003)

in their possiblities for movements to forge new political identities and alliances (Featherstone, 2003). Movements can also selectively present certain elements of their diverse membership to outside audiences to gain support: for example, in the Narmada struggle, the status of many (but by no means all) of those displaced by the dam as *adivasis* (or 'tribal' Indians) was used to build links with other indigenous people's environmental struggles globally (see Chapter 2).

Social movements use very varied forms of protest, but they have to play to their relative strengths. Given that, as in the case of APPO and the NBA, social movements in the Global South are often faced with fairly brutal state repression, tactics often have to be flexible and innovative, such as APPO's unofficial broadcasts from local radio stations. Non-violence is a frequent (but by no means universal)

BOX 6.9

Asamblea Popular de los Pueblos de Oaxaca (APPO), Mexico

The Popular Assembly of the Peoples of Oaxaca (APPO in its Spanish acronym) is an umbrella organization founded in the southern Mexican state of Oaxaca in June 2006. It developed after riot police attempted to remove striking teachers from the *zócalo* (central square) of Oaxaca City. Its membership includes the teachers' union, alongside organizations representing peasants, indigenous people and others. APPO is a non-violent, non-hierarchical organization deliberately attempting to practise a different kind of politics: its decisions are made following long debates and there is no one leader. APPO's members support a variety of causes, including opposition to state policies which favour urban upper-middle classes, foreign capital and tourists rather than the mass of the poor population. For example, rising land and housing prices are marginalizing thousands of poor families and the government is doing little to support them. APPO's campaigning has focused on the removal of the governor Ulises Ruiz Ortiz, who came to power in 2004 in what were widely believed to be fraudulent elections.

By August 2006, it was estimated that over 20 people had died at the hands of the police, and there had been numerous disappearances. APPO continued its campaign and organized regular 'mega-marches' with reported participation of between 50,000 and 500,000 people. In August 2006, a local radio station was taken over by a group of women who started broadcasting as 'Radio Cacerola' (Cooking Pot Radio). They were removed by the police, but APPO supporters took over other radio stations and continued to broadcast. In October 2006, Ulises Ruiz asked the federal government for assistance. This was prompted by the murder of an American, Brad Will (who had been filming events in Oaxaca to publicize them via the Internet), but also by the governor's inability to control the situation and the collapse of the key tourist industry. In late November, the federal police swept through the streets restoring the appearance of calm.

APPO continues to campaign for the governor's removal and for the release of those imprisoned for their involvement in the protests. On the first anniversary of the confrontation between the protestors and the federal police, marches were held in the city, graffiti adorned many of the key buildings (see Plate 6.7) and indigenous groups held a silent march involving

BOX 6.9 (*CONTINUED*)

BOX 6.9 (*CONTINUED*)

cleansing rituals at key sites. While APPO may not yet have achieved its aims, it provides an ongoing example of attempts in the Global South to create new political forms which challenge entrenched structures of economic and political capital.

(Sources: Adapted from Dalton, 2008; Martin and Gática, 2008; Mutersbaugh, 2008; Osorno and Meyer, 2007)

Plate 6.7 Graffiti painted during APPO rally, Oaxaca City, Mexico. The slogan reads 'Love it, care for it, defend it! The land is for everybody!'.
Credit: Katie Willis.

feature of protest within Southern social movements. In addition to its inherent value, it can also be powerful in subverting a key area of the state's strength, for when the state meets unarmed protesters with physical force, it exposes the authoritarian nature of its **power**. Mohandas K. Gandhi (Box 11.1) used this well in the mass mobilization against British **colonialism**, and various forms of non-violent protest such as *dharnas* (sit-ins) and protest marches have subsequently become part of the language of political activism in India.

The inventive use of space and scale by social movements is also often important to their success. It is no surprise that the teachers' protest that galvanized APPO was based in Oaxaca City's *zócalo*, as claiming central squares (or other key public spaces) can have symbolic importance. Well-networked social movements are often sensitive to the need not only to ground their protests in particular places, but also to 'jump scale' and transmit their messages to wider audiences. For example, in opposing industrial pollution in Durban, SDCEA (the South Durban Community Environmental Alliance) has built up its own base of evidence about air pollution in the heavily industrialized South Durban basin (see image used with Part 4 introduction). It has also been particularly effective in staging protests and environmental awareness-raising events that have engaged television and print journalists in the movement's activities in an ongoing way (Barnett, 2003; Scott and Barnett, forthcoming; Plate 6.8). This careful building up of both evidence and contacts with the media has been vitally important in allowing SDCEA to oppose South Africa's weak environmental legislation: its wider links have also helped it to restate its credentials as a group concerned with **environmental justice** (see Concept Box 6.2) and have ensured that it is not written off as a NIMBY (Not-In-My-Back-Yard) protest group.

CONCLUSIONS: RESHAPING THE STATE

This Chapter has argued that political **power** is never absolute, and is always open to contest (active or passive) from actors within and beyond the formal political system. This is both a cause for caution and optimism. The optimism stems from the fact that ordinary people in the Global South can and do challenge established patterns of rule in various ways. The modern state, for all its formal **power** and institutional reach, is not able to control all aspects of its citizens' lives; whether through individual and covert action or collective and open protest, citizens' actions can have dramatic effects on the nature of government within the South. As the brief look at social movements in the South has also shown, there are spaces for alternative political ideas to emerge, and these may quickly spread far beyond the areas in which they emerge. Some such as the *Narmada Bachao Andolan* draw on international support for their activities, and others take aim at wider aspects of a *global* **political economy** as their target for protest.

This optimism needs to be balanced with caution for two reasons. First, the political odds are stacked heavily against the most marginalized and poorest groups in the South: as shown in Boxes 6.3 and 6.9, even when national governments are formally committed to enlightened democratic procedures, these can often be undermined at the grassroots level by local elites that jealously defend their exist-ing power and influence. This does not mean that **democracy** is worthless or a sham in India, South Africa or elsewhere, but rather indicates that democratizing local patterns of rule is a slow process, and one in which the poorest are only

Plate 6.8 South Durban Community Environmental Alliance, South Africa. Credit: SDCEA.

CONCEPT BOX 6.2

Environmental justice

Environmental justice refers to both a socio-political movement, and a set of wider political and theoretical debates about how rights to the environment (and the rights of non-human nature) can be articulated and defended. The environmental justice movement in the USA has been concerned with the redistribution of environmental costs and benefits, challenging the spatial concentration of environmental 'bads' (such as pollution and locally unwanted land uses) in poorer neighbourhoods and among people of colour. As such, there are strong links to environmental activism within the Global South that approach the environment in terms of the resources and risks it provides to the poor. In both the Global North and South, proponents of environmental justice sometimes find themselves in conflict with more elite forms of environmentalism that stress nature conservation regardless of its impact on social justice.

likely to gain a significant voice through sustained struggle. This is an important caution to bear in mind when we turn to development strategies that aim to deliver '**good governance**' in Chapter 9. Despite many international development institutions' growing interest in improving the quality of **governance** by making rule more democratic, accountable and transparent across the Global South, these are changes that are difficult and very slow to implement. To do so does not just require changes to the behaviour of the state itself, but also sustained engagement with grassroots **power** structures.

Second, where **resistance** movements do emerge, the temptation is to celebrate these 'alternative' political programmes or patterns of rule. As this Chapter has aimed to illustrate, it would be wrong to do this uncritically. Social movements can be important sources of new political ideas and change in society (Box 6.9), but authentic local voices are not always democratic ones. Box 6.3 illustrated that *patriarchy* was an important part of dominant 'indigenous values' in rural South Africa (as it is in many other places across the South) and, as Mamdani's work (Box 6.2) has indicated, 'traditions' are themselves often shaped in response to the formal institutions of government. Chapter 11 will look at the growing interest in 'grassroots' and 'community-led' development over the past couple of decades; however, we should be realistic in our expectations of what these approaches may offer. In the Global South, just as in the North, 'local communities' are always diverse, and contain **power** differences and prejudices alongside their more positive aspects. Recognizing this is an important first step in building genuinely democratic community development.

 ## Review questions/activities

1   Think about your own everyday political geography: In what ways do you come in to contact with the state and the formal political system? In what ways are informal politics important in your daily life?

2   Look back at the case studies of people's political lives presented in Box 6.3, 6.5, 6.6 or 6.8 (if possible, find and read the original source materials). In what ways do formal and informal politics interact within these examples?

3   Look at the Friends of the River Narmada website (see below): How is the social movement being represented within this website? What issues are stressed, and in what ways is space/place important within the movement's activities?

SUGGESTED READINGS

Blom Hansen, T. and Stepputat, F. (eds) (2001) *States of Imagination: Ethnographic Explorations of the Postcolonial State*, Durham, NC: Duke University Press.
A collection of essays that considers state–society relationships across the Global South.

Corbridge, S., Williams, G., Srivastava, M. and Véron, R. (2005) *Seeing the State: Governance and Governmentality in India,* Cambridge: Cambridge University Press.
A study of state-society relationships, illustrated through detailed fieldwork in contemporary India.

Peet, R. and Watts, M. (eds) (2004) *Liberation Ecologies: Environment, Development and Social Movements* (second edition), London: Routledge.
Contains chapters on a range of different struggles for environmental justice across the Global South.

WEBSITES

www.groundwork.org.za The South Durban Community Environmental Alliance
The SDCEA is linked to *groundWork*, an environmental justice NGO working across Southern Africa.

http://lanic.utexas.edu/ Latin American Network Information Center, University of Texas
An excellent site which provides links to sources of information on a range of Latin American topics. There is a significant section on government and politics, including social movements.

http://www.narmada.org/ The Friends of the River Narmada
This website has background on the anti-Narmada protest, and further links to the activities of the *Narmada Bachao Andolan*.

7 Making a living

INTRODUCTION

Making a living is fundamental to people's survival across the Global South and also reflects the complex array of everyday practices that people engage in. This Chapter furthers some of the critical themes raised already (see Chapter 4 in particular) but changes the scale and focus of attention to the local level to explore how people are actually making a living in the Global South within the context of these broader processes. This Chapter makes four key arguments. The first is that ways of making a living in the Global South (and North) can only be fully appreciated if they are analysed holistically, recognizing their complexity, fluidity, illegality and informality. We discuss different ways of understanding making a living in the first part of the chapter, drawing on the 'whole economy model' and the concept of **livelihoods**, and illustrating these analytical concerns through the **informal economy**. The ways in which people across the Global South make a living are intimately tied to globalized economic processes and structures, which are constantly changing. Our second argument is that ways of making a living must be understood as historically contingent and never static. We explore this issue in relation to changing political and economic processes, and also in relation to changing state-based forms of employment, and through the example of **deagrarianization** and the internationalization of agriculture in the context of the rural Global South.

Our third argument is that ways of making a living are ridden with inequalities, both in terms of the structures which shape opportunities but also in terms of the skills, capital and opportunity that individuals possess. These inequalities are influenced by various factors including **identity** differences (see Chapter 5) and generally they point to uneven **power** relations. We explore the gendered inequalities evident within, and constitutive of, ways of making a living. We then examine the experiences of children and youth as a particular category of people who challenge simple analyses of inequality and the unequal exercise of **power**.

Our final argument is that analyses of how people survive must not overlook the **agency** of individuals by focusing purely on the broader structures and processes that shape their lives. Even in contexts of extreme powerlessness individuals express **agency**, and we explore this through an examination of collective approaches to work. This discussion points to the politics of work and the imbalance of **power** relations in work contexts. We conclude that in fact strategies to work collectively are severely constrained for many.

UNDERSTANDING THE WAYS IN WHICH PEOPLE MAKE A LIVING

There has always been a tension involved in trying to understand how people make a living. Mainstream neoclassical approaches (see Chapter 10) tend to assume that individuals' economic behaviour is driven solely by market forces, and thus overlooks the broader context and structures within which people function, as well as the specificities of individual actions. Structural approaches (like Marxism) generally focus on abstract processes and structures which shape economic relations, such as the economic system (e.g. **capitalism** or **communism**), labour markets and levels of unemployment. These approaches tend to assume that these broader structures determine economic behaviour and thus overlook the roles and actions of individuals. These more traditional approaches have thus been criticized for failing to take into account the multitude of factors which shape how people actually make a living, be they material factors (such as gifts between neighbours) or non-material factors (such as **power** relations as a function of one's **gender**) (see Friedman, 1992; Rigg, 2007).

Structural approaches have also been criticized for overlooking the **agency** and actions of people at a local level. In addition they (and also at times their critics) have tended to overlook specificity, difference, complexity and change over time. Jonathan Rigg argues for a holistic approach: 'if we are to understand everyday living then the scope of view must extend from the cultural to the economic, from the social to the political, from the present to the past, and from the local to the global' (Rigg, 2007: 42) (see Rigg's table 2.1 for a very useful comparison of agency-centred versus structure-centred approaches). With this holistic view in mind, Table 7.1 provides an overview of typical strategies used to understand the myriad of ways in which people make a living.

There have been various academic and policy-based outcomes of the broad criticisms of neoclassical and structural approaches and two key influential responses (but no means exclusive) are briefly discussed below. These are the 'whole economy model' and the **livelihoods** approach (and its variations). Their purposes and origins are quite distinct, but they provide far more flexible and holistic approaches to understanding ways of making a living, although they too are not free from criticism.

Table 7.1 Different strategies for understanding ways of making a living

Strategies which undermine a holistic understanding and thus risk oversimplification or the exclusion of individuals/actions	Strategies to achieve a more holistic understanding	Examples resulting from a holistic understanding
A focus on the household head as the significant income earner, usually male, and usually adult.	A focus on all household members whatever their age, gender, (dis)ability, presence or absence.	The value of female child labour
A focus on formally employed workers or formalized work arrangements	A focus on formally and informally employed workers and people working in formal and informal ways	The value of income generated from working as an unregistered taxi operator
A focus on legal activities or activities that secure taxation, or are state endorsed	A focus on legal, semi-legal and illegal activities that may or may not contribute to taxation, and may or may not be state endorsed.	The income generated by working as a sex worker or drug trafficker
A focus on paid activities or activities generating a monetary transfer usually between employer and employee or worker and customer	A focus on paid, unpaid and paid-in-kind activities, including transfers of goods and services between families, friends and neighbours	The value of food donations between family members
A focus on activities classified within national economic statistics	A focus on activities classified and also not classified or accounted for within national statistics	The value of income generated through the sale of barbequed corn on the street
A focus on individual activities	A focus on individual activities and also community or group work	The value of communal agricultural or sewing schemes
A focus on the work of individuals, neglecting the relational nature of work	A focus on the work of individuals but also on the relations between individuals as shaping experiences of making a living	The ways in which the value of children's work is related to the needs and working practices of their households

Table 7.1 (*continued*)

Table 7.1 (*continued*)

Strategies which undermine a holistic understanding and thus risk oversimplification or the exclusion of individuals/actions	Strategies to achieve a more holistic understanding	Examples resulting from a holistic understanding
An assumption that to make a living involves the performance of labour	Acknowledging that most forms of making a living require an input of labour but some are transfers, handouts or charity that are a function of a person's situation not labour product.	The importance of aid in situations of extreme poverty and the value of state pension transfers for all household members
A focus on ways of living that necessarily involve the market	A focus on market-based livelihoods as well as subsistence production	The value of growing food for household consumption
A focus on single ways of making a living, or the most lucrative or obvious	A focus on the most lucrative or obvious ways of making a living in combination with an understanding of the multiple livelihood strategies that people adopt	The combination of formal day work plus informal night work plus seasonal farm labour
A focus on productive labour	A focus on productive labour and also reproductive, communal and so-called 'non-productive' work (see Harrison, 2000)	The value of work conducted within the home to maintain the household
A focus on existing or current strategies	A focus on previous, existing and possible future strategies to living	Understanding a person's previous skilled employment in a context of current unemployment

The 'whole economy model'

John Friedmann's 1992 model stems from his critique of neoclassical models of economic activity (typically those used by governments in the formulation of national accounting statistics). He criticizes these approaches for providing 'a false conception of the economy' (1992: 44) and claims they leave much that occurs within poor countries (and households) invisible, precisely because of the narrow focus on market relations. In an effort to overcome these failings he reworked a model originally developed by the Vanier Institute of the Family and produced the 'whole economy model' (see Figure 7.1). The model begins with the household, rather than the individual, and explores a broad spectrum of ways of making a living from its perspective. Fundamentally the economy is divided into two overlapping parts, the subsistence economy and the economy of capital accumulation. Sociocultural factors are highlighted by including the sphere of **civil society** as a space where economic relations (not necessarily wage- or income-related) take place. In addition the role of domestic labour and the interconnections with **consumption** practices are illustrated here.

Friedmann's model can be criticized for suggesting that clear distinctions between the market and **civil society** are evident, and that formal, informal, domestic and communal work can so easily be separated. In addition relations of **power** and changes over time are not well represented, despite both being key to Friedmann's critique of mainstream approaches. But as Friedmann himself acknowledges, the process of modelling itself is always a simplification (1992: 44). Despite these

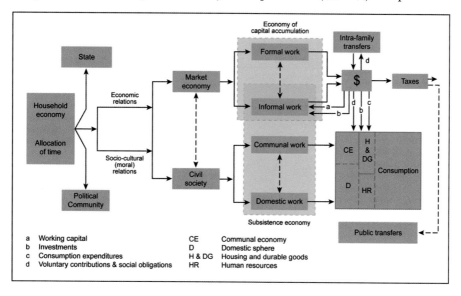

Figure 7.1 The whole economy model.
Source: Adapted from Friedmann (1992: 50).

concerns, this model provides a very useful overview of the various ways in which individuals within the Global South make a living as part of households, and its key contribution is forcing a wider examination of these multiple economic relations by looking beyond standard formal market relations.

The livelihoods approach

Ideas about **livelihoods** were developed by Robert Chambers in the mid-1980s and have been developed by Chambers, Gordon Conway and various others since this time (see Chambers and Conway, 1992; de Haan and Zoomers, 2005). The value of **livelihoods** approaches is debated within the social sciences, but their adoption by the Department for International Development (DFID) of the UK Government in the late 1990s as its framework for structuring policy responses and development assistance has given them global importance (see DFID, 1999; and see Figure 7.2). DFID defines **livelihoods** as:

> the capabilities, assets (including both material and social resources) and activities required for a means of living. A livelihood is sustainable when it can cope with and recover from stresses and shocks and maintain or enhance its capabilities and assets both now and in the future, while not undermining the natural resource base.

> (DFID, 1999)

The logic of the approach is outlined in Figure 7.2. People are understood to operate within a context of vulnerability, within which they have access to assets. The value of these assets is determined by 'the prevailing social, institutional and

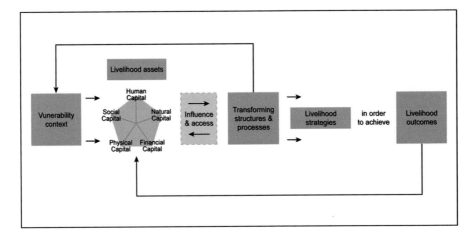

Figure 7.2 The DFID livelihoods framework.
Source: Adapted from DFID (1999).

organisational environment' (DFID, 1999). This environment shapes people's strategies as they pursue particular outcomes.

The **livelihoods** approach plugged a fundamental gap in the development industry (and academia). Specifically, it demanded, for the first time, that people were to be placed at the centre of development analyses and decision-making; furthermore it stressed the importance of processes of making a living rather than only examining the outcomes (in terms of wealth or **poverty**). The approach's impact and popularity is evident by its wide adoption and adaptation by national and international organizations (see Rigg, 2007: 30), yet within academic circles, it has experienced varying criticisms, and recommendations for its rethinking. Leo De Haan and Annelies Zoomers (2005: 32) call for an improved understanding of the non-economic and immaterial 'objectives of life' so as to improve the holistic value of the approach. Further, the actors involved are not homogenous, but rather they are at times conflicting and at other times cooperating. This forces a stronger recognition of **power** as a key factor shaping relations and outcomes, particularly in relation to the issue of access (de Haan and Zoomers, 2005: 36; see Rigg, 2007: Chapter 2 for a useful overview and critical discussion).

The 'whole economy' and the **livelihoods** approach both herald a sustained attempt to obtain a holistic understanding of a highly complex reality. They also place people at the centre of questions of 'development', and situate ways of 'surviving' within their structural and institutional contexts. We make use of these principles in our discussion below, and also highlight the **power** relations and inequalities embedded within ways of making a living. Finally, our discussion below maintains an acute awareness of the role of time and historical contingency in shaping social relations, power relations, work opportunities and institutional arrangements: we illustrate this through the dominance of informal ways of making a living.

Informal ways of making a living

Much of the work conducted within the Global South occurs within the informal economy (see Concept Box 7.1). In some African countries the sector contributes towards 30 per cent of income; between 45 and 60 per cent of non-agricultural GDP in countries where estimates exist; between 40 and 60 per cent of urban employment across the Global South; between 83 and 93 per cent of new jobs in Latin America, the Caribbean and Africa; and over 50 per cent of the non-agricultural labour force in most (but not all) countries of the Global South (Chen, 2001). Given this high level of labour absorption, the **informal economy**, particularly the informal production or service sector, is hailed by some, such as Hernando de Soto (1989), as a success: it is seen as evidence of **capitalism**'s ability to respond to the diverse needs of the market (such as taxi transportation) and of individuals' **agency** and entrepreneurial spirit. However the reliance of the state on the **informal economy** could lead to an abdication of the state's responsibilities, as informal actors provide services (e.g. housing) in the place of the state. In addition,

---

CONCEPT BOX 7.1

The informal economy

The **informal economy** is a broad term used to describe 'informal production units, but also informal employment in formal enterprises' (ILO, 2004: 45). The meaning of 'informal' is crucial here and refers to economic activity which takes place outside of formal controls and legislation, or through non-formalized relations of employment. This framework is important and illustrates recognition of the complexities of this type of economy. Historically the sector was termed the informal sector (in comparison with the formal sector), but it is now recognized that this dichotomy is too simplistic and that it overlooks employment relations which straddle both.

---

the **informal economy** is often highly unprofitable and exploitative for many working within it.

In reality much informal work occurs alongside, or in conjunction with, the formal economy and separating out the two sectors creates a false dichotomy (see Santos, 1979, for a discussion of the linkages and overlaps between the sectors). Traders sell food products and goods which are often purchased wholesale from formal sector companies, or they add value to goods, such as barbequing nuts and corn to sell as street food (see Plate 7.1). An important component of the informal economy is service related and this can range from hairdressing, to sex work, to car repair, to taxi transportation. These services are highly diverse in terms of their organization and can be offered by lone individuals or teams of managed workers. Criminal activities, such as the theft and sale of stolen goods, can also be described as contributing to the **informal economy**.

The **informal economy** is also characterized by its heterogeneity and its internal inequalities. A key source of inequality relates to **gender** differences. Women dominate numerically, particularly in relation to home-based work (see Plate 7.2) and street vending; indeed women make up a higher proportion of workers in the whole **informal economy** than men in most countries. For women across the Global South the **informal economy** is the primary source of employment. For example, in Benin, Mali and Chad it employs 95 per cent of female workers outside of the agricultural sector (Chen, 2001). Despite their numerical dominance, women tend to cluster in the least secure and most poorly paid parts of the economy. Fewer women are owner-operators (entrepreneurs) or paid employees of informal businesses: far more are self-employed or sub-contractors (Chen, 2001). Christian Rogerson (1996) describes these jobs as 'survivalist', as women in these roles earn very little, have limited security and possess limited prospects of long-term capital accumulation. In contrast, men are more likely to engage in 'entrepreneurial' work

Plate 7.1 Informal trading in Jos, Nigeria.
Credit: Hassan Sani.

where they may employ staff, and may earn a reasonable profit having the ability to grow as an investment. Gendered dualisms are only one explanation and differences may arise for a range of reasons, including training, education, wealth, race, ethnicity and obviously the nature of the market within which people are working – although arguably gendered distinctions do appear to be highly significant. Many men suffer within this economy too, often as a direct result of their **gender** and the expectations associated with being male (see Box 7.1).

The **informal economy** is typified by informal employment conditions (such as self-employment, unpaid workers in family enterprises, sub-contracted workers and temporary paid employees) (see Chen, 2001; Wilson, 1998), but these working conditions are also increasingly dominant within the **formal** sector across the Global South (and indeed in the Global North). Contract work occurs across the economic spectrum: service sectors (hotel cleaners, restaurant staff); manufacturing (home-based factory workers); extraction (casual mining labourers); public works (as temporary state employees); and health (temporary care workers, hospital cleaning staff), for example. Employment conditions can be highly exploitative, fragile and very poorly paid. Workers often lack job security, health and safety protection; they are excluded from additional employment benefits, and have very few rights

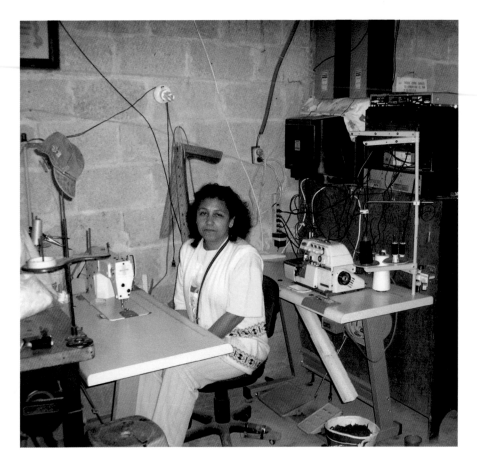

Plate 7.2 Ofelia at her home-based sewing workshop in Xalapa, Mexico.
Credit: Melanie Lombard.

and no possibility of unionization. In the case of public works contracts, workers are often paid below subsistence rates, despite such programmes being billed by national governments as solutions to long-term unemployment (McCord, 2005). This type of work is a good example of an insecure form of employment within a formal system. Crucial to understanding this form of work is seeing the links between making a living and global policy trends and changes, a key theme of the following section.

BOX 7.1

Male lobster divers in Honduras

The area of La Mosquitia on the Caribbean coast of Honduras has a long history of lobster harvesting through diving. Lobsters are primarily sold to the USA. The business of diving and capturing lobsters is entirely gendered, and diving is undertaken by men only. In this particular region the work is also the exclusive preserve of the indigenous Miskito community. There are around 9,000 Miskito divers, of whom almost half (4,200) are living with diving-related injuries because of the dangers involved. Furthermore, about 50 Miskito divers die annually from diving injuries. This form of informal employment is a tragic example of the inequities of local and global **commodity chains**, as the male divers earn on average US$2.5 per pound of lobster (minus 5 per cent of the total weight as it is assumed that the catch holds excess water). Divers have to pay about 80 cents of this US$2.5 to the young men who accompany them in small boats known as *cayucos*, alongside the large lobster boats managed by profiteering captains. The captains earn on average US$9 per pound, and the annual lobster business is worth around US$30 million dollars, the bulk of which is paid to the processors and the captains. The significance of this industry for the male divers and their broader community is evident in that 10 per cent of the total Mosquitia male population is involved in diving or steering small boats. As the supplies of lobsters continue to dwindle and pressure to provide for growing markets in the USA continues, the men are forced into diving to substantially deeper depths, from 40–60 feet to depths of 100–50 feet, with poor equipment. Culturally the move from working on the small boats to diving is viewed as a male rite of passage and this cultural pressure alongside weak national fishing policy leaves the men exposed to hazardous and often fatal employment.

(Sources: Adapted from Tassi, 2004; cited in Jones, 2006; Gollin, 2007)

WAYS OF MAKING A LIVING ARE HISTORICALLY CONTINGENT

Our second key argument in this Chapter is that life in the Global South is intimately connected to life across the Global North, and opportunities for making a living are explicitly shaped by changing socio-economic and political realities, as well as geophysical events (natural disasters, etc.). There are a range of key historical processes (see Chapters 3 and 4) which have directly shaped ways of making a living, a few of which we consider below, namely: **colonialism** and **decolonization**; the restructuring of the global economy and **TNC** investment strategies; the dominance of **neoliberalism** (including **SAPs**); rural restructuring (specifically changes to land tenure, the internationalization of agriculture and **deagrarianization**); and political, transport and financial systems which encourage labour migration and the growth of **remittance** economies. These broad processes, although often evident at the global scale, directly shape the ways in which making a living are actually experienced at the local level.

From colonialism to neoliberalism

The pre-colonial period reveals the practice of place-specific livelihood activities, for example, subsistence agriculture and the trading of goods such as beads, metal wares and food commodities. **Colonialism** dramatically altered economic practices across the colonies as local residents were brought in to newly established labour markets to work as miners, farm labourers and servants for the new ruling elite. Extraction and production were central to the national and imperial economies and formed the basis of much employment. (See Box 7.2.) These industries flourished

BOX 7.2

Changes over time: Zambian copper mine workers

Zambia's history of copper mining began in the early 1920s, and by 1930 the copper mines employed around 31,941 African mine workers. Following a global slump in the world economy in 1931 the level of employment in Zambian mining fell dramatically, revealing the sensitivity of extractive work to global markets. After this time many mines reopened and rates of production and employment grew steadily. The mines enjoyed a period of high growth from the postwar period onwards in particular with employment growing to 38,000 in the 1960s – shortly before independence in 1964. Mining companies provided amenities and housing for their employees and embarked on substantial house-building programmes for urban Zambians. Production peaked in 1969 and then declined due to global falls in price and a lack of investment.

Since the adoption of **neoliberal** policies in the 1990s in response to restructuring initiatives and international donor pressure, the Zambian government has privatized the majority of its state-owned mining companies.

BOX 7.2 (CONTINUED)

Privatization exposed Zambia to a range of vulnerabilities, such as in 2002 when key investor Anglo American withdrew in response to falling global prices. Unemployment followed and miners faced highly uncertain futures.

Other sources of investment (including UK company Vedanta) emerged, but the terms of trade and investment were not in Zambia's favour. In 2007 the price of copper on the global market rose again (see Figure 4.2 on global commodity prices) (similar to 1960s levels) but Zambia remains a very poor country – despite the significance of copper mining (which is estimated to generate three-quarters of the country's foreign exchange earnings). Zambian mine workers do not enjoy equal benefits and contract workers are particularly vulnerable as they lack unionized protection and benefits such as pensions and medical aid. The experiences of Zambia's mine workers illustrate the local impacts of the ongoing uneven economic relationships between countries of the Global North and the South.

(Sources: Adapted from ACTSA, *et al.*, 2007; Ferguson, 1990a, 1990b; Mbendi Information for Africa, 2007)

Plate 7.3 Mine workers at the Chimbashi pithead, Zambia.
Credit: David Rose www.davidrose.co/ChristianAid.

during colonial times as labour costs were kept very low and the post-World War II global economy had a steady need for cheap raw materials. Following independence many industries in former colonies were nationalized, but then as global recession set in during the 1970s many reverted to being privatized. This trend towards privatization and the decline of state subsidization was entrenched through the introduction of **SAPs** in the 1980s and 1990s (see Chapter 10). Many industries struggled in this new tighter **neoliberal** fiscal context and rates of unemployment rose. In many post-colonial economies today the service and financial sectors are contributing more and more to GDP; however their impact on employment has often been less impressive. Changing political and economic policy directly shapes people's capacity to earn a living in particular ways (see Box 7.2), especially in terms of job security and wage differentials.

The changing fortunes of public sector workers

Across the Global South, changes to public sector employment over the past 25 years have had a significant impact on the abilities of workers to make a living. Historically, and globally, work as a civil servant was considered (in comparison to the private sector) to be relatively secure, although it often limited earnings potential. Increasingly, the differences between these two sectors have diminished (see Box 7.3) as public sectors across the Global South take on a more **neoliberal** management style and as partnerships between the two sectors grow. Within newly independent countries in the 1960s and 1970s work within government sectors (such as administration, health, education and welfare) was a fundamental source of good employment for many. The jobs offered security, the promise of pension and medical benefits, minimum wage protection and the possibilities for promotion and training. This however has changed substantially, as much work has been privatized and outsourced: globally 15 million public sector jobs were lost in 1999–2001 (ILO, 2004).

BOX 7.3

Civil servants in Malawi

Malawi is an economically poor country that is highly dependent on agricultural production. Malawi adopted **SAPs** and as a result of this the organization of the civil service has been transformed. The **good governance** agenda of the World Bank (see Chapter 9) and related **New Public Management (NPM)** have encouraged the reformulation of employment and payment structures for civil servants. Some elite civil servants are benefiting – albeit in a context of

BOX 7.2 (*CONTINUED*)

greater insecurity – although many junior employees have not seen benefits.

Malawi changed the ways in which it paid elite civil servants in 2000. Previously, elite servants who occupied the highest grades (known as the superscale) earned no more than US$250 per month (the value of which can be compared to house rental prices in better areas of the capital Lilongwe which are between US$500 and US$800 per month). From 2000 onwards these civil servants were given the choice of either retiring or accepting employment with the government on a contractual basis. Contracts lasted for three years and were dependent upon performance. On the plus side salaries increased to US$3,000 per month in addition to substantial benefits and non-taxable allowances. However, celebration of these income rises was somewhat naive because in reality the rate of inflation and growing national economic crises devalue most of these improvements. The state was aiming to manage and improve the performance of its officers by offering more attractive remuneration, but this benefit appeared hand-in-hand with greater job insecurity and a constant fear of politically motivated-retrenchments.

(Source: Adapted from Anders, 2005)

Plate 7.4 Striking civil servants in 2007, South Africa.
Credit: PA Photos.

Explanations rely on the global **debt crisis** of the early 1980s and the subsequent introduction of **SAPs** (see Chapters 4 and 10). These measures did encourage an increased expenditure in particular sectors of the economy, however the funding of public sectors was largely undermined. Two sectors that were particularly damaged were health and education. For example, in Ghana, the percentage of the national GDP spent on education declined from 6 per cent in 1976 to 1 per cent in 1983. As a result, by the mid-1980s 'around 50 per cent of all teachers in primary and middle schools were untrained' (SAPRI / Ghana, 2001: 10). Twenty years later the impacts of **SAPs** are ongoing, with persistent clashes across the Global South unfolding (see Plate 7.4) as public sector workers are squeezed out of reasonably paid employment.

Nonetheless, the public sector in itself is extensive and diverse and certain categories of workers enjoy substantial privileges and wealth, and their lifestyles and **consumption** practices reflect this. Thus, for some the transition to a more **neoliberal** style of management has meant a rise in benefits and payment (see Box 7.3). However these are constantly under threat as success within the workplace is now monitored within **New Public Management (NPM)** (see Concept Box 9.3) structures (see Harrison, *et al.*, 2008). These structures facilitate a constant evaluation of work performance which directly shapes promotion and earnings potentials. The rise of NPM has also led to growth in inequality within more affluent working environments.

Deagrarianization and the internationalization of agriculture

In early 2008 the UN predicted that the world had entered a 10-year food shortage crisis, evidenced by rapidly rising prices for basic products such as grain and rice. Agricultural trends and food **consumption** practices are at the heart of this concern, and debates over how to respond to this dire prediction abound. The **Green Revolution** in agriculture (see Concept Box 7.2) was the response to a similar crisis in the 1960s, and once again farmers and policy makers are considering technological solutions, particularly the introduction of GM (genetically modified) crops, to intensify production. However, as was evident from the impacts of the Green Revolution (which led to further inequalities for some), realities of access and **power** relations will shape the equity benefits of any such changes. The food shortages of 2008 (and onwards) illustrate the importance of two key changes within agriculture, namely, **deagrarianization** and the impact of internationalization. Agricultural production has been pivotal to making a living for centuries and is one of the oldest **livelihoods** approaches (including both farming and pastoral activities). It has however changed significantly over recent years as a result of changing international pressures and trends within global food systems, and also shifts in rural sources of income generated through non-farm based activities.

Deagrarianization describes the process whereby ways of making a living in rural areas become dominated by non-agricultural activities, as well as rural to

---

CONCEPT BOX 7.2

Green Revolution

The Green Revolution refers to the technologies and policies aimed at rapidly increasing agricultural production particularly within areas of the Global South. Following the pioneering work of Norman Borlaug in the 1940s, strains of maize, wheat and other staple crops were selectively cross-bred to produce higher food grain yields in response to the addition of nitrogenous fertilizer. With the support of the Rockefeller and Ford Foundations, as well as indigenous research and investment, these new high-yielding varieties (HYVs) spread rapidly across many areas of the Global South from the 1960s, particularly where irrigation infrastructure allowed significant gains in output to be achieved. For these areas, dramatic increases in wheat, rice and maize production were achieved, but the revolution also raised concerns about the loss of biodiversity (as indigenous crop varieties and crop rotations were replaced with HYV monocultures), and growing rural inequality (as policies supporting the technology accelerated commercialization of agriculture). The original HYVs were cross-bred 'naturally'; today, gene-splicing techniques potentially offer a new generation of HYVs that could have drought-resistance and other qualities that would allow their wider use across areas bypassed by the twentieth-century Green Revolution.

---

urban migration (see Chapter 5). This involves income reorientation and occupational adjustment, as well as changing social identities and strategies of spatial relocation (Bryceson, 2002: 726). The causes of **deagrarianization** are multiple but can be linked to 'turning-point policies', such as **SAPs**, which have removed subsidies from farmers for crucial inputs such as seeds, pesticides and fertilizers (Bryceson, 2002: 727). Linking back to the global food shortage, Deborah Bryceson maintains however that farmers have continued to practise and invest in subsistence agriculture in an effort to maintain household food supplies. As food shortages escalate, the trend of **deagrarianization** is unlikely to be reversed, but may become more complicated and diverse as households seek to survive in new ways. Bearing this in mind, agriculture remains the greatest source of employment in most Asian countries and across the continent of Africa. The number of people whose work is agriculture-dependent (those who may not work directly on farms, but are reliant on the sector) is even higher with the ILO calculating this as 42 per cent of the world's population or 2.58 billion people in 2001 (ILO, 2004). Nonetheless, processes of **deagrarianization** illustrate the vulnerability of agriculture to change and the subsequent impacts on people's **livelihoods**.

The internationalization of agriculture is a second key trend shaping agriculture. This refers to the **globalization** of agricultural production, marketing and retail processes and consumption (Rigg, 2007) and arises from the dominance of local, national and international agricultural practices by global agri-business conglomerates. These businesses, such as the American giants Dole Food Company and Chiquita Brands International, dominate the global market for the distribution of particular products, in their case, bananas and pineapples among other fruits. Their immense corporate **power** means that they effectively control the environments within which agricultural production and marketing occur; overriding the political will of weaker governments (see Raynolds, 1997, on such processes in the Caribbean). Thus the growing interconnectedness of agricultural practices means that decision making at the local household level (in terms of crop choices, labour and associated investments and sale of land) is increasingly shaped by international markets, financial standards and agreements (WTO, EU, etc.) and **consumption** patterns. Global **consumption** trends are highly significant. Crop choices, in particular moves towards non-traditional crops for export, the use of pesticides and forms of production are directly influenced by changing **consumption** patterns in the Global North, but also importantly in the dominant countries of the Global South, namely, China and India. But **consumption** trends are not simply related to cultural values and changing lifestyles (see Chapter 8), they are also shaped by natural disasters and changing climatic conditions which leave regions of the world in need of particular products, such as wheat or fresh vegetables.

The realities of internationalization are, however, very varied at the local and national level. Singular analyses (see Watts and Goodman, 1997, for a critical discussion) and policy responses which aim to overturn this trend are therefore misplaced because for some in the Global South certain aspects of this trend have been beneficial. Gendered impacts are also varied: Laura Raynolds explains how women (and immigrant workers from Haiti) have suffered exclusion and exploitation from new labour management practices associated with the growth of export crops, particularly bananas (Raynolds, 1997: 92). International donor organizations and NGOs have often been at the forefront of challenging gendered discrimination, often in relation to the control of land. But as Ann Whitehead and Dzodzi Tsikata (2003) outline in Kenya, the policy outcomes of this are not always positive for their intended recipients either. This example illustrates the complexity of these processes on the ground, and also reveals how cultural factors constantly shape the outcomes of broader structural events. The production of inequalities through ways of making a living is clearly informed here by globalizing processes and local and national norms and social values, such as attitudes towards women. This particular issue is the focus of the following section.

INEQUALITIES OF WORK

A primary form of inequality in ways of making a living are between those that benefit and those that suffer through working practices. In general, evidence shows that many working lives across the Global South are characterized by insecurity, difficulty and even desperation. In contrast to this are the very wealthy and affluent within the Global South who have succeeded through particular means to achieve economic success. We do not focus on them much within this book, but their experiences must not be overlooked. Although they are an absolute minority (in numerical terms) they are hugely powerful and their actions and decisions shape the lives and experiences of many. In Chapter 8 we explore some of the **consumption** practices of the wealthy. These have an important impact on the social and spatial character of cities (through the building of shopping malls for example). The presence of wealthy individuals (and families) within the Global South also unsettles **representations** of the South as a site of homogenous **poverty**. All countries have business elites, but in states such as Argentina, Nigeria, China, India, South Africa and Brazil, particular (and different) political and economic histories have facilitated extreme differences in wealth between a minority and the vast poor majority. These inequalities are a function of the unevenness of **capitalism** and the impacts of the **globalization** of economic practices. Nonetheless business elites are also vulnerable to global trends (see Lever-Tracey, 2002) although the consequences are not usually life-threatening. For the non-elite residents of the Global South, vulnerability to broader global trends usually leads to a perpetuation of inequality and **poverty**. Questions arise thus about the longer term benefits of relying on market-led approaches to development, a theme which is explored within Chapter 10.

Ways of making a living shape and are shaped by social **identities**. These **identities** are also a key axis along which inequality is expressed, and **power** relations that structure working relationships (such as between factory manager and worker) contribute to experiences of inequality. In the Global South a range of inequalities are experienced at a variety of scales: here we use **gender** and age as illustrations, but other social **identities** and differences such as ethnicity and sexuality are also keys sites of inequality. Inequality conveys unfairness, where someone benefits while another does not. We illustrate here that inequalities are more ambiguous and complicated than this, and that inequalities cannot be mapped on to particular social **identities** or ways of making a living in straightforward ways. Women's working lives are generally far more unequal than men's across the Global South, but there is much inequality between men too. Working children experience inequalities in the workplace, but this can also be, in part, a site of freedom and responsibility. Furthermore, in the absence of children's work and income, the potential inequalities for their associated households could be far higher. We explore some of these ideas below.

BOX 7.4

Masculinity and work in South Africa

Young men living in the informal settlement of Cato Crest, in Durban, South Africa, have grown up in a context of high unemployment all of their lives. Durban was once a competitive manufacturing centre but a mixture of global shifts (the rise of cheaper sites for production in Asia) and national particularities (recessions, neoliberal solutions and Apartheid's legacy) has reduced the availability of work. Many of these men have never known formal employment, or the luxury of a permanent or skilled job. They have also had very poor education at the hands of the Apartheid education system. Most of these men identify themselves as Zulu, and are fiercely patriarchal. They view a man's role as household head as given. Thus, men's **identities** are closely shaped by their abilities to fulfil this role. The realities of unemployment in South Africa mean that these roles are severely undermined. The men feel demoralized as parents to their children, as they are unable to pay for necessary goods (food and school uniforms). Being unemployed changes the **power** balance within the household, and men feel lacking in authority in situations where their wives or partners are employed (albeit often informally and poorly paid). One male resident Philani explained: 'It is painful … to be unemployed if you are a man while your wife is working. … People are not respecting you as man. They used to call you all the bad words. Even your wife doesn't respect you as a man. She use to gossip with the neighbours about you' (Meth, 2007). Domestic violence is a common consequence of these experiences and countless women suffer at the hands of their emasculated partners. Analyses of masculinity also suggest that the male sense of self is highly vulnerable to failure, particularly in comparison with other seemingly successful men. Tensions between men over work and wealth in Cato Crest are at times expressed in xenophobic terms when the successful other men are foreign (from other African countries). Siboniso explained that 'The people who are coming from the neighboring countries many women of this area they interested to them because they have money' (see Plate 5.3 on xenophobia in South Africa). Work, status, **identity** and **power** are all interconnected.

(Source: Meth, 2007)

Gendered ways of making a living

Ways of making a living are inherently gendered (see Concept Box 5.3 and Boxes 7.1 and 7.2). They are gendered because cultural norms and values shape ideas about work, about what is appropriate or inappropriate work for men and women to do, and these values exert a powerful influence. For example, it is inappropriate in many Muslim countries for women to work outside of the home, and in particular for women to work in public places with male strangers, although strategies to overcome such norms are evident. For example, Deborah Bryceson notes that some women in Muslim countries use children to retail their merchandise so as to observe *purdah* restrictions (2002: 732). Nonetheless, because of the gendered nature of work, men and women experience changing economic or political processes (at a local, national or international scale) in differing ways, and these processes can in fact entrench gendered divisions and inequalities. Women's work, for example, has changed considerably over the past 30 years as opportunities for employment in expanding service sector industries, such as tourism, have grown phenomenally. These jobs however are often poorly paid and insecure. Many men, on the other hand, have lost employment in traditional male sectors such as mining and manufacturing. These broader changes have led to particular tensions between and among men and women whose **gender** roles are being redefined as a result of changing work opportunities. In the Global South (and North) work is central to men's **identity** (Morrell and Swart, 2005) and thus employment changes and unemployment are linked to crises of masculinity. Box 7.4 details the feelings and fears of unemployed men in the context of Cato Crest in South Africa, illustrating the connections between **identity** and employment.

Across the Global South, many women (like men) do work outside of the home in a broad range of environments, but women also take responsibility for the majority of **domestic labour**, unpaid work conducted within the home (Concept Box 7.3). Although feminist geographers (McDowell, 1998) have challenged the simplistic binary of public and private space, **domestic labour** can be considered a responsibility which keeps women within private space far more than their male counterparts (see Plate 7.5). But gendered inequalities over work are not so clear-cut, and neither are definitions of what is actually 'work'. Elizabeth Harrison (2000) argues (in a Zambian context) that male visiting practices and social meetings are often key to securing resources for households and by only focusing on so-called productive activities, these contributions will be overlooked.

Child workers

Ramya Subrahmanian describes the debates relating to child work (and education) as 'highly contested, (and) morally and ethically challenging' (2002: 403). Children and youth provide a substantial source of labour across the Global South. Table 7.2 details the three key areas of work that children are involved in, excluding **domestic labour**. This reality is the focus of much condemnation (particularly by

---

CONCEPT BOX 7.3

Domestic labour

Domestic labour is unpaid work usually performed in or around the home, including food preparation, cooking, cleaning, caring and maintaining the needs of household individuals, as well as maintaining the home itself. These tasks allow the household to thrive and continue, and in turn allows household members who may be wage earners to work. At a global scale, **domestic labour** is highly gendered, with women taking responsibility because it is viewed as an extension of their reproductive role.

Much **domestic labour** goes unnoticed and unregistered as it is considered a 'normal' part of a woman's day, despite efforts by feminist researchers to gain proper recognition for this work. The burdens of **domestic labour** are a fundamental point of inequality between men and women across the Global South as women often carry this burden along with other ways of making a living (see Table 8.4). In many households girl children are also burdened with **domestic labour**, either in the form of helping their mothers, or taking responsibility for tasks themselves. This is particularly a problem in households where adults are absent because of HIV infection or migration for employment.

Domestic labour must not be confused with paid domestic work where a household employs a person to assist them with household tasks – a job also commonly undertaken by women.

---

organizations in the global North such as the ILO) about the contravention of the rights of children.

However, researchers and practitioners working on this issue, while acknowledging and condemning the hideous nature of much child work (drug trafficking, prostitution, bonded labour, etc.) challenge the seemingly universal assumption that child work is necessarily and inherently bad (see Bourdillon, 2006; Nieuwenhuys, 2007). Their arguments are complex, but centre around questioning the nature of inequalities experienced by children in the context of their households. **Poverty** is a key explanation for why children work, but it is not the only factor. Pressures to compete within the global market, as a result of the neoliberalization of national economies, have forced many children into work (Lloyd-Evans, 2002). In addition the concern for child workers has largely focused on children working in industrial or productive settings (such as carpet making and match factories). This view has overlooked the far more common, but often invisible, work that is carried out, usually by girl children in the home in terms of **domestic labour**. Furthermore, the largest area where children are known to work is in agriculture, a sector which is

Plate 7.5 A woman cooking in her home, Yafa area, Yemen.
Credit: Nikky Wilson.

Table 7.2 Key sectors within which children work (excluding domestic labour)

Agricultural sector	Industry sector	Services sector
Agriculture	Mining	Wholesale and retail trade
Hunting	Quarrying	Restaurants and hotels
Forestry	Manufacturing	Transport
Fishing	Construction	Storage
	Public utilities (gas, electricity and water)	Communications
		Finance
		Insurance
		Real estate
		Business services
		Community work
		Social personal services
Percentage of total		
69	9	22

Source: Hagemann, *et al.*, 2006.

again often invisible to policy enforcers and the reach of global institutions (such as UNICEF). The International Labour Organization estimates that 132 million boys and girls aged 5 to 14 work within this sector, accounting for 70 per cent of all child work (ILO, 2008). Child workers can also, in contexts of extreme vulnerability, be a fundamental survival strategy for households: without their labour and income, some households would suffer starvation. A final argument centres around the loss of education when children are working, but as Subrahmanian explains, this area too is partly open to debate, as some children work to pay for their education, although engaging in both school and work is very challenging for children (2002).

Despite these debates there is a clear consensus that the worst forms of child labour (prostitution, bonded labour and working in hazardous industrial employment) should be abolished (Lloyd-Evans, 2002: 218) and the ILO has pursued this agenda through the Worst Forms of Child Labour Convention of 1999 (and subsequent recommendations). This convention details labour practices which are particularly harmful for children such as working underground, or under water, or working in an unhealthy environment (see Plate 7.6), despite which such conditions remain fairly common across the Global South (see Box 7.5). What is evident from this discussion, however, is that understanding inequality is never straightforward, whether it relates to **gender** or age-related practices. Vulnerability is also contained at times by the **agency** of individuals, although as we argue below this is often highly limited.

BOX 7.5

Child garbage pickers in Brazil

Garbage picking is a common activity across the Global South as a source of income generation for men, women and children. In Brazil, as elsewhere, children commonly accompany their parents and assist in the task of picking. Street children also engage in independent garbage picking. The types of pickings vary depending on the nature of the market and the type of waste site, but commonly children will pick for cardboard or metal products or clothing for resale. This type of work is intensely dangerous, in terms of the health and safety risks it poses regarding the exposure of children to dangerous waste products. Furthermore children garbage pickers often forgo their studies at school to assist their families in this enterprise and thus suffer disproportionately in terms of their educational achievements. Children are often found working through the night, making

BOX 7.5 (*CONTINUED*)

their ability to attend and concentrate at school near impossible. Their failure rates at school are high as a result of their absenteeism and serious fatigue. Garbage picking does however contribute to household incomes, or to the abilities of homeless children to make an independent living. For some the alternatives, such as prostitution or drug running, are far worse.

(Sources: Adapted from ILO, 2007; Pan American Health Organization (PAHO), 2007)

Plate 7.6 Child garbage pickers, Rio de Janeiro, Brazil.
Credit: © Still Pictures.

WORKING TOGETHER? USING COLLECTIVE ORGANIZATION TO IMPROVE WAYS OF MAKING A LIVING

This final section of the Chapter explores the politics of work by examining strategies of collective action by workers (see Plate 7.4) to further their interests. These practices illustrate the **power** relations embedded within working relationships (between employer and worker for example), but also workers' **agency** and the constraints upon this.

Millions of workers across the Global South have used the strategy of working together to combine their influence and **power** and to promote job security and work entitlements. The right for workers to associate allows for workers to join collectively, with trade unions being one common avenue. Trade unions are familiar (if somewhat declining) institutions in the Global North, and they are also significant in many parts of the Global South, alongside other organizations such as worker cooperatives and support organizations. More recently, global labour-management agreements are being developed by multinational organizations whose impact is felt across many countries of the Global South (ILO, 2004). In addition to this the rise of **corporate social responsibility** (see Chapter 10) shapes the working experiences of some workers at a global level.

The International Labour Organization enshrined the principle of freedom of association and the right to collective bargaining in their 1919 constitution. Ratification of this right has grown with the spread of **democracy**, but some countries still deliver very poorly in this regard. Furthermore, workers' abilities to form associations are limited within particular sectors, namely: the public sector, agriculture, **export processing zones (EPZs)**, migrant workers, domestic workers and **informal economy** workers (ILO, 2004: 4), all major employers within the Global South. Reasons for a lack of rights vary between these sectors, but usually workers are denied the freedom to 'associate' relating to legal controls or where existing labour legislation is not enforced. Where workers are allowed to associate, they often experience a restriction of their rights (ILO, 2004: 37–8). In EPZs restrictions on workers' collective action are often used to attract foreign business (see Chapter 4). In relation to domestic workers and workers within the **informal economy**, their lack of rights is often related to the illegal or informal nature of the employment relationship. Domestic workers are particularly hard to organize, because of the personal nature of the work and the fact that women often live in their employers' houses (ILO, 2004: 43). For women of the Global South this is a substantial problem: domestic work provides much employment (and is a significant source of **remittance** income), and female labourers predominate in both the **informal economy** and **EPZs** (ILO, 2004: 37). Thus restrictions in workers' rights are distinctly gendered, adding to a complex pattern of gendered employment practices at a global scale.

Aside from trade unions, workers can also organize in other ways to support the needs of their members. For those making a living in the **informal economy**, the growth over the past few decades of worker organizations has been central to their ability to achieve fair working relations with city authorities. Organizations vary in terms of their goals but commonly they work to obtain licences for business practice (e.g. trading licences), establish credit lending societies and demand better management and treatment from authorities, particularly in relation to energy, housing, water, road and transportation concerns. They also challenge bureaucratic corruption and ongoing harassment by authorities, making use of alternative forms of authority to ensure that their rights are upheld in court (ILO, 2004: 47). Examples of organizations that have formed to advance the needs of women

Plate 7.7 The 2006 International WIEGO Convention, Durban, South Africa.
Credit: Suzanne van Hook.

working from home as well as on the street as vendors are HomeNet, StreetNet, WIEGO (Women in Informal Employment: Globalizing and Organizing) (see Plate 7.7) and SEWA (Self Employed Women's Association). HomeNet's reach is substantial with affiliates in 130 countries and WIEGO had affiliates in 25 countries in 2001 (Chen, 2001: 81).

The significance of these organizations must be understood in the context of the lack of traditional trade union organization in these sectors. The activities of SEWU (Self Employed Women's Union) in South Africa illustrate the capabilities and limitations of such organizations (see Box 7.6: SEWU in Durban).

CONCLUSIONS

This Chapter used four related arguments to explore the complexities of ways of making a living across the Global South. First, we argued that the analytical framework for understanding how people make a living must adopt a flexible, holistic and fluid approach, or risk overlooking key forms of **livelihood**. The informal sector illustrated the value of adopting this approach and also revealed the multiple ways

BOX 7.6

SEWU in Durban, South Africa

The South African organization the Self Employed Women's Union (SEWU) was officially launched in 1994 and is modelled on the Self Employed Women's Association (SEWA) (based in India). SEWU has a constitution and a strong democratic structure and its membership is targeted at women working in the **informal economy** – and thus it concentrates on the poorer end of the **informal economy**. SEWU represents women involved in home-based production, small-scale farming, street trading and part-time domestic work. These areas cover activities such as craft and clothing making, trading vegetables and food stuffs and poultry and pig farming. The organization originally had a strong membership in relation to city street-trading but increasingly its rural membership has risen as more home-based workers join. This shift has affected the activities and focus of the organization.

For street traders, key successes were gained for women in Durban working in the local *muthi* (traditional medicine) market, such as improved facilities in 1994 and in 1995 the guarantee and provision of a semi-formal market known as the *muthi* market providing shelter, toilet and water facilities. Successful negotiations also led to the provision of storage facilities and nearby childcare provision. Previously, caring for children at the market had raised many problems as it is a site of much petty and serious criminal activity. A further additional achievement has been the provision of a nearby cheap overnight accommodation facility. This latter provision was absolutely crucial, as traders were unable to afford housing in surrounding townships and were often forced to sleep in unsafe and unhygienic spaces such as the basement garages of inner city buildings. In addition to this SEWU has been involved in policy formulation, engagement with national decision-making, work within international forums and also influencing policy within the ILO. In this case the practice of working together has served to reduce the inequalities experienced by women traders.

(Sources: Adapted from Devenish and Skinner, 2004;
Streetnet International, 2007)

in which informal and formal sectors intersect. Second, we explored the ways in which a historical perspective and an appreciation of the impact of global changes were central to understanding **livelihoods**. Global changes in agriculture are particularly relevant here. Inequalities are inherent within ways of making a living, as well as being shaped by them. We used two key axes of difference, namely, **gender** and age, to explore inequalities. Using a gendered approach, and that of age, should not result in an unsophisticated assumption of women's and children's marginalization, without examining the complex **power** dynamics between and within men and women, and children and their households. Finally, we explored the tradition of working together to effect changes in working conditions and we considered the difficulties of this strategy across various sectors. This discussion illustrates the politics of work, but also the politics of individual **agency** to act and challenge a situation.

 Review questions/activities

1   Assess the ways in which making a living is defined in different contexts, such as in the media and in government statistics (in different countries across the world). Ask how flexible and holistic these interpretations are, and then think about what other ways of making a living are being overlooked.

2   Consider the politics of work within your own household, throughout your childhood to now. What different working practices were you a part of, how did you think about these practices and how did they change (or not) over time? Try and step away from the specifics of your own history and see if you are able to identify broader structural processes which might have impinged on your household's livelihood, and also if any particular power relations were evident.

3   Using the commodity chain approach outlined in Chapter 4, identify a number of products in your home that have travelled from countries of the Global South to you (whether you are in the North or South). Consider how these products might have been manufactured or grown, marketed, packaged and transported. Reflect on the spaces within which these processes have occurred and the individuals who may have been involved. Consider the inequalities between various actors along this chain.

SUGGESTED READINGS

Chen, M. A. (2001) 'Women in the informal sector: a global picture, the global movement', *SAIS Review*, Vol. 21, No. 1, Winter–Spring: 71–82.
Chen provides a highly comprehensive overview of this global phenomenon and covers key issues concerning gender and inequality.

Rigg, J. (2007) *An Everyday Geography of the Global South*, London: Routledge.
This is a very well-written and clearly argued book which considers ways of making a living in several chapters. It provides a useful introduction to the livelihoods approach and also key references and sources.

WEBSITES

www.ilo.org International Labour Organization
This is the most comprehensive overview of international labour issues and provides a range of research papers and statistical databases covering work-related matters. Much of the website is in English but some country-specific information is in other languages (making it a little difficult for English readers to access). Nonetheless overall the site is user-friendly and informative.

www.livelihoods.org/info/info_guidancesheets.html
These web pages provide an excellent overview of DFID's livelihoods approach, providing detailed discussion of key concepts used, the history of the approach and supporting references and examples. Users of this material should be aware that this is only one interpretation of the livelihoods approach.

8 Ways of living

INTRODUCTION

The social and cultural practices of the peoples of the South have often been represented either as static and 'traditional', compared with the 'modern', 'progressive' and 'dynamic' North (Chapter 2), or at the mercy of the **cultural globalization** or **Westernization** juggernaut (Chapter 5), with its 'modern' ideas and commodities which undermine traditions and ways of living (Tomlinson, 1999). However, as has been argued throughout the book, both these **representations** are overly simplistic and aspatial. Not only are people in the South responding to new flows from elsewhere (including other places in the South), there is also a continuous process of reinvention and negotiation of **identities** in 'local' environments.

Such processes do not mean that 'tradition' is a redundant concept within the South; rather, that notions of 'tradition' are as **socially constructed** as concepts of 'modernity', in fact the two are clearly linked in a **relational** fashion. Eric Hobsbawm (1983) developed the concept of the 'invention of tradition' in relation to the ways in which 'traditions' were made up to perform particular purposes in a political context (see Concept Box 8.1). This notion of the instrumental use of the concept of 'tradition' will be discussed in this Chapter in three main ways; first, practices which claim to reassert 'tradition', most notably in relation to religious practices; second, the way in which ideas of 'tradition' are mobilized as a commodity; and finally, 'new' practices which draw on 'traditional' values. In all three cases, the ways in which this mobilization of 'tradition' reflects **power** relations and also how it fits into wider social, economic, spatial and political flows will be discussed.

The focus in this Chapter on cultural and social practices in the everyday lives of people in the Global South does not imply that politics and economics do not have social and cultural dimensions, or vice versa. Two processes which will be of particular importance in this Chapter are '**hybridization**' (see Concept Box 8.2) and '**transnationalism**' (see Concept Box 8.3). Both these concepts recognize the fluidity of **identities** and the mobility of ideas, people and things, but they also include

CONCEPT BOX 8.1

Invention of tradition

Eric Hobsawm (1983: 1) defines 'invented tradition' as 'a set of practices, normally governed by overtly or tacitly accepted rules and of a ritual or symbolic nature, which seek to inculcate certain values and norms of behaviour by repetition, which automatically implies continuity with the past'. Hobsbawm identifies three major reasons for the invention of tradition since the nineteenth century. First, attempts to create social cohesion or ideas around group membership; this has been particularly important in the creation of feelings of nationhood. Second, traditions may be invented as a way of establishing or legitimating particular institutions or hierarchies, such as between classes, races or genders. Finally, the invention of tradition can be a form of socialization through which beliefs and practices are inculcated.

Terence Ranger (1983) outlines how European powers invented traditions in Africa as a way of embedding colonial power and justifying relations of domination. These 'traditions' often drew on European interpretations of pre-existing cultural patterns, such as systems of tribal leadership which both the Germans and British used in their representations of and justifications for 'imperial monarchy' (see also Box 6.2 on Mahmood Mamdani's discussions of the invention of 'traditional authority' in colonial Africa). Educational establishments, religious missions and the armed forces also implemented practices which claimed to be 'traditional', but which had been developed as part of the colonial project. These included forms of dress, rituals, songs and naming practices, for example.

While such invented traditions came from 'outside' African cultures, some Africans drew on these new practices to justify their exercise of power. Ranger discusses how hierarchies of age, **gender** and ethnicity were reinforced by African claims of 'tradition' based on ideas and practices introduced by Europeans.

(Source: Adapted from Hobsbawm, 1983; Ranger, 1983)

the possibilities of **resistance** or **agency** on the part of marginalized groups. They are thus a challenge to some of the more pessimistic interpretations of 'cultural **globalization**' (see Chapter 5).

While mindful of the possibilities of **resistance** and change, we also discuss the obstacles which frame people's daily lives. For example, in many situations a lack of economic resources means that individuals, households or communities may have little, if any, choice in the ways they live. Similarly, other axes of marginalization

---

CONCEPT BOX 8.2

Hybridization

Hybridization is used to signify the coming together and mixing of two cultures, for example through migration or the global media. However, 'when have societies or "cultures" ever been so isolated, static and "authentic" as to not be hybrid in some fashion?' (Hodgson, 2001: 6). It is therefore important to recognize that the **hybridization** associated with processes of **globalization** since the mid-twentieth century is only a particular form of cultural mixing which has a much longer history.

CONCEPT BOX 8.3

Transnationalism

Transnationalism refers to economic, social and political processes which take place across international boundaries (Vertovec, 1999). Unlike 'globalization', it is important to recognize the continued importance of the **nation-state** within **transnationalism**, because without national borders there can be no 'transnational'. A key aspect of **transnationalism** is that the cross-border processes are ongoing. In the case of migration, for example, this would mean international migrants keeping in contact with friends or relatives 'at home' through visits, telephone calls, emails (Wilding, 2006) or **remittances**, for example. Within the ideas of **transnationalism** is the possibilities which come from being 'in-between', or in 'liminal space'. For some authors (see e.g. Bhabha, 1994), this provides opportunities for individuals or groups to challenge existing inequalities, while for others (see, Mitchell, 1997), such emancipatory readings are overly optimistic.

---

such as **gender**, sexuality or ethnicity may limit choice. However, as Chandra Tolpade Mohanty (1991) warns in the context of 'Third World women', it is vital not to read off particular forms of disadvantage from one social characteristic.

To help examine these issues, this Chapter focuses on three main groupings of social and cultural practice: home-making, food and clothing **consumption**, and leisure and play, and the ways in which different spaces are used or created through these practices. There is a particular focus on **consumption** within these practices; what, how and where people consume is not just a matter of chance, rather it reflects people's **identities**, social norms and spatial practices. Ideas about **cultural**

globalization often present the global spread of clothing brands and fast food as a one-way process, which the 'local' environments and communities in the Global South cannot resist. However, in this chapter, we will consider the ways in which these everyday practices are shaped by, and in turn shape, broader processes in the Global South. The Chapter does not aim to provide a comprehensive discussion of everyday life in the Global South, but rather to highlight key dimensions and processes.

HOME-MAKING

The concept of 'home' refers not only to the physical space where people live, but also a social space, ranging in scale from the social relations with co-residents, to ideas of the **nation-state** as 'home' (as in 'home country') (Blunt and Dowling, 2006). This section focuses on the home for three main reasons. First, it is a highly significant space for the vast majority of people, both because of the amount of time that is spent there, and because of the emotions and relationships which are experienced within the home. It is important to remember, however, that for some people, the idea of 'home' is highly contested and a lack of a formal structure means that where one sleeps (for example, on the street or in an informal shack) does not have homely connections (Meth, 2003). Second, domestic **material culture** provides an insight into the manner in which flows of ideas, commodities and people, as described in Chapter 5, are grounded in particular spaces. Third, examining the social relations of the home allows us to consider people's everyday lives and the ways in which kinship, marriage, **gender** relations and ideas of 'the family' are both reinforced and challenged by processes such as **urbanization**, **transnationalism** and **globalization**.

House construction and domestic material culture

Before long distance travel was common and populations relied on resources from the local area for subsistence, the style of house building was based on the natural resources available. In addition, populations made sure that houses were built to provide protection from local weather conditions and that they allowed for cultural norms to be reproduced, such as, for example, compounds with different huts for polygamous or extended households in parts of rural Sub-Saharan Africa. However, as communities have become exposed to different forms of living, house style may have changed or may differ depending on location and purpose. Plates 8.1 and 8.2 illustrate two contrasting house styles belonging to Sakhile Shangase (a pseudonym) in South Africa, both making use of locally sourced materials. The first is his rural home compound in Ndwedwe (made from mud and thatching), and the second his shack in Cato Crest (Mayville), Durban (made from scrap pieces of wood). Sakhile used these drawings to highlight the differences between housing types, and the

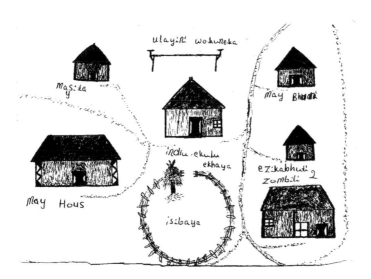

Plate 8.1 Sakhile's drawing of his rural homestead in Ndwedwe, South Africa. The Zulu words Sakhile used in his picture are as follows: 'ulayini wokuneka' – washing line; 'isibaya' – corral for cattle; 'maysita' – my sister's house; 'may hous' – my house; 'may bhuti' – my brother's house; 'indlu enkulu ekhaya' – big house which is used to talk to the ancestors; 'ezikabhuti lezi' – two houses for my other brother.
Credit: Sakhile Shangase/Paula Meth.

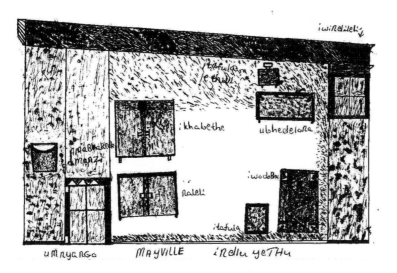

Plate 8.2 Sakhile's drawing of his urban shack in Cato Crest, South Africa. The Zulu words Sakhile used in his picture are as follows: 'umnyango' – door; 'indlu yethu' – is our house; 'iwindi leli' – this is a window; 'naleli' – also a window; 'itafulu lethu' – is our table; 'eMayville indlu yethu' – this is our house at Mayville.
Credit: Sakhile Shangase/Paula Meth.

impacts of their differing designs, on his experience of safety and privacy, specifically in terms of his ability to practise (private) sexual intercourse with his partner (Meth, 2007).

One of the key processes which has broken the link between the natural environment and house building, is **urbanization** (see Chapter 5). The rapid growth of urban areas in the Global South has not been accompanied by a concomitant growth in quality housing. Neither the government nor private companies have been willing or able to provide accommodation for the vast majority of new urban residents. This has meant that most migrants and low-income urban dwellers have been forced to rely on informal housing. This may be, for example, renting a room in a subdivided city-centre building, or constructing a dwelling on a piece of illegally occupied land (Davis, 2006; Drakakis-Smith, 2000). In the latter cases, residents will use whatever material is available to construct a dwelling, often using corrugated iron or cardboard if they cannot afford bricks (see Plate 8.3).

While informal housing is not necessarily of poor quality, and over time residents may be able to improve their houses through small investments (Mangin, 1967) millions of people in the cities of the Global South are living in unsanitary and overcrowded conditions (see Plate 5.1 of Lagos). UN-Habitat (2003) would

Plate 8.3 Informal housing, Cali, Colombia.
Credit: Melanie Lombard.

classify these as 'slums'. In 2005, UN-Habitat estimated that approximately one billion people lived in slums, most of which were located in the cities of the Global South (see Table 8.1). Such poor living conditions have severe implications for the health and life chances of the residents (Chapter 5).

A house is both a place to live and also an investment for the future (de Soto, 2000). For this reason, residents will often seek to improve their houses if they have money available. The appearance of the house is an outward symbol of how the house's residents are doing economically and can be used as a signifier of status. For example, **remittances** from overseas family members or migrants elsewhere in the country can be used to improve the house. This is sometimes interpreted as 'unproductive investment' as it is not being used directly for income generation, but it can create jobs indirectly through the purchase of building materials and construction labour, and if it improves the health of the residents then it could have knock-on benefits in the future. Hein de Haas (2006) outlines the ways in which **remittances** to the Toghda Valley, Morocco, are used to improve houses. Rather than building traditional-style houses, migrants' families use this income to build what are seen as more 'modern' houses. Plate 8.4 shows a new development in Sonjiang near Shanghai, China, which recreates an image of an English country town. Prospective residents are attracted by the idea of living in this kind of setting as it suggests a particular awareness of the world and engagement with things beyond China.

This desire to be seen to be 'modern' and to use the domestic space as a way of claiming status and expressing **identity** has been discussed in many parts of the world. The focus on domestic **material culture** (Concept Box 8.4) provides insights into the **identities** of household members because the domestic space is

Table 8.1 Slums by global region, 2001

	Urban population (%)	Slum population as % of urban
Northern Africa	52	28.2
Sub-Saharan Africa	34.6	71.9
Latin America & Caribbean	75.8	31.9
Eastern Asia	39.1	36.4
Eastern Asia (excluding China)	77.1	25.4
South-Central Asia	30	58
South-Eastern Asia	38.3	28
Western Asia	64.9	33,1
Oceania (excluding New Zealand & Australia)	26.7	24.1
WORLD	47.7	31.6

Source: Adapted from UN-Habitat, 2003: 27, table 3.

Plate 8.4 Thames Town, Sonjiang, China.
Credit: Andrew Marton.

often the only space where people have **power** to change their living environment. Of course, as outlined below, not everyone in households necessarily has the same **power** to make such decisions.

Domestic **material culture** provides a useful example of how **transnational** and hybridized practices can be played out in the Global South. Objects within houses may reflect the ways in which the household members have direct links with people and places elsewhere, for example, stereo systems, fridges or washing machines brought or sent by migrants from cities or overseas. Alternatively, domestic objects may be chosen to distinguish the residents from people they see as more 'backward' or 'traditional'. Thus, locally produced wooden furniture may be rejected in favour of imported soft furnishings. However, such furnishings may be presented alongside other 'global' objects, such as posters of American popstars or European footballers, while in another part of the room there is a religious icon or shrine. Domestic spaces may therefore reflect the intersection of a range of commodity flows.

Plate 8.5 shows Srei Mon in her one-roomed house which is built on stilts on squatter land along the river in Siem Reap, Cambodia. She moved to Siem Reap from Phnom Penh in 2000 and works selling palm juice on a stall outside her

---

CONCEPT BOX 8.4

Material culture

Material culture is the term used to describe the objects which people use and how this reflects wider cultural practices and meanings. Anthropologists and archaeologists have long had an interest in material culture as a way of understanding past societies or particular cultures. Within Geography, the study of **material culture** dates largely from the early twentieth century when the Berkeley School of Cultural Geography led by Carl Sauer examined how landscapes across the USA varied. They focused on differences in house or barn design, for example, as a way of identifying and understanding differences in local cultural practices, as well as recognizing how certain practices spread.

The study of **material culture** is now applied to an array of places and objects. There has also been a recognition that objects are not just a reflection of existing cultural practices, but that they also contribute to the construction of social relationships and understandings of the world. With increasing migration flows, the movement of commodities and ideas around the world, looking at **material culture** has been a key dimension of work examining **cultural globalization**.

(Sources: Adapted from Anderson and Tolia-Kelly, 2004; Crang, 2005; Jackson, 2000)

---

house. She was deserted by her husband and was left to bring up her two teenage daughters, one of whom is still at school, and the other sells soup. The three of them live together in this house and they are constantly threatened with eviction and find it hard to cope with the upkeep of the building. The picture shows how Srei Mon and her daughters have furnished and decorated their house; the stereo system takes pride of place, but there are also family photographs, a curtain with pictures of Winnie the Pooh and a variety of plastic goods, including flowers. There are also advertisements for ABC Stout, which is very popular in Cambodia, but is produced by Asia Pacific Breweries Limited, a company with breweries all over the region and which is listed on the Singapore stock exchange. Srei Mon's possessions reflect her income, but they also show how home decoration and **material culture** encompass local, regional and global flows of ideas and products.

Plate 8.5 Srei Mon in her home in Siem Reap, Cambodia.
Credit: Katherine Brickell.

Household formation

'Home' can refer not just to the dwelling where we live, but the people that live there and the social relations that bind them together and make people feel safe. This idea of 'home' as a place of **social reproduction** and care is part of the concept of 'household'. While 'households' may be difficult to identify on the ground, in theory they are domestic units of production and **consumption**. This implies that members share activities such as cooking, as well as pooling of income or subsistence production although, as the next section shows, such practices are not always present. Household members may be kin, but they do not have to be (see, for example, Box 8.1).

Cultural norms about household formation vary over time and space, and are often intertwined with changing economic circumstances. For example, large households have been seen as important in subsistence agricultural economies when labour is needed. However, smaller, nuclear households may be more acceptable in urban settings or when alternative forms of income generation become available. This may lead to a decline in fertility as parents decide to limit the number of children they have, although there are many other reasons for falling birth rates (see Table 8.2).

BOX 8.1

Mema's house, Mexico City

Mema is an AIDS educator and gay rights activist in Ciudad Nezahualcóyotl, a low-income district on the outskirts of Mexico City. His home provides a sanctuary for a range of largely young people from the local area, many of whom are gay (although they would not necessarily use that term) and have experienced problems with their families because of this.

The house provides a safe space for people to chat, listen to music, have sex and drink. Everyone is welcome to stay for meals as well. Some young people live there if they have been thrown out of their parents' homes, although in some cases, family reconciliation is possible after a period of time. For the transvestites (*vestidas*), Mema's house also allows them to share make-up and dressing tips and to try out new 'looks'. The house can also be used for earning money through hairdressing or prostitution.

Mema is called *abuelita* (grandmother) by many of the young people and is certainly head of this unconventional household. As head, he sets and enforces a number of rules that members must abide by, such as sharing housework, not stealing and limiting alcohol and drug use. Condom use is also compulsory for anyone having sex on the premises. Anybody not abiding by these rules is asked to leave.

While nobody in Mema's house is related by blood or marriage, the occupants can be viewed as a household with Mema as the head. Not everyone in the house can contribute money, but everyone helps out with housework and cooking. There are also relationships of care and concern within the household, as well as a set of rules which Mema enforces. These bonds between residents and the widely held feelings that they are all members of one unit, mean that Mema's house can be regarded as a household.

(Source: Adapted from Prieur, 1998)

There are also household structures which are particularly associated with some parts of the Global South. For example, three-generational households (grandparents, parents and children) are sometimes represented as being an 'Asian' form of family (see Stivens, 2006, for an overview). Reflecting cultural norms about 'filial piety' – the expectation that children will respect, look after and care for their parents as they get older – three-generational households have been common in many parts of Asia. However, with **urbanization** and changing social norms,

Table 8.2 Reasons for lower or higher population growth rates

Reasons for lower growth	Reasons for higher growth
Access to family planning and health care	Death rates have fallen but birth rates have remained the same or grown
Low fertility rates (e.g. Germany)	Active government policy to increase fertility
Improved educational status (particularly of women)	Low level of available contraception
Improved political status (particularly of women)	Low educational status of women
Government policy on fertility and family planning (e.g. China)	The country has a larger proportion of its population of child-bearing age
Improved access to employment	Religious or cultural reasons for low use of contraception or family planning
Improved government services for the elderly, sick or unemployed	High in-migration and low out-migration
Devastation of war, famine or disease (such as HIV/AIDS in South Africa)	

sometimes due to **discourses** around 'modern families' portrayed on television, nuclear households have become more common.

In some places, shifts in household structures and practices have led to calls for the reassertion of 'traditional' norms and values because of perceived crises in 'family values' or morals (particularly sexual). For example, the massive rise in forms of evangelical Christianity in Latin America and Sub-Saharan Africa (see Chapter 5) have not been accomplished through a call for a return to the 'traditional' practices of indigenous peoples, but rather for a reassertion of the 'natural' order of respect, clear **gender** divisions of labour and sober behaviour (see also the discussion of 'Asian values' in Chapter 2). While Christianity may have been part of the process of '**modernization**' during the colonial period, in the early twenty-first century, certain forms of the religion are constructing a new way of acting in contrast to the perceived ills of **modernity**.

Despite some general trends towards smaller households over time, it is important to recognize that changes are not linear. While small nuclear households are sometimes presented as 'modern', it is not the case that households in either the North or South are becoming more homogenous. In fact, household diversity is growing. This may be because of changing legal or cultural norms about divorce or women-headed households (Chant, 1997; 2007), or the need to expand or reduce the size or composition of the household due to economic crisis (see Chapter 7 on household survival strategies and income generation). Research in urban areas throughout the Global South in the 1980s and 1990s suggested that household

Table 8.3 Shifts in household structure, Querétaro, Mexico

Household type	1982–83	1986
Nuclear	40% (8)	30% (6)
One-parent	25% (5)	10% (2)
Male-headed extended	25% (5)	30% (6)
Female-headed extended	10% (2)	30% (6)
TOTAL	100% (20)	100% (20)

Source: Adapted from Chant, 1991: table 4.11.

extension (merging households) was a strategy adopted by some low-income residents to cope with the economic effects of crisis and **structural adjustment policies** (see Chapter 10). Sylvia Chant (1991) did longitudinal work in Querétaro, Mexico, in 1982–83 and 1986 which allowed her to investigate how household structures change over time. Between 1982 and 1986 many households had undergone a process of extension (see Table 8.3). In some, but not all cases, this was a response to the economic crisis and the need to maximize income-generating opportunities and minimize expenditure.

Increasing levels of international migration have led to what have been termed 'transnational families' (Parreñas, 2001) or the practice of 'global householding' (Douglass, 2006). This has usually been associated with labour migration (see Chapter 7) where individual household members have moved abroad to find work. In some cases, this is associated with household-based work elsewhere. The idea of global care chains has been developed to describe how migrants (usually women) have moved to care for the elderly or children of middle-class households, leaving their own children to be cared for by others in their region of origin (Ehrenreich and Hochschild, 2002; Hondagneu-Sotelo and Avila, 1997). Other reasons for the development of transnational households have, however, become increasingly common. For example, parents who can afford it may decide to send children overseas for education, particularly to English-speaking countries, in order to gain both human and **cultural capital** which will place them at an advantage in the global labour market. In such cases, children are usually accompanied by a parent, almost invariably the mother (Lee and Koo, 2006; Waters, 2006). As demographic changes mean populations are ageing (see Chapter 5), transnational households may become increasingly common in the Global South. Retirement migration has been a feature of some parts of the world for some time, for example retirement communities in Florida, or the migration of British retirees to the Mediterranean (O'Reilly, 2000). However, some parts of the Global South are also becoming chosen destinations (see Box 8.2).

BOX 8.2

Japanese retirees in Chiang Mai, Thailand

Over 1,000 Japanese retirees live in Chiang Mai, northern Thailand. As the Japanese population ages (21 per cent were over 65 in 2005) and pension and welfare provision declines due to economic recession, many Japanese people have been considering ways of providing for themselves in old age. From the early 1990s onwards, the Thai government has sought to attract 'long-stay' Japanese visitors through changes in the visa regulations and joint ventures with Japanese companies to provide quality accommodation.

As can be seen from the quotations below, Japanese retirees have been attracted to Chiang Mai for a range of reasons. Not only is the cost of living much lower, but there is a large pool of labour for caring and nursing jobs, as well as the possibility of finding a wife.

'I can afford to hire a live-in maid or nurse to look after me here. I would much prefer to be looked after by someone else than my own family members.'

(Mr Tanimura)

'I lost my job before retirement. ... I somehow need to eke out a living with limited savings until I start receiving my pension. I came here because the cost of living here is low.'

(Anonymous man)

'Although I have a house and land it is not easy to find a wife for someone like me who graduated with only high school education in Japan. ... But I really have to find a wife now. That is why I came to Chiang Mai. I need to find someone who can look after my mother.'

(Mr Miichi)

The presence of the Japanese retirees in Chiang Mai has provided economic opportunities, but it also creates new forms of social activity within the city, as services (newspapers, Japanese Association and restaurants) are established for the newcomers.

(Source: Adapted from Toyota, 2006)

Domestic social divisions

Feminist researchers (see e.g. Dwyer and Bruce, 1988) have been particularly keen to examine the social processes operating within households. Images of households as harmonious and equitable social institutions fail to recognize divisions and the operation of **power** relations within them.

As units of production and reproduction, often containing individuals of different ages and genders, the operation of households often relies on a **division of labour**. This division both reflects and constitutes wider social expectations, for example, regarding the roles of women and men in society. The domestic sphere or private sphere has often been considered the 'woman's realm', while the street, workplace and political environment or public sphere is constructed as masculine space (see Chapter 7). Such distinctions have been based on women's socially constructed primary role as wife and mother.

The distinction may be enforced through particular cultural practices, such as *purdah*, where women of reproductive age may only move outside the house if accompanied by a male relative (Bose, 2007). The control of women's mobility may, however, be achieved through less formally sanctioned practices, such as gossip when women are viewed as being 'out of place'. Similar social disapproval may also be received by men who are seen to transgress norms of masculinity, for example by doing housework (Gutmann 1996).

Clear distinctions between 'women's space' and 'men's space' are both spatially and temporally variable. For example, seclusion or *purdah* as a cultural practice associated with some forms of Hinduism and Islam often has particular class dimensions. The ability of a husband to maintain his wife within the home means that he has sufficient wealth to support the household himself, without relying on women's wages. Thus 'class **identity**', as other forms of identity, can be seen as **performative** (see Concept Box 6.1).

While **normative** ideas about a women's place being 'in the home' are prevalent in many parts of the world, in reality women working outside the home in the Global South is nothing new. For example, both men and women have contributed to subsistence agriculture throughout the Global South for centuries. However, since the 1970s, economic crises and economic restructuring have been associated with increasing levels of female labour force participation (see Chapters 4 and 7). These shifts have both contributed to and reflect shifting social norms regarding the role of women in society. Such shifts in the public sphere have sometimes been associated with growing **power** for women within the home, although this is not an automatic outcome of paid employment (McClenaghan, 1997) (see Box 7.4 on masculinity and work in South Africa). In addition, while in some cases men have taken on greater roles in domestic tasks, in the vast majority of households women's **domestic labour** burden remains much higher than that of men. Work in non-market activities means working for no remuneration, for example, in subsistence farming or housework in your own home. Such activities are dominated by women and constitute a large percentage of women's work time on an average day (see Table 8.4).

FOOD AND CLOTHING

An analysis of the **consumption** of food and clothing provides an excellent way of examining some of the broader debates about **cultural globalization** and processes of increasing social inequality (see Chapter 5). Both food and clothing are classified as 'basic needs' for human populations, but they also represent key ways of demonstrating **identity** and status once basic survival requirements have been met.

Food consumption

Food is vital for human survival. However, for the vast majority of the world's population, food is also 'packed with social, cultural and symbolic meanings. Every mouthful, every meal, can tell us something about our selves, and about our place in the world' (Bell and Valentine, 1997: 3). While the bulk of this section will deal with eating practices and **identity**, it is important to recognize that for millions of people, the majority of whom live in the Global South, getting food is a daily struggle. This may be because they have insufficient money to pay for food, or subsistence farmers may have to deal with unreliable rainfall, soil erosion or limited inputs to improve crop yields. For example, in Mulanje, Malawi, subsistence

Table 8.4 Gender, work burden and time allocation

	Total work time (minutes per day)			% work time spent on non-market activities	
	Women	Men	Women's work time as % men's	Women	Men
Urban areas					
Colombia (1983)	399	356	112	76	23
Indonesia (1992)	398	366	109	65	14
Kenya (1986)	590	572	103	59	33
Venezuela (1983)	440	416	106	70	13
Rural areas					
Bangladesh (1990)	545	496	110	65	30
Kenya (1988)	676	500	135	58	24

Source: Adapted from UNDP, 2004: table 28.

farmers often intercrop to maximize their yields, diversify their food sources and protect the soil. Plate 8.6 shows intercropping using maize, beans and pumpkins. However, problems with the subsidized government fertilizer distribution scheme have meant that some farmers have been unable to obtain the subsidized fertilizer to which they are entitled.

As outlined in Chapter 2, familiar images of the Global South, particularly Africa, are of malnourished and starving children. While these images may reinforce Western notions of African peoples as passive victims, food insecurity and its effects are a reality in some parts of the continent and elsewhere in the world, including parts of the Global North (see Table 8.5). The global rising costs of food in 2007–08 due to growing demand, especially from China, the use of land for biofuels rather than food crops and the increasing costs of fertilizers due to high oil prices, look set to continue. This will have a particularly harsh effect on populations who are already struggling to meet their basic food needs.

Food **consumption** choices can reflect economic and social processes and change over time. In economic terms, food prices may vary depending on environmental factors, such as drought, or political decisions, such as cuts in food subsidies associated with **structural adjustment policies** (see Chapter 10). For individuals

Plate 8.6 Intercropping in Mulanje, Malawi.
Credit: Katie Willis.

Table 8.5 Regional patterns of undernourishment

Region	1969–71 (%)	2002–04 (%)
DEVELOPING WORLD	37	17
Asia & Pacific	41	16
East Asia	45	12
Oceania	24	12
South East Asia	39	12
South Asia	37	21
Latin America & Caribbean	20	10
North America	12	5
Central America	30	19
Caribbean	25	21
South America	20	9
Near East & North Africa	23	9
Near East	21	12
North Africa	27	4
Sub-Saharan Africa	36	33
Central Africa	30	57
East Africa	44	40
Southern Africa	34	39
West Africa	31	15
DEVELOPED WORLD	n/a	<2.5
Countries in transition	n/a	6
Commonwealth of Independent States	n/a	7
Baltic States	n/a	<2.5
Eastern Europe	n/a	3
Industrialized countries	n/a	<2.5
WORLD	n/a	14

Source: Adapted from Food and Agriculture Organization (FAO), 2007.

and households with limited economic resources, rising food prices clearly have implications for food **consumption**. However, food choices are also made within particular cultural contexts. As Pierre Bourdieu (1984) outlined, **consumption** practices are a way of conferring or claiming status. This may be because they are associated with 'modern' or 'foreign' influences, but it may also be an attempt to distance the consumer from the survival practices of the poor (see Box 8.3).

The impact of 'foreign' influences on food **consumption** patterns in the Global South has been one of the key dimensions of work on **cultural globalization**. In

BOX 8.3

Changing patterns of food consumption in Mwanza, Tanzania

Mwanza, on the shores of Lake Victoria in northern Tanzania, had a population of 277,000 in 1992. Karen Coen Flynn's (2005) study of food **consumption** in the city in the mid-1990s provides an excellent example of how economic, environmental, cultural, social and political factors intertwine to explain food **consumption** patterns.

The staple food crops in the area were sorghum, maize and rice. The latter two had only become popular following increased cultivation in the surrounding area. Maize had been introduced by the British in the 1950s as a way of reducing dependence on imports from neighbouring Kenya, while rice was originally developed as a cash crop in the 1970s for export to Kenya and Zaire, but was now produced purely for domestic **consumption**. Maize and rice were popular in urban areas as they were viewed as cleaner and not as a 'poor person's' food. Thus, food **consumption** was a way of claiming status. However, economic realities were also present, with rice being viewed as better to store than maize, as well as cooking more quickly, so reducing the need for water and charcoal.

This need to ensure food availability at times of economic crisis or food shortages due to drought is summed up by Mama Shukuru, a mother of five children from a middle-income household:

> The most expensive food is wheat flour. It is used by bakeries, restaurants and other businesses. I usually do not buy it. I buy rice, potatoes, *ugali* [cooked, mashed maize], fish and kidney beans because these are the least expensive foods to get. The cheapest food is cassava flour and I buy it so that we will not go hungry when money is in short supply.
>
> (Flynn, 2005: 59)

As well as locally produced food, local residents could also buy imported goods, such as baked beans from the UK and peanut butter from Kenya. The diversity of ethnic groups in the town, including a number of 'African' ethnic groups, as well as Asians of different ethnicities, was also reflected in food **consumption**.

(Source: Adapted from Flynn, 2005)

particular, McDonald's and Coca-Cola have been highlighted as symbols of cultural destruction and the homogenization of cultures worldwide. Such products have certainly become more available in many parts of the Global South. Their popularity is partly related to local interpretations of these commodities as 'modern' and therefore desirable as a way of claiming status. However, in-depth examination of how and where these products are consumed shows that there is not always a simple adoption of commodities and **consumption** practices. Rather than a linear process of crude cultural imposition, individuals and groups within the Global South have the **agency** to reshape the meanings and practices associated with these goods (see Miller, 1998; and Box 8.4).

As well as these commercial settings for food **consumption**, rituals around food preparation and **consumption** are vital for reinforcing particular social ideas around family and community, either on a day-to-day basis or for special festive occasions. For example, as discussed earlier in this chapter, the sharing of food production and **consumption** is sometimes used as an indicator of household membership. As Monica Janowski (2007: 5) states, '[t]he consumption of everyday food is one of the most important everyday arenas in which rigid rules about how things should be done are often apparent, although they are often unspoken or only partially explicit'. These rules may include specifications about who should do the food preparation (Devasahayam, 2005), what foods are suitable for particular times or people, and who should be served or eat first (see Plate 7.5 of a Yemeni woman preparing food).

Clothing

Just as the **consumption** of food and drink is embedded in particular social, economic and political processes operating at a range of scales, so too is clothing a reflection of how individual **consumption** practices reflect **identities** potentially combining influences from a range of sources (Kothari and Laurie, 2005). Examining dress can also help disrupt the unhelpful dichotomies of 'traditional'/'modern', 'local'/'global' and 'indigenous'/'foreign'.

What people wear is often the key way in which they present their **identity**. As with food, changing clothing practices have been interpreted as an indication of the imposition of external norms through the exercise of **power**. For example, European colonial accounts in both Latin America and Africa used the indigenous populations' relative lack of clothing as a sign of savagery and lack of civilization (see also Chapter 2). The imposition of external norms, as well as the demonstration effect, which associated high status with particular forms of dress, meant that new forms of appropriate clothing were adopted (see Hay, 2004, on the case of western Kenya).

The increasingly widespread adoption of jeans, sports clothing, trainers/ sneakers in the Global South, particularly among young people, could be interpreted in a similar way for a globalized era, with the association of particular forms of 'modern' lifestyle circulating through television, pop music, films and

BOX 8.4

Fast food in Beijing

American-style fast food has been a feature of the Beijing landscape since the late 1980s. A Kentucky Fried Chicken branch opened in 1987, while McDonald's followed in 1992.

Being able to consume in these outlets became a sign of **modernity** and status. Eating in McDonald's and KFC was a very different experience to that of existing restaurants. Not only was the food different, but the space was clean and brightly lit, staff were friendly and service was quick. For example, Yunxiang Yan (2000: 211) quotes one customer in a Beijing McDonald's who said 'The Big Mac doesn't taste great; but the experience of eating in this place makes me feel good. Sometimes I even imagine that I am sitting in a restaurant in New York or Paris.' Customers were not going to these restaurants because they necessarily liked the food, but because of what being in these spaces represented. Visitors from the Chinese countryside would sometimes go to KFC or McDonald's and would take the food and drink packaging back to their villages as tourist souvenirs.

Parents, keen to give their one child 'modern' experiences that would help them in later life, would also save up to go to KFC or McDonald's. For many families, such expenditure was significant. In 1994, an ordinary worker in Beijing made 500 RMB (US$60) per month, while a McDonald's value meal cost 17 RMB (US$2.10). The introduction of children's birthday parties at McDonald's and KFC also represented a change in cultural norms regarding birthday celebrations, and fitted in with the growing focus of family life on the nuclear family.

The introduction of KFC and McDonald's has not, however, been a wholesale transference of the US experience to Beijing. KFC, for example, used a chicken mascot called 'Chicky' alongside the Colonel Sanders logo because it was felt to be more appropriate for the Chinese market and marketing promotions are linked closely to school activities in the local area. Rather than being 'fast food' outlets, both KFC and McDonald's have become places in which to 'hang out'. In response to the arrival of American fast food chains in Beijing, local variations, such as Ronghuaji (Glorious China Chicken) have developed, drawing on the business model and appearance of KFC and McDonald's. Such restaurants are cheaper than the 'foreign' version and may add a 'Chinese' twist, such as claiming to use 'sacred recipes' so the food has medicinal properties.

(Sources: Adapted from Lozada, 2000; Watson, 2000; Yan, 2000)

advertising, often originating from the Global North (Klein, 2001; Wright, 2006). The two young people in the forefront of Plate 8.7 are wearing jeans and T-shirts, whereas in the background you can see a man wearing more traditional Nigerian robes and a woman wearing a brightly coloured and patterned wrap skirt. Despite seemingly contrasting in fashion styles, in all cases the clothes are likely to be factory made, so the distinction between 'traditional' and 'modern' is blurred. For rural–urban migrants, returning to their home village wearing 'modern' clothing is a very immediate indicator of success, demonstrating both economic achievements and an ability to navigate in an unfamiliar environment (Hay, 2004), although these interpretations may vary greatly between male and female migrants (Gaetano, 2008; and see below).

It is important, however, to recognize the ways in which the **consumption** of clothing by individuals within the Global South can be the outcome of very complex understandings of desirable and appropriate attire for different social settings. Jean Allman (2004: 6), in her introduction to an edited volume on the politics of fashion in Africa, makes the strong claim that 'fashion may be a language spoken everywhere. It was, and remains, profoundly local, [and] deeply vernacular'. Thus, there is a need to understand how particular forms of dress are understood in particular contexts, rather than assuming homogeneity. The example of fashion in Mumbai and London which is outlined in Box 8.5 demonstrate that places in the Global South are not merely recipients of Northern tastes which are viewed as superior, but rather that flows can be reversed and it is Mumbai fashion which is interpreted as being much more cutting-edge and modern (see also Walton-Roberts and Pratt, 2005) on similar sentiments within the **transnational** circuits between India and Canada.

Plate 8.7 Diverse fashions in Lagos, Nigeria.
Credit: Muyiwa Agunbiade.

Because clothing **consumption** is an inherently embodied practice, issues of **gender** and dress have often been of great concern as women's position as bearers of 'tradition' (Yuval-Davis, 1997) has involved attempts to ensure women's modesty in both appearance and behaviour. However, the interpretation of women's choice of clothing, for example, wearing trousers, rather than skirts, or wearing

BOX 8.5

Fashions in London and Mumbai

Using material from focus groups in London and Mumbai, Peter Jackson, Nicola Thomas and Claire Dwyer highlight the need to deconstruct the 'traditional'/'modern' binary and the way in which this has been imposed on an 'East'/'West' distinction.

While British-Asians in London and young women in Mumbai all wore jeans, T-shirts and other 'Western' clothes, they also wore what they sometimes identified as 'Indian' fashion. For both groups, it was India that was the centre for such fashion, particularly Mumbai and the associated Bollywood film industry with its high-profile stars. The following two quotations sum up this reversing of the usual flows of fashionable and desirable clothing:

'We never knew what the fashions were in Pakistan and India, and suddenly we're five years behind them, we're wearing something they wore five years ago and you go over there and you feel "oh my God, I'm out of fashion."'
        (London focus group participant, quoted in Jackson, *et al.*, 2007: 913)

'In fashion, we are not actually alike, in fashion they [in Britain] are behind. … For Indian clothes they are very much behind. They don't get all the Indian fashion clothes there. Like in London, you won't get the latest. They are actually not aware of what is going on here.'
        (Mumbai focus group participant, quoted in Jackson, *et al.*, 2007: 913).

For many Mumbai participants, certain branded goods, such as Levi jeans, were coveted, but when it came to 'Indian' fashion, it was the lack of branding which gave it authenticity and cachet. British-Asians in London who wanted to be in fashion or required a special outfit for a wedding said that they would usually try and get clothes sent from India as they would be more fashionable and also be very well made.

(Source: Adapted from Jackson, *et al.*, 2007)

tighter fitting clothes or short skirts, cannot be easily slotted into a discussion of women being 'modern', rather than 'traditional', or trying to use their clothing to assert their independence from 'traditional' framings. For example, Karen Hansen (2004) discusses women wearing miniskirts in Zambia, highlighting how at the time of independence in 1964 politicians sometimes interpreted this form of clothing as being a challenge to the development of a 'national culture' as it represented 'foreign' ideas and morals regarding the covering of the female body. By the 1990s, this interpretation had been abandoned and replaced with **discourses** around modesty and the need for women to keep covered up to avoid inflaming male passions. Women were attacked in the streets for wearing miniskirts, and female members of parliament were also criticized for such attire. Young women in Zambia, aware of such **discourses** and practices, were able to negotiate their dress by ensuring that they wore suitable clothing for different situations, particularly outside the home.

It is in areas around clothing, particularly women's clothing, that debates around 'tradition' are often most marked. Responses to rapid change have sometimes led individuals, communities or governments to attempt a reassertion of tradition as a bulwark against undesirable or frightening activities, particularly if these practices are viewed as coming from 'outside'. The most obvious recent example of this would be the Taliban regime in Afghanistan which banned girls' education and forced women to wear burqas which completely cover the body and head, including the face (Kandiyoti, 2007). Such actions cannot be, however, interpreted as a wholesale return to some previous way of life. The very act of reasserting these practices is done in response to new practices and influences. As Edward Said (1978) argued in *Orientalism*, national **identities** are constructed through engagement with 'the other', not in isolation (see Chapter 2). In addition, just because individuals behave in a way which may be viewed as 'traditional', it does not mean that their reasons for adopting such practices are the same as they were in the past. Debates around Muslim women and the covering of the head, for example, demonstrate the complexities of interpreting clothing choices and the importance of recognizing women's **agency** in such decisions (see Box 8.6).

LEISURE AND PLAY

Making a living is a key element of life for the vast majority of people in the Global South (see Chapter 7), however, there are also opportunities for leisure and play. Both these concepts are **social constructions**, varying over time and space, but in all cases they provide an additional insight into how people's lives in the Global South are framed by, contribute to and contest global flows of ideas and commodities. How and where certain leisure practices are performed can help construct individuals' and group **identities** (see e.g. Visser, 2008, on white gay men's leisure spaces in Bloemfontein, South Africa).

Shopping is a way of obtaining commodities to meet basic needs, but it is also

BOX 8.6

Different interpretations of 'the veil'

The covering of the head and hair by some Muslim women has often been used as an indication of Islam's 'backwardness' and 'tradition' compared with 'Western' women's freedom and liberation to wear what they want. However, such interpretations are built on **Eurocentric** ideas of **modernity** which do not consider the possibilities of Islamic modernities (see Chapter 2). Within the Global South, some regimes, for example Iran under Mohammad Reza Shah Pahlavi (1953–79), encourage women to remove their veils as a sign of **modernization**.

Viewing the 'veil' or *hijab* as inherently oppressive, fails to acknowledge the diversity of practices within Islam (see Box 5.7 for an overview of the main groupings within Islam) and the potential for women's **agency** in choosing their attire. There is also very little discussion of how the interpretation of Muslim texts and teachings affects constructions of Islamic masculinities (Ouzgane, 2006).

For many Muslims, the covering of a woman's head is not required to be a 'good Muslim', although some places and practices would require it, for example when in a mosque. Even among those who advocate head-covering, what this involves can vary from a small headscarf covering the head and hair, to a *burqa* which covers the entire body, including the face. This diversity reflects different interpretations of the Qur'an and other holy texts.

Wearing a veil may be enforced by law, with very severe punishments for non-compliance (as in Afghanistan under the Taliban), but in most Islamic societies and communities wearing the veil is not legally enforced. This does not mean that women make free choices to veil; in some cases social disapproval (expressed in violent ways in some cases) will leave women with little choice.

In other contexts, for example among some university students in Bangladesh (Rozario, 2006), women choose to wear a headscarf to ease their public movements, for example, being allowed to go to university. This is therefore a form of strategic decision. Historically, women sometimes chose to wear the veil as a political statement against **colonialism** as in Algeria in the 1950s. In the early twenty-first century, women have increasingly chosen to cover their heads as a public demonstration of their faith in the context of the 'War on Terror' which is interpreted by some as being an anti-Islam campaign.

BOX 8.6 (CONTINUED)

> **BOX 8.6 (*CONTINUED*)**
>
> Public displays of faith or 'public piety' as Lara Deeb (2006) terms it, may also be expressions of an individual's **identity** and belief system.
>
> In practice, the reasons given for wearing the veil and the choice of head-covering cannot be separated into neat categories. This brief discussion has demonstrated, however, the need to move away from Northern interpretations of women's dress to consider the role of individual **agency** in clothing decisions.
>
> (Sources: Adapted from Bulbeck, 1998; Deeb, 2006; Rozario, 2006)

associated with forms of leisure and the performances of class **identity**. The history of middle-class shopping spaces in Cairo is relatively long, with colonial influenced department stores emerging in the nineteenth and twentieth centuries to serve wealthy Egyptian and foreign customers. Recently constructed malls, such as the City Stars mall (see Plate 8.8) consume vast tracts of urban land and offer a dazzling variety of services and experiences for consumers, namely 150,000 square metres of shopping space with 550 shops. Other mall developments in Cairo have formed part of government plans to gentrify parts of the city, including the removal of *ashwaiyyat* (slums) to make way for these new urban developments (Abaza, 2001: 108). Malls within Egypt provide middle-class consumers with a 'safe' space – safe from crime, free from pollution and away from the heat and noise of the street. This clearly discriminates against the economically poor and young people, who have limited purchasing power (Vanderbeck and Johnson, 2000). Also, health clubs and golf clubs are not only the preserve of elite expatriates in the Global South, as they are becoming more common sites for middle-class leisure activities.

Leisure and relaxation does not have to come at such a high economic price. The home and public space can be used and enjoyed for free, although access and behaviour in such spaces can still be subject to surveillance, either by public authorities or by family members. For millions of low-income migrant workers, certain public spaces have become key areas in which to spend their limited leisure time, mixing with fellow migrants and reclaiming the spaces from which they are usually excluded because of economic or social pressures (see Yeoh and Huang, 2000, on Indonesian and Filipino migrants in Singapore). Young people, as in the Global North, may often depend on public spaces for free leisure, usually 'hanging about' with friends. Such activities may be viewed with distrust by adults and the authorities, but they are vital in providing an opportunity for young people to meet outside the confines of school or home.

The separation of 'work' and 'play' is also problematic because of the type of jobs undertaken by many people in the Global South (see Chapter 7). Informal sector

Plate 8.8 City Stars shopping mall, Cairo, Egypt.
Credit: Steve Connelly.

activities around the home may be combined with rest or entertainment, while for children, work and play are frequently difficult to disentangle (see Box 8.7).

Community or national celebrations may also be a part of **identity** construction for individuals and groups, but in some places these practices may also have changed meanings and outcomes. 'Tradition' can be mobilized for economic reasons. Authenticity can command a high price in a world where many consumers feel

---

BOX 8.7

Young people, work and play in rural Sudan

Cindi Katz's (2004) study of children's lives in the village of Howa (a pseudonym) in Central Eastern Arabic-speaking Sudan, really demonstrates the importance of children's work to household survival (see Chapter 7 in this book) and also how work and play are inseparable in children's daily lives.

Children in Howa are expected to work from an early age. This may be by collecting water, running errands to the shop, weeding plots or herding sheep and goats, for example. When doing this work, children introduce elements of play into their activities. For example, when boys are out herding, they may play games such as *shedduck,* which is play-fighting where participants have to hop with one leg behind them. By combining games with overseeing the herds, boys can make their work more enjoyable and the time pass more quickly.

During play, Katz also observes how work and important environmental knowledge are included. Children use scrap metal and material to make dolls, tractors and layout houses, fields or shops. They then use these to act out domestic life and agricultural cycles; developing knowledge that both helps them in their current work and will also contribute to their work in the future. Their awareness of how trade and wage labour operates is exemplified in their acting out of payments for 'crops' using money made out of broken pieces of china.

The kind of work children engage in has changed over time. For example, the introduction of a major irrigation-fed agricultural project in 1971 increased children's workloads because of the need for weeding and planting of new crops. These changes are reflected in the games they play. Despite the increased workload putting pressures on their time, children are still able to combine their work with enjoyable games and fun.

(Source: Adapted from Katz, 2004)

they are at the mercy of faceless corporate entities, and the homogenization of experiences (Cole, 2007). Tourism in some places has expanded due to the use of 'tradition' either as a marketing tool or as a way to package activities for visitors. This in no way implies that the individuals involved in activities such as handicraft production, dancing, music-making and cooking are merely performing actions for the tourist dollar (although this may be case) but rather, that particular forms of local practice can be reframed and presented in a way that can be of economic benefit, as well as potentially contributing to greater understanding between 'locals' and 'visitors' (see Chapter 4 on global tourism) (Box 8.8).

John Urry (2002) coined the term 'the tourist gaze' to describe how places are consumed and experienced visually by visitors. For tourists seeking an 'authentic' experience, how places look, how people dress and what rituals are performed are important factors in creating feelings of authenticity, as shown in the discussion of tourism in Bali in Box 8.8. However, as in the case of the Bali dances, 'authenticity' and 'tradition' may not be indicators of long-standing practices, but are rather particular performances created not only for the tourists, but also part of an assertion of cultural pride. Stephen Coleman and Mike Crang (2002) argue that seeking for the 'authentic' in tourism implies a very fixed idea of bounded places; something which, as we have argued throughout the book, fails to recognize long-standing interactions and connections at a range of scales. While 'authenticity' is usually mobilized in contrast to '**modernity**', grounded experiences of tourism in places like Bali demonstrate the untenable nature of such distinctions.

CONCLUSIONS

In this Chapter we have used the themes of home-making, food and clothing **consumption**, and leisure and play to examine how people in the Global South negotiate and challenge the global flows of ideas and commodities which are sometimes represented as unstoppable waves of **Westernization**. The Chapter has demonstrated that local practices still vary and that even when people are receiving the same products, they often interpret and use them in very different ways. Forms of 'local' **consumption** are often transformed by such processes, but they are not automatically eradicated (Jackson, 2004).

Second, this Chapter has highlighted how social and cultural practices originating in particular localities in the Global South have become part of global flows. We focused on Indian fashion, but we could have discussed forms of music or food as other key commodities that are consumed far from their places of origin, and, in turn, are reinterpreted. This again challenges ideas of **modernization** coming from the Global North.

Finally, this Chapter has again stressed the **agency** of people in the Global South. The discussion of shopping centres in Cairo demonstrated the consumption choices that can be made by middle-class Egyptians, but leisure, fun and enjoyment

BOX 8.8

Tourism spectacles in Bali

The island of Bali in Indonesia is often used as an example of an archetypal paradise island, with beautiful beaches and a friendly and welcoming population. The ability to see traditional culture, particularly in the form of dance and music, is also used as a selling point for the island.

These **representations** of Bali and the mobilization of 'tradition' are not new, rather they go back to the 1920s and 1930s when tourism to Bali began to develop. Bali was seen as a place where Asia and the Pacific came together and this meeting of cultures created particular forms of dance. However, many of these dances, such as *kecak* and *barong,* were not 'traditional'; rather they had developed as hybrids following the arrival of the Dutch colonists on the island. Similarly, the concept of a 'Balinese culture' had not existed prior to **colonialism**.

In the 1970s, tourism was identified as a key development strategy under Indonesia's President Suharto. Bali was crucial in this strategy and cultural authenticity was used as a key marketing tool and this continues to the present day. Tourist accommodation incorporates features of 'traditional' Balinese architecture and tourist guides are often encouraged to wear 'traditional' clothing so that visitors can experience the authentic Bali.

Musical and dance performances are often laid on for tourists. Shinji Yamashita (2003) describes the history of many of these dances including the *Ramayana* ballet which was created in 1961 by a government minister specifically for tourist audiences. The need for government permits for dance performances aimed at tourists has meant that dance groups have become more standardized. Most tourists watching the performances have no idea of the history of the dances, reading them as 'authentic' because the costumes, music and choreography match their expectations of Balinese dance and they are marketed as 'traditional'.

Yamashita argues that rather than causing an eradication of 'real' Balinese culture, the supposed commodification of Balinese dance for sale to tourists can actually be associated with an assertion of pride in local culture. A similar conclusion has been made on the neighbouring Indonesian island of Flores (Cole, 2007). Practices may change in the light of increased foreign visitors, but this is not necessarily a dilution of culture; similarly pre-existing practices are likely to be hybridized and have changed over time, rather than being long-standing traditions which must be upheld.

(Sources: Adapted from Cole, 2007; Yamashita, 2003)

are not purely the domain of the economically well-off; the examples in this chapter, such as the children in Sudan, contrasts very strongly with frequent **representations** of people in the Global South as desperate, sad and passive with no joy in their lives (see Chapter 2). The discussions of poor housing and food shortages mentioned in this Chapter highlight the challenges which millions of people face on a daily basis, but this does not mean that their lives are defined purely by these preoccupations.

 ## Review questions/activities

1 Choose a practice from the Global South which you would regard as 'traditional', e.g. a form of dress, style of music. Research the history of this practice and how it has changed or been maintained over time. What do these processes tell you about the distinction between 'traditional'/'modern'?

2 Select five objects from your home that are important to you. What is the history of these objects, how did they end up in your possession and what do they reveal about your identity?

3 How far would you agree with the assertion that the world is *not* becoming more homogenous in cultural terms?

SUGGESTED READINGS

Flynn, K. C. (2005) *Food, Culture and Survival in an African City*, Basingstoke: Macmillan.
A fascinating account of food consumption practices across class, gender and ethnic groups in urban Tanzania.

Schech, S. and Haggis, J. (2000) *Culture and Development: A Critical Introduction*, Oxford: Blackwell.
A very clearly written account of the centrality of culture to processes of development.

Skelton, T. and Allen, T. (eds) (1999) *Culture and Global Change*, London: Routledge.
An excellent, wide-ranging collection which considers the ways that supposedly 'global' processes are experienced and challenged at a 'local' scale.

WEBSITES

www.fao.org UN Food and Agriculture Organization
A very useful source of information about food production, prices and food shortages worldwide.

www.unhabitat.org UN-Habitat site
This is the homepage of the UN Human Settlements Programme and is full of useful statistics, as well as policies to improve living conditions for poor urban dwellers.

PART FOUR

Making a difference

SAPREF oil refinery, Durban
Source: Glyn Williams.

In this part, we turn to explicit strategies to 'develop' the countries and peoples of the Global South. As we argued in Chapter 2, development has been an important way in which the Global South has been represented, especially from the mid-twentieth century onwards. The extent to which particular areas of the South have 'caught up' with the Global North has often been key to whether Southern governments, particular policies or the international development community more broadly have been seen to be 'successful'. As a result, one common way of telling the story of development is to try to identify a series of dominant ideas or strategies that have emerged since the 1950s, and to explain the evolution from one to another over the decades that have followed.

We take a rather different approach here, partly because there will always be many examples of Southern development practice that *do not* fit easily within such an overarching history. Also, because different development ideas are always coming into and falling out of fashion, it is useful to take a slightly more distanced perspective. As a result, our underlying question in this section is: *What has inspired different attempts to 'develop' the Global South, and what effects do these interventions have?* In different places and at different times, academics and policy makers have made powerful arguments that the state, the market or people themselves (as communities and holders of valuable indigenous knowledge) can or should take the lead in delivering development. We therefore look in turn at ideas of state-led development (Chapter 9), market-led development (Chapter 10) and people-centred development (Chapter 11) to see the different views contained within each of what a better future for the Global South might look like, and how it could be achieved.

Each idea has a long history that pre-dates the international focus on developing the Global South that has emerged since the collapse of European empires at the end of World War II. Ideas of state-led development drew directly on the Global North's response to the Great Depression in the 1930s (and less directly on the experience of industrialization in the Soviet Union) and those of market-led development have a far longer lineage, dating back to Adam Smith's work in the late eighteenth century. One frequently cited inspiration for community-led development has been Gandhi's ideas, which emphasized autonomous development based around India's 'village republics', but these can also be seen as part of a far wider re-evaluation of indigenous knowledge and values.

Presented in their own terms, each idea has considerable attraction, offering the hope of a world in which there is a more rational planning of development (Chapter 9), a more efficient use of resources (Chapter 10) or greater autonomy and room for the expression of alternative perspectives of development at the local level (Chapter 11). Each also has its downsides, however: state-led development can lead not only to inefficiency, but also to authoritarian **power** and its abuse, while unfettered markets can have the effect of concentrating resources in the hands of the already powerful, exacerbating inequality and **poverty**. Community-led development may not have unleashed suffering on the scale of either of the other two, and it would be difficult to point to a direct equivalent to the mass displacements caused by massive state-led infrastructure projects, or the sudden shock to **livelihoods** of **Structural Adjustment Programmes**. This does not mean, however, that calls for 'grassroots' development are wholly innocent or without cost. One important argument of Chapter 11 is that the coalitions of interest behind calls for community-led development need to be examined carefully, for they can support highly iniquitous 'traditional values' locally, or be used to mask important shifts in **power** relations (such as those between states and citizens) at other spatial scales.

Rather than seeing any single perspective as being better or worse, we highlight the possibilities, tensions and limitations inherent in all three. Our intention in doing this is to raise critical questions that need to be asked about *all* strategies to develop the Global South. Any development intervention inevitably changes **power** relations: relationships of authority are altered, resources are redistributed, and more subtly **power** is expressed differently within institutions and everyday practices as particular development strategies support certain values and norms over others. Looking critically at state-, market- and community-led development helps to provide a framework for analysing what might be gained and lost through each. We argue that it is only by asking such critical questions of past and present development interventions that the context-specific search for more socially just and environmentally sustainable development alternatives can proceed.

9 Governing development

THE STATE'S ROLE WITHIN DEVELOPMENT

In Chapters 4 and 7, we looked at the ways in which Southern governments are located within an international political system, and at the **power** the state has in shaping the everyday lives of people in the South. From what we have seen thus far, Southern governments may seem to be unlikely candidates as agents of 'development': externally, they are often constrained by unequal international geopolitical relationships, and internally their relationships with their own citizens are often far from perfect. Nevertheless, within the theory and practice of international development, two different arguments for focusing attention on the state have been put forward over the past half-century. The first, important in the decades following World War II and the **decolonization** of much of Asia and Africa, was that national governments in the South were uniquely placed to push forwards an agenda of economic and social **modernization** that would allow their countries to 'catch up' with the West. The next section of the Chapter introduces this idea of the **developmental state**, looks at examples of its impact in practice and considers the criticisms that have been made of them.

In contrast to this role as direct agents of development the second, and more recent, way in which Southern states have been of interest is as institutions in need of reform. International development agencies have paid growing attention since the early 1990s to the quality of **governance** in the South, and have argued that development is often held back by states that are performing badly. As a result, there has been significant debate over what '**good governance**' is, and how the international community can induce programmes of state reform in the Global South. We look at these debates, and the successes and failures of the **good governance** agenda in practice, in the second half of this chapter.

Underlying the ideas of a **developmental state** and of **good governance** are sets of assumptions and aspirations about the role the state can and should play in changing society. Regardless of their ideologies and forms of government, all states exercise **sovereignty** over their national territories and populations, and this gives them two quite unique features. First, even though their authority may be challenged by both regimes of international **governance** (Chapter 3) and economic **globalization** (Chapter 4), they retain the **power** to shape important elements of the context in which national development might take place through their choices over national policies, laws, tax regimes, and so on. Second, they are institutions that claim to represent the collective will of their people. Whether these claims are based on universal electoral mandates (the post-Apartheid state in South Africa in 1994), appeals to the divine right to rule of a monarch (in Bhutan or Lesotho) or the ideology of a national party (in China), they are always contested to some degree. However, no other types of institutions – whether corporations, international development institutions or NGOs – can make equivalent claims to be representative at this scale. Together, this **power** and the claim to legitimate authority can make states very powerful agents of development.

This Chapter looks at the state as a key agent in the reshaping of Southern economies and societies, but also raises wider questions about who has the right to govern development. As we show below, the ambitions of high-modernist developmental states have often led them to act in an authoritarian fashion, often ignoring the interests of precisely those parts of their own populations that are most in need of development. At the same time, the idea that states of the Global South should be asked to conform to the '**good governance** agenda' of international development agencies is itself problematic, and can erode elements of national **sovereignty**. As this Chapter will indicate, governing development remains highly contested both in theory and in practice.

THE RISE, FALL AND RETURN OF THE DEVELOPMENTAL STATE

During the period of rapid **decolonization** of the Global South following World War II, the idea that national governments should actively direct social and economic change was a central theme in development thinking. This idea had gained the status of international common sense for a number of reasons. The first was the process of **decolonization** itself: independence often followed a period of popular struggle, which was itself important in mobilizing a sense of national **identity** and a desire for a dramatic transformation of people's lives. Many leaders of newly emergent nations in the South could therefore legitimately claim that they had a mandate to undertake sweeping social and economic change, and that state-led development offered the most effective way to respond to these expectations.

Second, the experiences of Europe and the USA immediately before **decolonization** suggested that the economic and technical knowledge to bring about this transformation of Southern economies was available. The 1929 stock market crash and the subsequent Great Depression had thrown millions of people out of work (*national* unemployment rates exceeded 20 per cent in the USA and the

UK, with local peaks still higher), suggesting that private markets needed a degree of government regulation to avoid damaging cycles of boom and bust. President F. D. Roosevelt's 'New Deal', a raft of policy measures to respond to the effects of the Depression in the USA, also indicated that the state could be proactive in re-starting economic growth through its own spending on public works programmes. The experience of government control of large sections of the economy within the war effort itself, and the rise of Keynesian economics (see Concept Box 9.1) in the North over the postwar period, had further helped to cement the idea that the state could and should act to plan and coordinate economic activity in the interest of the citizens. The Tennessee Valley Authority's (TVA) impact on a population impoverished by the Depression (see Box 9.1) was a concrete example of some of these ideas in practice. For international development thinkers and Southern political leaders alike, the TVA seemed to offer a model of integrated, state-led development that could be replicated across the South, fulfilling expectations of national progress that had been released by independence.

BOX 9.1

The Tennessee Valley Authority

To counter the effects of the Great Depression in one of the poorest parts of the USA, President Franklin D. Roosevelt set up the Tennessee Valley Authority – a state-run development agency charged with transforming a massive catchment area of over 40,000 square miles. The building of dams and power stations through massive public works programmes directly provided much-needed work to thousands of unemployed labourers, but the TVA's ambitions went far wider. Cheap hydro-electric power was to be the means for modernizing the entire valley: the TVA's activities encompassed the setting up of fertilizer factories, rural electricity cooperatives, demonstration farms and agricultural extension services, and even entire planned model communities. In several respects, the TVA prefigured the aspirations of many development programmes in the Global South. There was great ambition in its plans, and contemporary reports on its activities expressed not only wonder at its practical achievements, but also a wider faith in the possibility of state-led planning:

> In its program for flood and navigation control, for land reclamation, and for cheap electric light and power the TVA is substituting order and design for

BOX 9.1 (CONTINUED)

BOX 9.1 (*CONTINUED*)

haphazard, unplanned, and unintegrated development. Through its social and educational activities it is bringing to this region a consciousness of its own rich natural and human resources.

(Federal Writers' Project, 1939)

The TVA's Rural Electricity Administration promoted the benefits of electrification in ways that were to have echoes in many newly independent Southern countries (Plate 9.1): new technology would deliver modern, efficient ways of living, and allow a 'backward' section of the USA's population to catch up with more fortunate citizens elsewhere.

This was a vision driven through by the TVA's board and its technical experts, but it was not without controversy. The TVA faced legal challenges from both the private electricity suppliers it supplanted and from local African-American farmers, who were excluded from its demonstrations of improved farming methods or living in its model planned communities. Thousands of people in the upper reaches of the valley were displaced when the dams flooded their communities and no financial compensation was offered to tenant farmers, the valley's poorest people. In some instances then, the TVA was exacerbating the very problems of social and economic marginalization its 'modernizing' vision hoped to solve.

Plate 9.1 TVA advertising on the benefits of rural electrification, USA.
Credit: Franklin D. Roosevelt Library.

(Source: Adapted from *TVA: Electricity for All*, and original source materials archived at the New Deal Network (http://newdeal.feri.org/index.htm))

CONCEPT BOX 9.1

Keynesian economic management

John Maynard Keynes (1883–1946) was important in developing a school of economic thought that challenged the **laissez-faire** idea that an unfettered market would produce the most efficient economic outcomes. Writing against the background of the Great Depression, he argued that government should have a more interventionist role in the economy, in particular by regulating demand for goods and services. During an economic downturn, governments could intervene to stimulate public demand by lowering interest rates, and investing public money in the provision of public infrastructure. The latter not only delivers new productive capacity, but by providing jobs and incomes at a time of potential growing unemployment it shored up people's spending power, in turn securing a market for other goods in the economy (and the jobs of *their* producers), ultimately generating an overall increase in demand several times greater than the government's own investment. Critics of Keynes's analysis have argued that regardless of any benefits of such demand-management during times of economic crisis, the longer-term danger is that it encourages *permanent* increases in state spending, and a growing public sector that ultimately crowds out both the private sector and **civil society**.

Finally, the economies of many Southern countries at independence were still largely based around primary industries, such as agriculture and mineral extraction, an economic structure that reflected colonial needs for the South to act as a source of raw materials and as a market for European manufactured goods rather than the needs of these nations themselves. Radical economic change to reflect the new political realities of independence thus seemed logical. Diversification of the economy, and in particular the promotion of manufacturing industry, could deliver greater national self-sufficiency than had been possible for many countries under colonial rule. Equally importantly, it also offered the hope that the benefits of technologically-led rapid **modernization**, would put countries of the South on a path to 'catch up' with the industrialized nations (see Box 9.2; see also the discussion of President Truman's inauguration speech in Chapter 2).

Projects like those of the TVA were virtually impossible for the private sector to conduct in the Global South. Owing to their massive scale, the scarcity of capital and the long-term nature of the investment required, government would have to take the lead within these activities. Many Southern governments, supported by international **aid**, did undertake such 'mega' development projects – examples include dam-building programmes in Colombia (one of the World Bank's first projects, dating back to 1949) and India (see Box 9.3). At their most successful, projects such as these provided key infrastructure that enabled future growth; at their worst, however, they were overambitious or inefficient failures. James Scott (1998) discusses the case of Brasília, the city brought into being by the populist president Juscelino Kubitschek in the late 1950s. Built from scratch on an empty site deep in the interior of the country, Brasília was to be the administrative capital of Brazil. It was to embody the high **modernist** architectural ideas of Le Corbusier and its dramatic built environment was designed to change city dwellers' use of space, making them efficient modern citizens appropriate for a 'new Brazil'. By 1980, however, the planned city had reached less than half its intended size, it was ringed by informal settlements and it had acted as little more than a distraction from pressing urban problems elsewhere in the country (such as those of Rio's *favelas*, see Box 6.8).

BOX 9.2

Modernization theory

The hopes of newly independent countries and the North's experience of state intervention in the economy both came together within **modernization** theory, the newly emergent field of Development Studies' dominant perspective in the 1950s and 1960s. The central problem here was how to achieve rapid transformation towards modern, industrialized societies, and for Walt W. Rostow (1960) the answer was relatively simple: learn from the experience of the North. Looking at global economic history, he claimed to identify five key 'stages of growth' through which all countries had to pass to transform from static traditional societies to a 'developed' age of high mass consumption. Crucial for him was the stage of 'take-off' – where a local modernizing elite saw the potential of industrialization, and concentrated a growing proportion of savings and investment in this sector. Once a high savings ratio was established, the high-productivity of industrialization would become self-sustaining, eventually leading to a diversified economic base and growth that 'trickled down' from leading sectors and regions to create a fully modernized society.

What was attractive in Rostow's analysis was that it appeared possible to compress this period of take-off (Figure 9.1), so that a process that had taken over a hundred years in eighteenth-century Britain could be completed within a generation in the Global South. It also offered the North a role: foreign investment in the 'modernizing' sectors of the economy could substitute for indigenous savings (Lewis, 1955), and accelerate the process of 'take off'.

BOX 9.2 (*CONTINUED*)

There are numerous criticisms of **modernization** theory, not least that it uncritically holds up a Western model of high mass consumption as a universal goal, and sees 'traditional society' in the South as something that is irrelevant to this transformation (or still worse a barrier to be removed – Hoselitz, 1952). It is also rather geographically naive: it gives the impression that **modernization** is a process that nations go through individually and in virtual isolation – with the possible exception of benign outside investment. The existing concentrations of wealth in a fundamentally international economy (Chapter 4) clearly shape the South's possibilities for development, but they get scant attention within a theory that portrays Northern countries as historical models to be followed, rather than current and powerful competitors with interests of their own.

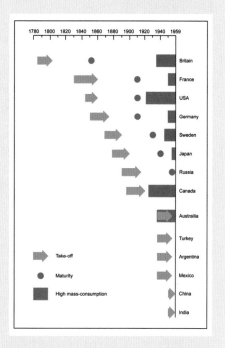

(Sources: Adapted from Hoselitz, 1952; Lewis, 1955; Rostow, 1960)

Figure 9.1 Walt W. Rostow's stages of growth.
Source: Adapted from Rostow (1960: xii).

The experiences of European postwar reconstruction and, in particular, the rapid industrialization of Japan all suggested that the state had a role to play in development that went beyond individual projects: to manage the economy to actively direct growth. The lesson here seemed to be that the state needed to mobilize, guide or even replace private enterprise if the South was to 'catch up' with the industrially developed nations. For many international development thinkers and Southern political leaders alike, there was a degree of faith in the ability of technical experts (among the national bureaucratic elite or international advisers) to plan the economy centrally to deliver the sustained growth that would be in the 'national interest' of the countries involved. In practice, this involved Southern governments intervening in the economy through a variety of measures. Some

BOX 9.3

Dams and development: the 'temples of modern India'

In 1949, the Narmada Valley in central India was first identified as a site for a TVA-style combined irrigation and hydro project, and in 1961 Prime Minister Jawaharlal Nehru (who once described dams as the 'temples of modern India') inaugurated the Sardar Sarovar Project (SSP). With a planned height of 138 metres this was to be the largest of 30 large dams to be built in the valley, along with a further 135 medium-sized and 3,000 small dams. In India, which is highly dependent on monsoon rain for its main cereal crops, large-scale irrigation projects have seemed to offer an 'obvious' engineering solution to its problems of slow agricultural growth for over 150 years (Chapman, 2002). For their supporters, the costs of building them – financially, but also in terms of loss of **livelihoods** and habitats in the submergence zone – have been justified by claims that the development they deliver is in the greater national interest.

Opposition to the SSP, led since 1985 by social scientist turned activist Medha Patkar, has uncovered some of the dangers of mega-projects undertaken by the state. First, there are profound questions about the state's ability to plan accurately such projects. An independent assessment of India's dam-building record has argued that it has consistently exaggerated benefits and minimized estimated costs of large dams, and the long-term financial viability of projects like the SSP is in doubt (Rangachari, *et al.*, 2000). A proper environmental impact assessment of the project has never been completed. Second, the state's attitude to those affected by the dam has shown a degree of callousness. Official figures of 40,000 displaced people may underestimate the true figure by a factor of ten (Routledge, 2003), and the failure to compensate properly and resettle even this officially recognized group has been at the core of the legal battle to halt construction. Patkar's *Narmada Bachao Andolan* (Save the Narmada Movement) has coordinated mass protests and raised international opposition to the SSP: in return, its protesters have been subject to heavy-handed government repression, including beatings and illegal detention of its members.

Whilst Nehru himself may have later spoken against 'the disease of gigantism' within Indian development projects, political interests have ensured that construction continued, even after the World Bank withdrew financial support for the SSP in 1993. In June 2008, almost half a century after Nehru inaugurated the SSP, the dam stands at over 120 metres: construction

BOX 9.3 (*CONTINUED*)

continues, metre by legally contested metre, as do the protests and the displacements.

(Sources: Adapted from Chapman, 2002; Rangachari, *et al.*, 2000; Roy, 1999; Routledge, 2003)

Plate 9.2 A boat on the flooded Narmada Valley, India.
Credit: Amita Baviskar.

of these were directly aimed at fledgling national manufacturing industry itself, such as limiting the import of foreign products (see Concept Box 9.2 **Import-substitution industrialization** (ISI)), using industrial licensing and regional policy to coordinate growth between sectors and across space, or even government takeover of key industrial activities like electricity generation or steel production through nationalization. Others aimed to boost industrial growth by shaping national economic parameters: government control of the banking sector could direct national savings towards growth areas; control of agricultural markets could ensure a reliable and cheap supply of food to industrializing areas; and control of foreign exchange rates could help to promote the export of a country's manufactured goods abroad. In some cases, governmental guidance of the economy was able to deliver dramatic and sustained growth that exceeded even the expectations of **modernization** theory, and regimes such as that of Taiwan and South Korea (Box 9.4) were celebrated as '**developmental states**' that other Southern governments should emulate.

When states act to initiate large-scale development projects, such as those in the Narmada Valley, or to direct the economy, as in South Korea, this involves both potential benefits as well as substantial risks. One issue here is the potentially negative impact of individual government policies and projects, such as **import-substitution industrialization**, delivering 'second rate' products. Plate 9.3 shows the Hindustan Ambassador, a model based around 1950s technology that dominated the Indian car market until the 1990s through protection from foreign competition (it still has a loyal fan-base, partly due to its robustness and ability to deal with rural roads). Potentially more significant are the dangers of concentrating **power** and resources in the hands of political elites and technical experts. In short,

---

CONCEPT BOX 9.2

Import-substitution industrialization

Import-substitution industrialization (ISI) is a development strategy that was a widely accepted part of mainstream development thinking in the 1960s, and derived from the work of the Economic Commission for Latin America (ECLA), based in Santiago, Chile. According to this viewpoint, the state has an important role in fostering inward-looking economic development, and in particular in promoting domestic industry. By progressively replacing imported manufactured goods for domestically produced ones, Southern economies would increase the breadth and sophistication of their industrial base, move away from dependency on primary production alone and capture more of the added value to be gained within the manufacturing sector. A government could foster this drive towards industrialization by placing high import tariffs on foreign manufactured goods, thus protecting its own fledgling manufacturing sector from international competition until it matured, and by intervening to redistribute wealth (such as through land reform). This second measure would ensure that a broader section of the population had some measure of disposable income, and thus expanded the domestic market for the manufactured goods the country could now produce for itself. Although formalized as an economic model through ECLA's work, similar strategies were widespread across the Global South, and in India they were only completely abandoned at the start of the 1990s.

state-led development was intended to overcome the difficulties Southern countries faced in achieving economic development, but did so by replacing the logic of the market with the logic of political control. Alongside advantages such as longer-term planning, greater coordination of activities across sectors and realizing economies of scale, this can bring limitations and problems of its own. The 'dark side' of this political logic should lead us to ask critical questions about *all* development activities where the state is a major player.

One important question is whether the state is effective in delivering development outcomes that are appropriate for the people and areas concerned. James Scott (1998) has argued that, in general, centralizing **power** in the hands of the state can lead to the production of standardized, simplified 'solutions' to development problems that ignore local diversity

Plate 9.3 Import-substitution industrialization in practice: the Hindustan Ambassador.
Credit: Glyn Williams.

BOX 9.4

South Korea's 'Tiger Economy' and state-led development

South Korea's history made it an unlikely candidate for rapid development: its agrarian economy had been subject to authoritarian Japanese colonial rule (1905–45), political upheaval and war with North Korea (1950–53), before finally gaining some stability under Park Chung-hee's military rule (1961–79). By the late twentieth century, however, it had become one of Asia's leading 'Tiger Economies': over the 1960s and 1970s its economy grew at an incredible average rate of over 8 per cent per annum, and it developed a global presence in the ship-building, electronics and automobile industries. Today, it is one of the Global South's

BOX 9.4 (*CONTINUED*)

BOX 9.4 (*CONTINUED*)

richest nations, with the world's twelfth largest economy and a per capita income equivalent to Portugal (World Bank, 2005).

The Korean state actively shaped this economic success through economic policies that have changed significantly over time. Under Park Chung-hee's leadership, South Korea largely abandoned its earlier policies of **import substitution industrialization** (ISI) in favour of export-led growth. Incentives – such as low-interest loans – were offered to companies with good export performance, and to those industries that the government thought had the highest growth potential. These strategies made government-managed growth particularly effective, and turned some of Korea's family-controlled business conglomerates (*chaebols*), such as Samsung and Hyundai, into globally recognized brands. The 1970s saw a partial return to ISI, before greater economic liberalization in the 1980s. Throughout these policy changes, the state maintained its active role in economic management, and its close relationship with private business. Its firm control ensured that business interests followed government policy, rather than vice versa: by both supporting the *chaebols* and yet encouraging competition among them it maintained its credentials as a '**developmental state**' (Evans, 1989).

There are, however, two important question marks over this remarkable economic record, and its replication elsewhere. First, a strong civil service, the close links between the business houses and government, and an effective (if brutal) police force had been carefully built up during the Japanese colonial period, along with significant investments in infrastructure, agricultural productivity and mass primary education. Park Chung-hee did not therefore create a **developmental state** overnight through policy change alone – it had deep historical roots. Second, the South Korean economic miracle was bought at a high cost to many of its citizens. Industrial labourers were tightly controlled within a network of employer–employee 'clubs' to boost productivity and squash dissent, force was used against political opponents and to suppress civil unrest, and democratic elections for the Presidency were only restored in 1987.

(Sources: Adapted from Kohli, 1994; Evans, 1989; Wade, 1990, 2000; Cumings, 1998; World Bank, 2007b)

and local knowledge (as happened in the collectivization of Tanzania's agriculture: Box 6.4) When this is coupled with the desire of political leaders to impress their supporters or the international community with a symbol of their **power** or '**modernity**', such as the building of Brasília, the chances of failure on a grand scale will be magnified. These related dangers of excessive central control have led to calls for 'alternative' forms of development that challenge standardization and involve community-based planning (see Chapter 11) or decentralized market-led development approaches (see Chapter 10).

A second question concerns the degree to which the actions of the technocrats and political elites guiding the development process can be trusted to act in good faith, rather than out of narrow self-interest. Peter Evans (1989) and others have argued that effective **developmental states** like South Korea maintain a degree of 'embedded autonomy'. Their autonomy – holding **power** above and beyond particular interest groups – enables them to plan for the long term, free from immediate social pressure; while their close links – or embeddedness – with leading industrialists allows their policies to be targeted and effective. But if key policy and investment decisions are focused in the hands of high-level bureaucrats, and their actions are removed from public scrutiny, the possibility of them abusing their position always exists. The same tightly knit relationships between government and industry that allowed planners to have a dramatic impact on the development of the Korean economy also led to claims of widespread corruption in the 1990s. One quantitative estimate of the scale of bureaucratic corruption in India in the 1960s was that the 'rents' earned from public office were equivalent to 7.3 per cent of the national income (Krueger, 1974); a staggering figure, if accurate, and one that stresses the need to keep development 'experts' under public scrutiny at all times.

A third question revolves around whose interests are represented within state-led development. States usually claim to be operating in 'the national interest', but as the example of the Narmada Valley projects shows, these claims are often highly contested in practice. No pattern of political representation is perfect, and even within parliamentary democracies within the Global South, there are dangers that the state's developmental plans and visions may not reflect the full range of citizens' voices – particularly those of the politically marginalized.

Finally, questions need to be asked about the spatial scale at which economic transformation can be managed. Many of the policies undertaken by countries such as South Korea were aimed at controlling the economy at the *national* level. While this was both logical and possible until the 1970s, the extent to which states could do the same today has been undermined in several respects through economic **globalization**. Internationally integrated markets, rapid movement of capital and free-floating exchange rates (see Chapter 4) have all eroded some of the sources of government control over the economy that aided the industrialization of the 'Tiger Economies' of Asia (see Chapter 10 on the Asian Miracle). Countries attempting the same economic policies today would face pressure from international markets, especially if they had small domestic economies and were reliant on foreign capital to kick-start growth. China does, however, act as an important exception, with the state-controlled China Development Bank investing in massive infrastructure projects such as the building of Shanghai Pudong International Airport and the Three Gorges Dam, for which it provided the vast majority of the US$30 billion of the construction costs (China Development Bank, 2008). The Pudong Economic Development Zone across the Huangpu River from Shanghai has also been transformed from agricultural land in the 1980s to a thriving commercial centre (see Plate 9.4). China's sheer size and its long history of strong centralized control means that its development path will be difficult for others to emulate (see Boxes 3.6 and 4.5 on Chinese involvement in Africa), but it does show that even within a globalized world possibilities still exist for some countries to engage in state-directed development.

All of these questions suggest that there are definite limits to the state's ability to direct development, and make the optimism that characterized much of the development theory and the national plans of newly independent countries of the 1950s and 1960s seem misplaced. Certainly, the rapid transformations to industrialization across the Global South that were predicted have been hard to achieve in practice. Even where **developmental states** have been 'successful' in these terms, it is important to ask whose interests they have served, and at what social and environmental costs. This does not mean that the state has become a minor actor in development practice over subsequent years: donor-funded and state-led 'mega projects' are still being rolled out (such as the Lesotho Highlands Water Project), and recent attempts have been made to reassert a greater degree of state control over economic development in some countries, such as in Venezuela under Hugo Chávez (see Box 2.5). But from the early 1990s onwards there has been a new emphasis in development thinking on how states affect development, led by the drive by the World Bank and others to achieve '**good governance**' in the Global South.

Plate 9.4 Pudong New District, Shanghai, China.
Credit: Katie Willis.

REFORMING THE STATE

Recent interest in **good governance** has usually started from the assumption that states in the Global South need to improve the quality of their rule. The reasons for this are not hard to find: in most Southern countries, state-led development had not led to the dramatic transformations hoped for in the 1950s and 1960s, and there were many examples of inefficient or corrupt intervention in the economy. According to **neoliberal** ideas of market-led development that became globally important over the 1980s (see Chapter 10), states in the South were trying to do too much, and doing it too badly. Furthermore, some Southern regimes, such as Pol Pot's Cambodia (1975–79) (see Chapter 2), had presided over or actively undertaken crimes against humanity so severe that their domestic governance arrangements seemed to be a legitimate area for international involvement. More recently, **good governance** has been linked with the **Millennium Development Goals** (see Table 5.2): reform programmes that make government more accessible and responsive to poor people are seen as an important part of achieving global **poverty**-alleviation targets (World Bank, 2000). There are therefore strong arguments for setting global **governance** standards on both developmental and humanitarian grounds, and for making improvements in **governance** a central concern of international development agencies. But what does **'good' governance** comprise of? Box 9.5 provides one important set of definitions – that of the World Bank.

Setting these global standards is not a simple matter, and any definition will always reflect its authors' values to some degree. In the case of the World Bank's criteria, this relationship between a particular definition of **good governance** and the values underpinning it is of immense practical importance to citizens in the Global South for at least two reasons. First, since 1997 the World Bank has been aiming to provide a reliable statistical measure for these six criteria by creating its Worldwide Governance Indicators. These indicators are based on perception data drawn from a variety of sources (including the country analysts of major multilateral development agencies, and international NGOs such as Amnesty International), which are combined to produce a composite measure which, in theory at least, allows the detailed comparison of the quality of **governance** between countries and over time (Figure 9.2). These aggregate measures cannot hope to reflect the

BOX 9.5

What is 'good' governance?

States across the world – whether democratic or authoritarian – have always used their own values to reflect on the quality of their governance. Ruling justly and well in accordance with local norms is important in ensuring the support of their citizens (maintaining legitimacy) and thereby retaining **power**. The more recent drive for **good governance** however is rather different: it aims to produce a degree of *global* consensus over what constitutes good and

BOX 9.5 (*CONTINUED*)

BOX 9.5 (*CONTINUED*)

just rule, and then ensure all states are moving towards these ideals. The World Bank (Kaufmann, *et al.*, 2006: 4) identifies six dimensions of **good governance**:

1 *Voice and accountability*: the extent to which a country's citizens are able to participate in selecting their government, as well as freedom of expression, freedom of association, and free media
2 *Political stability and absence of violence*: perceptions of the likelihood that the government will be destabilized or overthrown by unconstitutional or violent means, including political violence and terrorism
3 *Government effectiveness*: the quality of public services, the quality of the civil service and the degree of its independence from political pressures, the quality of policy formulation and implementation, and the credibility of the government's commitment to such policies
4 *Regulatory quality*: the ability of the government to formulate and implement sound policies and regulations that permit and promote private sector development
5 *Rule of law*: the extent to which agents have confidence in and abide by the rules of society, and in particular the quality of contract enforcement, the police, and the courts, as well as the likelihood of crime and violence
6 *Control of corruption*: the extent to which public **power** is exercised for private gain, including both petty and grand forms of corruption, as well as 'capture' of the state by elites and private interests.

All six may seem to be globally applicable aims, but a close examination reveals that hiding within these 'universal' values is a particular way of defining 'good' rule. Western-style democratic models are perhaps implicit in the definition of *government effectiveness*: other states may not separate the civil service from political interests in this way, but that does not mean that their rule is automatically flawed. More importantly, the definition of *regulatory quality* assumes that states should be acting to 'promote private sector development'. This ignores both other ways of organizing the economy (such as through state ownership) and other aims of regulation (such as promoting an equitable society or promoting environmental sustainability) that citizens in different contexts may see as more appropriate to their needs.

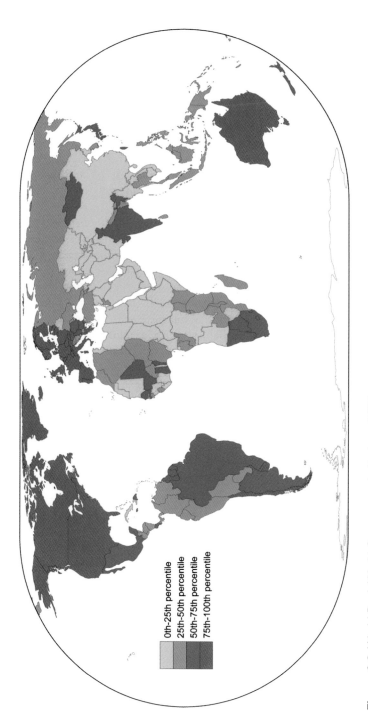

Figure 9.2 World Bank World Governance Indicators, 2005.
Source: Adapted from World Bank (2008). Map data © Maps in Minutes™ (1996).

Note: the World Bank's World Governance Indicators aim to measure each of the dimensions of good governance described in Box 9.5 and provide an overall composite measure (mapped here); countries in the higher percentiles having 'better' governance.

0th–25th percentile
25th–50th percentile
50th–75th percentile
75th–100th percentile

diversity of **governance** quality on the ground: for example, India is a federal nation with significant internal variations in how government acts across its states, and how these actions are felt by its 1.1 billion people, but it is represented through a single set of national measures. Despite their limitations, these statistics have powerful, real-world effects. Being labelled as an 'unstable' or 'corrupt' country not only may deter private investment, it can also affect the sources of international **aid** a country receives. The Millennium Challenge Corporation (MCC) was established in January 2004 and distributes a significant amount of the USA's **aid** budget (over US$2bn in its first two years of operation). It specifically targets its development assistance to countries that show signs of '**good governance**', and will only provide aid to those that already score well on the World Bank's **governance** indicators, among other measures (MCC, 2007).

The second reason that the World Bank's definitions are important is that they are based on a set of values and practices that the Bank believes Southern governments *should* be striving towards. Over recent years, government reform has become a significant theme of the Bank's own development projects: it ran over 600 **governance**-related programmes in 95 countries between 1998 and 2000 (Nanda, 2006). The details of the Bank's **good governance** agenda have evolved since the term was first popularized at the end of the 1980s, but the overall direction of these programmes has been consistent. In the Bank's vision, a 'reformed' state is one where government provides the supporting conditions for sustainable, market-led economic growth. It is also one where the state should strictly rein its ambitions in to match its administrative and fiscal capabilities (Table 9.1) – it should concentrate on 'getting the basics right' first. This is a far different role for government than that of *directing* the economy envisaged in the early postwar decades, and one underpinned by **neoliberal** values. Reform is to be achieved in three main ways: by changing the internal rules of government (in particular, by strengthening the checks and balances on governments' executive **power**, and improving the transparency of their actions); by opening the state's own operations up to competitive

Table 9.1 Redefining the role of the State

Functions of the State	Addressing market failure			Improving equity
Minimal functions	Providing pure public goods			Protecting the poor
Intermediate functions	Addressing externalities	Regulate monopoly	Overcoming imperfect information	Providing social insurance
Activist functions	Coordinating private activity			Redistribution

Source: Adapted from World Bank, 1997: 27, table 1.

pressure; and by seeking partnerships (with NGOs, community organizations or the private sector) to deliver key government services.

Governance reform has often aimed to change dramatically both the day-to-day practices of civil servants and also the overall financial discipline of the state. With regards to the former, ideas of **New Public Management** (see Concept Box 9.3) have been implemented within the Global South (see Box 7.3 on civil servants in Malawi), but alongside this there have been efforts to learn from concrete experiences of improvements in public service delivery emerging from within the Global South itself. Judith Tendler's work on health service reform in Ceará, Brazil (Tendler, 1997) gives details of one such example. Here a new body of 'barefoot' health workers was created: by employing these people within their own communities and giving them strong recognition within the community for their work, the state managed to create a committed and dynamic workforce. The more general hope is that a combination of personal incentives (such as merit-based promotion) and commitment to collective goals will change the ways in which

---

CONCEPT BOX 9.3

New Public Management

New Public Management (NPM) refers to a broad sweep of public sector organizational change across the Global North and South from the 1980s, aimed at creating more financially efficient and change-oriented public services. Policy details vary greatly between contexts, but **New Public Management** has been associated with the break up of large public sector units (often contracting out elements of their work to the private sector) and the introduction of market-based mechanisms within them, including competition between public sector agencies and performance incentives for individual employees. Ewan Ferlie and his co-writers (Ferlie, *et al.*, 1996) also note that these shifts are accompanied by wider political and economic changes, including weakening public sector unions, declining autonomy of professionals or technical experts vis-à-vis a growing cadre of general managers and growing **power** of appointed (rather than democratically elected) executive directors running public services. Although their work was based on health service reform in the UK, similar contextual factors have also been important in many places in the South. The outcomes for the general public are ambiguous at best, even though the public are often invoked as the 'clients' for whom change is being implemented. Declining local democratic oversight of public services is one concern; when combined with various forms of cost recovery, this can mean that poorer people are increasingly excluded from public service provision in many parts of the South.

public sector employees work: they will have a sense of ownership of their jobs, the respect of their 'clients' among the public and a stake in improving the capacity of their own organizations. As will be clear from Chapter 6, however, effecting these changes on the ground can be a difficult process.

To ensure that states in the Global South follow the World Bank's ideas for internal reform, debt-relief programmes from the international community are commonly dependent upon them signing up to elements of the '**good governance agenda**'. For example, countries participating for relief within the World Bank's Highly Indebted Poor Country (HIPC) initiative have had to commit to a number of external controls, including a Medium Term Expenditure Framework agreed with the Bank. This effectively places a cap on government spending on a ministry-by-ministry basis, with an agreed formula of expenditure based on performance and delivery. The intention is to improve government financial accountability, but this also gives the Bank a fairly significant oversight and control over the public finances of the countries concerned. By applying this financial pressure from above, and seeking to galvanize the potential enthusiasm and energy of public sector employees from below, the intention is that faceless, sprawling bureaucracies will be made in to 'leaner', goal-oriented institutions that are more responsive to the public. As Box 9.6 shows, this process has had mixed results in practice, even in some of the states the World Bank considers its 'success stories'.

BOX 9.6

Implementing good governance in Uganda

Since 2001, the World Bank has linked its aims of improving governance and **poverty** relief in the Global South through the **Poverty Reduction Strategy Paper (PRSP)** approach. PRSPs are country-specific evaluations of the policies required to move people out of **poverty**, produced through partnership between Southern governments and **civil society** actors, with oversight from the Bank itself. Conducting a Bank-approved **PRSP** is important in delivering financial support: it is an entry condition for debt relief under the HIPC Initiative, and has also become linked to the dispersal of the World Bank's Social Funds (which have been designed to offset some of the pain of **SAPs** for the poorest). This financial support, however, comes with conditions: countries are 'locked in' not only to ministry-specific spending priorities and caps, but also to

BOX 9.6 (CONTINUED)

BOX 9.6 (*CONTINUED*)

new accountability procedures that feed information on service delivery performance back up from local to national government, and ultimately to international donors. In this way, international spending of **poverty** relief is intended to drive countries towards **governance** reform, with local 'change agents' (particularly finance ministries) taking responsibility for spreading 'good practice' more widely.

Uganda has had a long series of **governance** reform initiatives, some funded by the World Bank, and as one of the first countries to gain HIPC status it is a good test case for this strategy of leveraging **governance** reform 'from above'. David Craig and Doug Porter (2003) suggest that the initial results have been rather mixed. **Aid** conditionality and improved accountability procedures mean that new money has reached local areas for provision of schools, health services and water supply. At the same time, those same conditions have put the moves Uganda made over the 1990s towards decentralized and empowered local government into complete reverse. Money is spent 'correctly', but this has reduced local government to a mere implementer of centralized, donor-approved plans, and made it dependent upon the continued flow of external **aid** (Harrison, 2005). The knock-on effects are that local government has decreasing incentives to raise revenue, pursue its own initiatives or build links to its own constituents. The technical detail of conditional lending is improving financial accountability 'upwards', but eroding the political accountability of local government 'downwards' to its own people. In doing so, it is undermining one of the foundations of lasting '**good governance**'.

(Sources: Adapted from Craig and Porter, 2003; Harrison, 2004, 2005)

Questioning the good governance agenda

Good governance may therefore seem like a praiseworthy goal in theory, but it is one that is always going to be contentious in practice. As such, the recent international-led push to 'improve' the **governance** of Southern countries raises important questions about this as a direction for intentional development. The first is how practical and possible is it to achieve results within the time frames of the World Bank and other donors? **Good governance** programmes are based on a degree of faith that states and their practices are reasonably malleable: if the right forms of financial discipline and incentive structures are provided, the state will be capable of fairly swift reform. As shown in Box 9.6, there can be particular problems with this in practice, as **governance** reform begins to produce unintended consequences for the relationships between states and their societies. It is possible that as the number and detail of **governance** interventions proliferates, these might actually be increasing the chance of their own failure: there have certainly been calls to simplify a reform agenda that outpaces Southern governments' capacity to manage both reform and the conflicts it produces (Grindle, 2004). More generally, as the discussion of South Korea's '**developmental state**' (Box 9.4) has shown, changes in state effectiveness have deep historical roots, and the 'good' institutional practices that have been established in the North often took generations to establish (Chang, 2002). For the international donor community to imagine that Southern countries will produce similar change within a decade or so of intentional reforms is at best optimistic thinking; at worst it can lead to dangerously misplaced ambition. The experiences of trying to effect post 9/11 'regime change' in Iraq and Afghanistan have clearly shown that is far easier to remove existing **governance** structures than it is to replace them with stable new ones built around Northern norms.

A second related question, above and beyond the practicality of reform, is that of whose ideas and values of **'good' governance** will be implemented. As was noted earlier (Box 9.5), there is an underlying set of **neoliberal** assumptions about the direction reform should be heading in, and the sorts of relationships between states, societies and markets that it should support. Although much of the agenda for **governance** reform may at first sight seem to be concerned with 'managerial' or 'technical' issues, it is important to remember that behind (or embedded within) these are a set of real *political* choices to be made about the roles and objectives of government, and supporting market-led growth will not necessarily be the priority of all Southern regimes. For those countries not willing to agree with some of the starting assumptions of the **good governance** agenda, the different forms and conditions of Chinese international development assistance (which has fewer strings attached) may therefore be an increasingly attractive option (Box 3.6 on Chinese involvement in Sub-Saharan Africa).

Even if the **neoliberal** roots of dominant approaches to **'good governance'** are ignored, problems remain about the spatial focus of their analysis. Just like ideas of **modernization** theory in the 1950s and 1960s, there is perhaps too much attention placed on the national scale such that current problems of **governance** are seen as a result of internal political and institutional factors, and the international context in which 'bad' **governance** emerges is hidden from view. A more geographically informed analysis would recognize that **governance** problems are not self-contained within national boundaries, and in particular that earlier rounds of World Bank and IMF reform through **SAPs** (Chapter 4 and Chapter 10) have often been responsible for eroding the very same state capacities that today's **good governance** programmes are looking to build up (Harrison, 2005). There is some recognition of these problems within Northern donor agencies. The UK Government's Department for International Development (DFID) has argued that *international* problems – such as money laundering, lack of corporate responsibility among **TNCs**, the global trade in conventional weapons and illicit sales of natural resources – are an important part of the production of *internal* governance

BOX 9.7

Good governance from the bottom up: Porto Alegre's participatory budget

Since the end of military rule in 1988, a number of Brazil's municipalities have experimented with forms of participatory budgeting. Most famous amongst them is Porto Alegre, a city of over 1.3 million people, where dramatic changes to local **governance** started in 1989 after the Brazilian Workers' Party (*Partido dos Trabalhadores*, or PT) won the mayoral elections. The PT had a mandate to redirect government priorities towards the poor and to improve popular participation in government, which it decided to implement by overhauling the process for setting the city's budget. The city was divided into 16 districts, within which a series of general assemblies were held annually where the city's budget was explained and debated, and where participants elected their own representatives to District Budget Forums. These assemblies were backed up with neighbourhood meetings where local spending priorities were debated; thematic forums to debate key issues such as health policy; and a city-wide Municipal Budget Council (drawn up of delegates elected from the District Budget Forums). Together, these produced significant opportunities for direct involvement in government, and a 1994 survey indicated that 8.3 per cent of the city's population had been involved at some stage of the participatory budget process (Matthaeus, 1995, cited in Souza, 2001).

The potential benefits for city **governance** are great: participation in setting the budget makes one of the most technical (and corruption-prone) activities of government open to popular scrutiny, and thus offers a people-centred mechanism for ensuring financial accountability, in contrast with that pursued by the World Bank in Uganda (Box 9.6). It has also allowed city spending to be redirected to match local needs: investment in water supply and sewerage were initially voted as priority issues, correcting the administration's assumption that public transport was most important for the poor. Questions about its success inevitably remain: the budget meetings only deal with part of the city's budget, and while over 40 per cent of delegates are drawn from the poorer sections of the city's population, participation levels among the very poorest are lower, as is women's representation in upper-level meetings. Nevertheless, most assessments of Porto Alegre's experiment with participatory budgeting agree that it has been successful in replacing client

BOX 9.7 (*CONTINUED*)

BOX 9.7 (CONTINUED)

and authoritarian city politics with less corrupt and more democratic practices, and as such it has become a model for good urban **governance** adopted by other cities in Brazil and beyond.

(Sources: Adapted from Abers, 1998; Baiocchi, 2001; de Sousa Santos, 1998; Souza, 2001)

---

problems in the Global South (DFID, 2006). As such it is important to improve *global* **governance** structures instead of simply assuming that reform of public bureaucracies of the South will be effective in isolation (Williams, 2009).

There is, however, an important difference between criticizing the particular **good governance** agenda followed by the World Bank, and abandoning the idea of reforming the state as a developmental aim altogether. If the **good governance** agenda has emerged in part because of the horrendous actions of past regimes such as Pol Pot's in Cambodia, more recent cases such as Robert Mugabe's brutal retention of **power** in Zimbabwe show that there is still a more general justification for the internal actions of states being legitimate areas of international concern. As examples in this Chapter and Chapter 6 have shown, the quality of government in far more 'normal' circumstances also has very important implications for the lives of citizens across the Global South. But here, it is far less clear that what the Global South needs is to follow a detailed programme of prescriptive changes from international donor agencies. Merilee Grindle (2004) has argued that a far better approach is to try to learn from successful examples of **governance** reform in the Global South, to ask why these have emerged and how they can be reproduced elsewhere. Porto Alegre's innovative processes of participatory budgeting at the municipal level provide one such success story (Box 9.7) that has been taken up in other Southern cities. It is perhaps here through processes of 'bottom up' and mutual learning from practice rather than the top-down imposition of a Northern-dominated agenda where the greatest potential for **governance** reform lies.

CONCLUSIONS

This Chapter has looked at two very different approaches to the question of how development should be governed, and to what ends. In the mid-twentieth century, there was much optimism in both the North and South that state-led development could deliver rapid **modernization** of entire countries, with the benefits of industrialization 'trickling down' to all. The negative aspects of this vision – grandiose projects that are often of questionable economic benefit, and that are undertaken at great environmental and/or social costs – are easy to see in retrospect, and are themselves a result of excessive faith in the capacity of bureaucratic and technical elites within Southern governments to deliver planned change 'in the national interest'. As the human costs of such projects as the Narmada Valley and others like it across the Global South add up, it is important to question whether the state should be left to define what 'development' is, and how it is to be implemented.

Programmes of internationally sponsored **governance** reform have in some ways tried to correct for the excesses of the **developmental state**, bringing in new structures to ensure governments' financial accountability, and the latest ideas of **New Public Management**. The promise here is that citizens are offered a new way of 'seeing the state', experiencing professional and publicly accountable civil servants instead of high-handed or corrupt officialdom (see Chapter 6). As the example of Uganda has shown, there are practical difficulties in implementing such schemes 'from above': donors may be able to lever changes in government practices out of Southern countries, but they may do so at the cost of removing the accountability of the state to its own people. There are also important questions to be asked about the underlying **neoliberal** development vision that is being supported through the World Bank's **good governance** agenda, and whether the attempt to buy poorer countries' compliance with this agenda is a justifiable erosion of their national **sovereignty**.

The problems of governing development may thus be easy to spot, but solutions are rather more difficult to find. South Korea has achieved the rapid economic growth that many Southern countries aspire to, but this was not simply the result of the adoption of a set of 'correct' development policies that others could copy. Its transformation relied on deep-seated investments in the state's institutions, and these have also had their authoritarian side-effects. If Porto Alegre emerges as the 'success story' of this chapter, its most important impacts have not been dramatic material improvements in the lives of its citizens. There have been some gains here, but they are modest ones, achieved by the better targeting of a limited budget towards poorer areas and people. Rather, its success is to be found in the changing relationships between its citizens and local government, where a greater degree of participation and democratic oversight of the actions of the local state have been achieved. Judgements about which is more important – an economically successful government, or one that is more publicly accountable – cannot be made in the abstract. These contrasting examples do however serve as a reminder that the aims of development are always contested, and as a result the question of who *should* govern development will always be one that is open to both theoretical and political debate.

Review questions/activities

1 One argument made for state-led development is that national governments have the authority to direct development in accordance with a country's national interest: look back at the case studies within this Chapter to evaluate this claim.

2 Look through a recent World Development Report (from the World Bank website, listed below) and a recent Human Development Report (from the UNDP website, listed below). What roles does each organization see the state as playing within development practice today?

3 Use the World Bank's Governance and Anti Corruption web pages (listed below) to look at good governance indicators for any region or country within the Global South. What aspects of governance there do you think they manage to represent, and what gets overlooked?

SUGGESTED READINGS

Roy, A. (1999) *The Cost of Living,* London: Harper Collins.
This is an account of what can happen when state-led development is pursued unchecked by a high-handed government. Novelist Arundati Roy presents an emotive but thought-provoking attack of India's nuclear programme and development in the Narmada Valley.

Ferguson, J. (1990c) *The Anti-Politics Machine: 'Development', Depoliticization and Bureaucratic Power in Lesotho,* Cambridge: Cambridge University Press.
This is a more in-depth academic account than Roy's book, but is highly readable and raises some equally profound questions about the dangers of state-administered (and internationally supported) development programmes.

Tendler, J. (1997) *Good Governance in the Tropics,* Baltimore, MD: Johns Hopkins University Press.
To balance any possible anti-government bias in the previous two, here's a well-grounded account of intelligent, and effective, state-led development intervention in Brazil.

WEBSITES

www.mca.gov Millennium Challenge Corporation
> This website provides an interesting example of how good governance indicators can have very real effects on a country's ability to access some forms of aid.

www.undp.org United Nations Development Programme
> The UNDP's annual *Human Development Reports* provide a rather different review of development practice, often suggesting a more active role for the state and the public sector within the Global South.

www.worldbank.org World Bank
> The World Bank's *World Development Report* provides an annual review of what it thinks constitutes 'good' development practice (and the appropriate role of the state within this).

www.worldbank.org/wbi/governance
> The World Bank's *Governance and Anti-Corruption* webpages include the database of the Bank's good governance indicators (searchable by country, year and topic) and a range of background papers and discussion on their definition and use.

10 Market-led development

INTRODUCTION

As in Chapter 9, the focus of this Chapter is on strategies to achieve 'development'. The overall theme is on market-led development, but as with discussions of the state in the previous chapter, there have been variations over time and between places in how this approach has been implemented. Markets have a part to play in state-led development strategies, but in this Chapter we will concentrate on development theories and policies which advocate the operation of the market with few state controls, that is a 'free market' system.

Current official international development policy is based on the market being the key actor in **poverty** alleviation, economic growth and improvements in the quality of life. While this ideological stance seems **hegemonic** at the start of the twenty-first century, this Chapter outlines the rise of market-led development policies and how their implementation and effects have varied spatially. A focus on market forces in development means that the private sector (from large companies to very small enterprises), as well as individual workers and consumers, are at the heart of economic decision making.

The focus on market forces does not mean that the state, **civil society** organizations, communities or households do not have parts to play (see also Chapters 9 and 11), rather that the emphasis of the approaches outlined in this Chapter is on market-based solutions to development 'problems'. This Chapter will outline how the focus on market approaches has framed and in some cases encompassed NGOs and individuals as contributing to non-state-led development. It will also stress that market-led development does not mean the absence of the state, but rather a transformation in the state's role from provider (of goods and services) to an institution which enables companies, organizations and individuals to make their own decisions.

MARKETS AND FREE TRADE

Market-based approaches are based on the premise that the most equitable and efficient way of distributing resources (such as food, land, housing, services and labour) is the market. This means that the price of a good, or the wage levels of a worker, will be determined by the relationship between supply and demand (see Figure 10.1). If the supply goes up then the price will go down, but if a good becomes scarcer, the price is likely to increase. For example, in Figure 10.1, demand curve 1 (D_1) is the pattern of demand for a certain good at a particular time and place. As the price goes up (along the y axis) the demand for the good goes down (x axis). For suppliers (i.e. companies) the supply curve (S_1) shows that the higher the price the more they will produce. Where the two curves meet (E_1) is the quantity that is actually produced (Q_1) and the price that is actually charged (P_1).

Different goods and services will have different demand and supply curves with some being steeper than others. There can also be shifts over time in the nature of the supply and demand curves because of changes in factors other than price. The shape of the curves does not change, but their position on the axes does. For example, if household income goes up, then the demand curve for certain

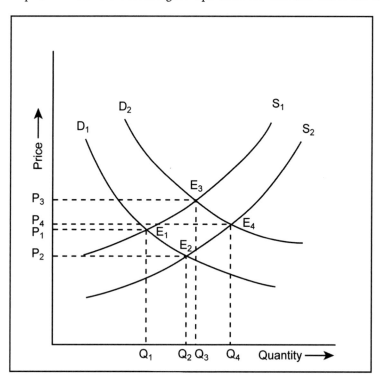

Figure 10.1 Supply and demand curves.

goods may shift to the right (D$_2$) meaning that at higher prices consumers are willing to buy more than they did when incomes were lower. Shifts in the supply curve may occur because of technological changes or fluctuations in input costs.

The same principle will hold for wages, with the demand curve representing the demand from employers for staff and the supply curve the supply of workers. Wage levels will be at the point where the curves cross. Such a free market system, it is argued, will allocate scarce resources most efficiently as it will encourage producers to switch production to goods which are in demand and workers to develop skills in areas where they could command higher wages or to move to a location where their skills are in greater demand. The obstacles to such shifts will be discussed later in this chapter. The system also assumes that everyone is able to participate in the market as a worker and a consumer.

Such free market ideas (sometimes called **laissez-faire** approaches) are most associated with Adam Smith whose book *An Inquiry into the Nature and Causes of the Wealth of Nations* (1979 [1776]) called for reduced government intervention in economic affairs, free trade and a division of labour. Ideas about **free trade** were also promoted by the economist David Ricardo who developed the concept of 'comparative advantage'. He argued that if countries specialized in producing the goods in which their resource base gave them an advantage relative to other countries, then overall production would increase. **Free trade** was vital to allow all nations to have access to the goods that they required to progress (see Figure 10.2).

Figure 10.2 is a simple representation of the ideas of absolute and comparative advantage. In all four scenarios there are two countries (A and B). The production figures are the amounts of wheat or textiles which can be produced in each country by one unit of production (e.g. a certain number of workers or a particular amount of land). Scenarios A and B show a situation of absolute advantage where each

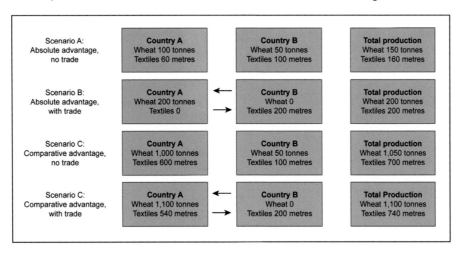

Figure 10.2 Relative and comparative advantage.

country can produce one product more efficiently than the other. In this case, trade allows Country A to specialize in wheat and Country B to specialize in textiles (Scenario B), and for them to import their requirements from the other country. Overall production increases.

In a situation of comparative advantage, which is what Ricardo identified, even when one country has no efficiency advantages over another, trade will help increase total production. In Scenario C, Country A produces both wheat and textiles more efficiently than Country B. If trade is possible, then by switching 10 per cent of productive capacity to the commodity in which it has a comparative advantage, then total production increases. Rather than trying to produce wheat relatively inefficiently, it makes sense for Country B to stop wheat production and specialize in textile production. By switching 10 per cent of its productive capacity from textiles to wheat, Country A will also experience benefits.

These basic principles of free market and **free trade** operation, outlined in the late eighteenth and early nineteenth centuries, have been used as a basis for national and international economic policies in a range of situations ever since. The rest of this Chapter will consider how the market has been mobilized as a route to development, particularly in the Global South, and the impacts of such policies on the peoples of Africa, Asia, Latin America and the Caribbean.

COLONIALISM

In Chapters 3 and 4, we outlined the ways in which the Global South became incorporated into a global political and economic system. The role of private companies, most notably the British and Dutch East India Companies, was highlighted. While such private sector activity was aimed primarily at profit-maximization for the companies themselves and also contributions to coffers of European governments, it also had effects on the development of infrastructure and economic activities in particular places in the Global South (see Box 10.1).

'DEVELOPMENT' AS AN INTERNATIONAL PROJECT IN THE POSTWAR WORLD

In the post-World War II period, 'big D Development' (Hart, 2001) became the focus of international policy in the 'First World' (see Chapters 2 and 9). Using the model of European or US economic development as a basis for development interventions in the Global South, meant a focus on industrialization, **urbanization** and greater use of scientific principles in agricultural production. Such ideas can be generally termed '**modernization**' and demonstrates a Northern-centric approach to development, whereby countries or peoples are defined by their distance from Northern ideals, rather than in relation to their own priorities or worldviews

BOX 10.1

Infrastructure development in Malaya

During the colonial period, part of what is now peninsular Malaysia was controlled by the Portuguese, the Dutch and then from the early nineteenth century, the British. The Straits Settlements, including Penang, Malacca and Singapore, grew up as key ports and trading sites on the routes between Europe and Asia (see Figure 4.1). Agricultural production developed around these ports and others on the west side of the peninsula as migrants settled in the area.

The development of infrastructure, including transport routes, hospitals, schools, post offices, banks and other indicators of European **modernity**, was driven by two main forces: first, the demand for infrastructure to support economic development, and second, the need for improved communications and transport to facilitate administrative control as British colonial rule spread across the peninsula. Infrastructure development was therefore partly a response to market forces and private sector demands, and partly politically motivated.

Economic development in Malaya in the nineteenth and early twentieth centuries was driven by the ports and trade, but also by the expansion of primary production. At first the focus was on coffee and sugar production on private estates near the ports, but later tin mining boomed, driven by an influx of Chinese miners. Centres of tin production, such as Ipoh, required good transport routes to move the tin to the ports. While rivers were originally used, the development of roads and railways greatly improved the efficiency. In the late nineteenth and early twentieth centuries rubber production became much more important as the demand for rubber expanded with the growth of the car industry in Europe and North America. Again, the processing and transport of rubber for export required new forms of infrastructure.

By the late 1950s when the Malay Federation gained independence, Thomas Leinbach (1972, 1975) calculates that almost the whole peninsula had been incorporated into modern transport, education, health and legal networks. However, there were key nodes of **modernization**, most notably Kuala Lumpur and Ipoh. The central and eastern parts of the peninsula had fewer services as they had been incorporated less intensely into the economic expansion driven by rubber, tin and trade.

(Sources: Adapted from Leinbach, 1972, 1975)

(Escobar, 1995; see also Chapter 9). While World War II is often used as a turning point in international development efforts, it is important to remember that for many parts of the Global South, independence did not arrive until much later. Thus, many of these **modernization** policies were implemented under colonial regimes and were then continued as part of development assistance to independent nations (Kothari, 2006).

As outlined in Rostow's model (see Box 9.2), the focus on external intervention to achieve **modernization** meant that the postwar period was characterized by increasing flows of **aid** (see Concept Box 10.1) to governments of the Global South. These **aid** flows were particularly significant in relation to large infrastructure projects, such as dams and roads (see Chapter 9). The role of national governments in the Global South was viewed as key in directing development. This was because of the perceived limited capacity for private sector or community action. However, while the state's role was viewed as very important, **aid** and support from Western nations clearly relied on Southern governments generally following a capitalist, rather than communist, development path. Thus the balance between market-led development and state-led development was somewhat more equal than Adam Smith and other neoclassical economists would advocate (see Chapter 9 for elaboration on the role of the '**developmental state**' in the postwar period). During the **Cold War**, **aid** was used as a way for the West and the Soviet Union to gain support from Southern countries (see Chapter 3). Cuba and China

---

CONCEPT BOX 10.1

Aid

Aid refers to the transfer of resources to promote improvements in economic development and social welfare. The resources transferred can be money (either as a grant, or as a loan with interest rates lower than those available on the free market); goods (such as seeds, technical equipment and medical supplies); or human resources and expertise as part of training and education programmes. **Aid** can be provided at times of emergency in the wake of a natural disaster or civil war, for example, or as longer-term development **aid** which seeks to achieve development which is sustainable. **Aid** that is transferred between governments (either bilaterally between individual national governments or multilaterally through organizations such as the European Union or United Nations) is called 'Official Development Assistance' (ODA). Large development organizations, like Oxfam and Christian Aid, are also **aid** providers, and smaller **NGOs** can also channel **aid** as well as being recipients (see Chapter 11).

have also been important providers of **aid** to fellow countries in the Global South (see Box 2.3 on Cuban health diplomacy and Box 3.6 on Chinese activities in Africa).

NEOLIBERALISM

In the late 1960s and 1970s, a succession of global economic crises led some researchers and theorists to reconsider the role of the state in development, both in the Global North and the Global South. Rather than being a positive actor in promoting economic growth and therefore, so the argument went, social development, economies where the state played a significant role and the national economy was relatively closed to foreign investment were experiencing sluggish growth. A common comparison was between the supposedly market-friendly, open economies of East Asia and the closed, protectionist economies of Latin America (Balassa, 1971). The grouping of ideas promoting greater market freedoms and a limited role for the state was termed '**neoliberalism**' (see Concept Box 4.3).

According to these **neoliberal** ideas, the state's role is 'to favour strong individual private property rights, the rule of law, and the institutions of freely functioning markets and free trade' (Harvey, 2007: 64). This focus on the rights of the individual (both person and company) to act as they see fit, resonates strongly with Adam Smith's arguments in the *Wealth of Nations*. **Neoliberal** theorists (such as Milton Friedman and Friedrich von Hayek) believe that by freeing up the market, economic growth will result, with positive **trickle-down** effects to benefit all sectors of society. How this works in practice, however, is not always as the theorists envisaged, particularly in relation to enforcing private property rights.

Although some of the East Asian economies had been praised for achieving economic growth, and continued to be fêted by the World Bank as examples to the rest of the Global South until the economic crash of 1997 (see Box 10.2) (Plate 10.1), it was Chile that provided the first example in the world of **neoliberal** political and economic principles in practice (see Box 10.3). The perceived economic success of the Chilean **neoliberal** experiment fed into the adoption of such policies in countries of the Global North, most notably the UK under Margaret Thatcher from 1979 and the USA under Ronald Reagan from 1980.

While **neoliberalism** is often presented as a coherent set of ahistorical and aspatial ideas, in reality it is important to recognize that the practices of **neoliberalism** can vary widely (Barnett, 2005). In discussing the ways in which **neoliberal** policies are implemented, experienced and contested, it has therefore become more usual to talk about '**neoliberalization**', stressing the process of **neoliberalism**'s adoption (England and Ward, 2007; Peck and Tickell, 2002). In the rest of the chapter, the focus will be on examining how **neoliberal** ideas became **hegemonic** in international development practice in the 1980s and 1990s, and remain key in shaping particular forms of development intervention. The remaining sections

BOX 10.2

'The East Asian Miracle'

In 1993, the World Bank published its report *The East Asian Miracle* which compared the experiences of East Asian countries with those of Latin America and Sub-Saharan Africa in the period 1965–89. The main conclusions of the report were that the East Asian countries had been able to achieve high levels of economic growth (over 4 per cent per annum average growth in GDP p.c.) without increasing levels of economic inequality. These countries were classified into what has been termed a 'flying geese' pattern, with Japan at the head, followed by the 'Newly Industrializing Countries/Economies' (NICs or NIEs) of Hong Kong, Singapore, South Korea and Taiwan. The next group of countries included Thailand, Malaysia and the Philippines, while economies such as Vietnam and Indonesia were in the next grouping.

The report's authors ascribed this success to the openness of the East Asian economies to foreign investment and also to limited state involvement in economic activities. Based on this interpretation, they urged other parts of the Global South to adopt these policies to achieve economic development without social inequalities. These representations of 'Asian' economic success relative to Africa and Latin America reflects other forms of representation which make distinctions between parts of the Global South (see Chapter 2).

Following its publication, the report was criticized on a range of grounds. For example, the very concept of an 'East Asian Miracle' was challenged, partly because it implied that all countries in East Asia were the same and had followed similar patterns, and partly because the concept of a 'miracle' implies something which cannot be explained, but according to the World Bank the success was due to very obvious policies. The interpretation of the state's role in the economic success of the different East Asian countries was also challenged. Rather than being free-market led, a number of researchers highlighted the role of state intervention, for example, in protecting particular domestic industries (see Chapter 9 on the **developmental state** in South Korea).

In 1997, economies across the East Asian region suffered tremendous economic problems following the devaluation of the Thai baht. Investor confidence in the region plummeted, and investors withdrew their money leading to severe economic consequences. The freedom to move money freely around the global economy had been, and still is, a key tenet of neoliberal policies, but in this situation it led to economic collapse in some

BOX 10.2 (*CONTINUED*)

of the countries with concomitant negative impacts on the lives of the population.

In response to the East Asian economic crisis, the World Bank argued that it was a result of financial mismanagement and what was termed '**cronyism**' in the region. Greater transparency of regulation was called for to ensure that such catastrophic events could not happen again.

(Sources: Bird and Milne, 1999; Rigg, 2003; Wade, 1999; World Bank, 1993)

Plate 10.1 Financial district, Singapore.
Credit: Madeleine Dobson.

BOX 10.3

Chile and the 'Chicago Boys'

General Augusto Pinochet came to power in Chile in 1973 following the overthrow of the democratically elected left-wing government led by Salvador Allende. The economic policies adopted by the Pinochet regime after about 1975 were developed largely by a group of Chilean economists who had been trained in the Economics Department at the University of Chicago. Because of this they became known as the 'Chicago Boys' even though not all economists associated with the Pinochet regime had been trained in Chicago.

Pinochet's government sought to implement policies that were identified as 'technocratic' rather than 'political', the argument being that problems should be identified and then solved in a rational and neutral manner. This focus on technocratic solutions to 'development problems' is now widespread within the international financial institutions (see e.g. Mawdsley and Rigg, 2002, 2003, on the World Bank). This approach did not (and does not) admit that the construction of 'problems' and so-called 'rational solutions' is political in that it depends on the exercise of **power** (see Chapter 2 on **discourse** of development).

In the 1950s, economics students from the Catholic University in Santiago, Chile, were able to study at the University of Chicago thanks to a grant scheme that was partly funded by the Ford and Rockefeller Foundations. The US Government was keen to promote such exchanges as a route to challenging what were seen as dangerous left-wing ideologies in Latin America. Between 1955 and 1963, 30 economists studied in Chicago and staff from Chicago lectured in Santiago. A key figure in the Chicago Economics Department at that time was Milton Friedman who greatly influenced the thinking of the Chilean students and their adoption of **neoliberalism** as the most appropriate route to development.

On gaining positions in Pinochet's government, the 'Chicago Boys' were able to implement these **neoliberal** ideas, including large-scale privatization, opening up the Chilean economy to foreign investment and promoting export-oriented industries rather than **import-substitution** enterprises. Such policies were implemented throughout Pinochet's regime (1973–90) although the more extreme forms of free market policies were reduced following recession at the start of the 1980s. The ability to implement the **neoliberal** reforms and control dissent was strongly linked to the authoritarian nature of the political regime, extreme oppression and state violence.

(Sources: Adapted from Harvey, 2007; Huneeus, 2000; Silva, 1991)

will, however, highlight the spatialities of **neoliberalization** in the Global South and will mention some ways in which these processes are being resisted (see Smith, *et al.*, 2008), although this theme will be examined in more detail in Chapter 11.

Debt crisis

Neoliberal policies had begun to be implemented in Chile and parts of the Global North in the 1970s, but it was in the 1980s that processes of **neoliberalization** became more prominent in most parts of the Global South (and later in the transitional economies of Eastern Europe and former Soviet Union). The catalyst for these processes was the '**debt crisis**' which unfolded following the Mexican government's announcement in 1982 that it would not be able to meet its public debt repayments, although the crisis had a much longer history (Corbridge, 1993b; Griffith-Jones, 1988; Nafzinger, 1993).

During the 1960s and early 1970s, many Southern governments, keen to finance '**modernization**' projects such as infrastructure expansion and industrial development, borrowed money from commercial banks (see Chapter 9). Given the relatively low levels of interest rates and the promising economic growth prospects, this appeared to be a reasonable course of action for most states. Banks, meanwhile, had significant funds to lend, particularly from oil-producing nations, which resulted in millions of '**Petrodollars**' (see Chapter 4) being deposited in banks and available for lending. This circulation seemed, at least for a short time, to be a strategy with few disadvantages.

In the 1970s, severe oil price rises had very negative impacts on the economies of non-oil-producing nations. As a key resource for economic development, oil imports were a drain on foreign currency reserves and increases in oil prices had knock-on effects on prices throughout the economy, for example in transport, power generation and food. These oil price rises, combined with rises in interest rates and recession in the global economy, brought many indebted countries to the brink of economic collapse. Given the levels of **debt servicing** compared to export earnings (see Table 10.1), it was only a matter of time before payment defaults resulted.

Structural Adjustment Policies

In the context of high levels of national debt and an inability to meet debt repayments, national governments in the Global South were forced to turn to the IMF and World Bank (see Chapter 4) for assistance. Private banks would no longer lend to indebted governments, but governments still had massive expenditure commitments. Further funding from the IMF and World Bank was conditional on governments adopting what were termed '**Structural Adjustment Policies or Programmes**' (SAPs). The fundamental principles of **SAPs** were modelled on **neoliberal** ideas of reducing the role of the state and increasing the influence of the market. With very few exceptions, by the mid-1980s, most countries in the Global South had no alternative but to adopt **SAPs** (Stewart, 1995).

Table 10.1 External debt, 1982–85

Country	Total external debt US$bn	Total debt servicing as % of export earnings from goods & services
Debt servicing>50% export earnings		
Argentina	51	83.0
Mexico	91	52.1
Debt servicing>40–50% export earnings		
Bolivia	5	49.5
Chile	20	48.6
Côte d'Ivoire	10	44.7
Uganda	1	43.1
Madagascar	2	42.8
Costa Rica	4	41.0
Jamaica	4	40.5
Debt servicing>30–40% export earnings		
Malawi	1	38.9
Nigeria	19	38.7
Brazil	106	38.6
Niger	1	33.7
Morocco	16	33.0
Ecuador	9	33.0
Philippines	27	32.5
Mozambique	3	31.6
Debt servicing>20–30% export earnings		
Togo	1	27.3
Zaire (now Democratic Republic of Congo)	6	25.7
Mauritania	2	25.3
Venezuela	35	25.0
Honduras	3	24.9
Senegal	3	22.2

Source: Adapted from Corbridge, 1993.

SAPs forced governments to implement policies to reduce expenditure, such as through reducing subsidies, privatization and the size of the public sector workforce (see Chapter 7), as well as increasing income through improved tax systems and opening the economy to foreign investment through the removal or lowering of tariff barriers and devaluing the currency (Mohan, *et al.*, 2000). Such policies have helped expand and deepen forms of economic **globalization** and the incorporation of countries in the Global South into the global economy (see Chapter 4).

While **SAPs** certainly allowed governments access to vital resources and economic stabilization usually ensued eventually, **SAPs** were frequently associated with increasing inequalities both socially and spatially. For the poorest people, cuts in basic food subsidies and the implementation of 'cost recovery' programmes within the health system (see below), meant that the daily struggles to meet basic needs were made even more difficult (Konadu-Agyemang, 2000; Stewart, 1995) (see Box 10.4). For some, migration for work, either nationally or internationally, was the solution (see Chapter 4 on **remittances**), while for millions of others, increasing workloads and reduced household expenditures were the strategies adopted. These actions clearly had an implication on the standard and quality of life for the majority of people in the Global South and led to calls for 'adjustment with a human face' (Cornia, *et al.*, 1987). **SAPs** were also criticized for being '**gender**-blind' in that they implicitly relied on increases in women's productive and reproductive labour (Elson, 1991). There were also concerns that stop-gap measures would have longer term implications on both household standards of living and national development, for example, through children's non-attendance at school, or levels of ill health (Moser, 1992).

It was not just the economically poor who were negatively affected by **SAPs**; for many middle-class households, cutbacks in the public sector workforce hit them very hard (see Chapter 7). Entrepreneurs who had previously been protected from foreign competition by tariff barriers or quotas found themselves facing competition from foreign imports, leading to many domestic industries closing down, particularly in sectors such as textiles.

SAPs were a one-size-fits-all approach to market-led development. While it could be argued that governments in the Global South had the choices of whether to accept them or not, in reality, there was no choice to be made as there were no feasible alternatives. As Richard Peet (2007) argues, **SAPs** represent a reinforcement of existing global **power** relations, with the Global North again reasserting its **power** over countries and peoples of the Global South. A recognition of the negative aspects of **SAPs**, particularly in social terms and in their lack of sensitivity to different national situations, has been addressed, at least rhetorically, in **Poverty Reduction Strategy Papers** (PRSPs) which are discussed below.

World Trade Organization

As outlined earlier in this chapter, the operation of a free market system relies on appropriate state regulation, enforcement of the rule of law and respect for individual property rights. Within individual countries, this can be exercised through legislation and regulatory bodies (as well as more repressive measures as were seen in the Chilean case, for example). As there is no global government, for

BOX 10.4

Social effects of SAPs in Peru and Zimbabwe

Both Peru and Zimbabwe implemented stabilization and adjustment policies in 1990, but they were implemented in very different contexts. In Peru, the policies were introduced by the new president Alberto Fujimori who took office in July 1990 at a time when inflation rates were about 3,000 per cent, there were high levels of unemployment/underemployment and **poverty** levels in metropolitan Lima had increased from 16.9 per cent in 1985/6 to 44.3 per cent in 1990. The so-called 'Fuji-Shock' was implemented in August 1990 and an IMF-approved SAP was agreed in 1991.

In contrast, the Zimbabwean government agreed a SAP with the IMF in 1990, not in circumstances of extreme crisis but in an attempt to provide an impetus to economic growth. Continuing conflict with neighbouring Mozambique was using up significant levels of government expenditure and the previous focus on small-scale agriculture and tariff barriers to protect domestic industry had resulted in low levels of economic growth and industrialization.

Despite these differences, the social impacts of the SAPs in both Peru and Zimbabwe were similar. In both cases, the government attempted to provide safety nets and support to protect the poorest households, but they had little effect. In Lima, the number of people living in **poverty** (according to a 'basket of goods' **poverty** line) rose from 7–8 million in August 1990 to 12 million in December 1990. This reflected the ending of food subsidies, relaxation of price controls and public sector job cuts, among other processes. Expenditure on health care among poorest families in Lima fell by over 85 per cent during this same 6-month period. Women were particularly badly affected because of the need to earn an income, find food for their families and also provide care for family members who would previously have sought medical attention.

In Zimbabwe, similar trends were apparent. Real wages fell as legal minimum wage regulations were liberalized. Food subsidies on maize meal, bread and sugar were removed causing the price of bread to increase by 40 per cent and that of sugar to rise by 50 per cent. The price of maize meal remained stable however; the deregulation of domestic maize meal production was part of the SAP and had the effect of increasing the number of small maize-meal providers, thus keeping the price stable. The introduction

---

BOX 10.4 (*CONTINUED*)

of user fees in the health sector, is also thought to have resulted in reductions in the number of women giving birth in medical facilities and rises in maternal mortality from less than 80 deaths per 100,000 births in 1990 to over 110 in 1993.

The negative impacts of SAPs on large sectors of the population are apparent. However, it is important to consider the pre-existing conditions (e.g. the widespread economic problems in Peru) and coexisting factors (e.g. drought in Zimbabwe in 1991–92 and the rise in HIV/AIDS infections) when assessing the role of SAPs.

(Sources: Tanski, 1994; Marquette, 1997)

---

a global free market, a different form of regulatory body is required. Since 1995 this role has been played by the World Trade Organization (WTO) which as of June 2008 had 152 members (WTO, 2008).

Prior to the setting up of the WTO, the General Agreement on Tariffs and Trade (GATT) was responsible for promoting **free trade** between nations. It was set up as one of the **Bretton Woods** organizations (see Chapter 4) and operated through eight rounds of negotiations which involved member countries discussing and agreeing rules for international trade. This included discussions on which sectors of the economy should be covered by these rules.

Whereas GATT did not have any **power** to enforce these rules, the WTO does. Members of the WTO agree to rules such as not providing preferential treatment to goods or services from particular countries (unless there is a formal free trade agreement). If there are disputes between WTO members, the WTO will act as a neutral arbitrator using the agreed rules. While the organization promotes liberalized trade, there are situations where restrictions on trade are allowed under WTO rules, such as situations of consumer protection or disease control (WTO, 2008). However, overall, the WTO is committed to freeing trade from state controls as summarized in the following statement:

> [T]he temptation to ward off the challenge of competitive imports is always present. And richer governments are more likely to yield to the siren call of **protectionism**, for short term political gain – through subsidies, complicated red tape, and hiding behind legitimate policy objectives such as environmental preservation or consumer protection as an excuse to protect producers. Protection ultimately leads to bloated, inefficient producers supplying consumers with outdated, unattractive products. In the end, factories close and jobs are lost despite the protection and subsidies. If other governments around the world pursue the same policies, markets contract and world economic activity

is reduced. One of the objectives that governments bring to WTO negotiations is to prevent such a self-defeating and destructive drift into **protectionism**.

(WTO, 2008)

While the aim to promote economic growth and social well-being is certainly laudable and these policies can contribute to **poverty** alleviation in the Global South (see below), the WTO has been criticized on two main grounds: first, for undermining fragile domestic economies with demands for liberalization of trade (Payne, 2006) and second, for acting in the interests of the countries of the Global North, rather than as a neutral actor (Peet, 2007). Protests against the WTO throughout the world (for example, in Seattle in 1999) demonstrate the depth of feeling against certain WTO policies and the perceived lack of transparency in its activities.

FAIR TRADE, ETHICAL CONSUMERISM AND CORPORATE SOCIAL RESPONSIBILITY

Trade as a route to improving the living conditions of individuals and communities, particularly in the Global South, has been taken onboard not just by international organizations adopting a **neoliberal** agenda, but also by social justice movements. These range from international **NGOs** such as Oxfam, to small, locally based groups. In particular, these organizations have campaigned for more equitable access to markets in the Global North and against what they see as unfair subsidies for Northern farmers and other producers (Oxfam, 2007). They argue that by being able to sell their goods overseas, producers in the Global South would be able to generate income and foreign exchange earnings that would benefit both themselves and their countries more broadly. Such campaigns involve lobbying the WTO to enforce its rules regarding fair market access (see Chapter 2 on cooperation within the South at WTO meetings).

However, wider liberalization of trade and better access to markets would not necessarily lead to better living conditions for producers. This is because of the way in which trading systems are structured, with small producers of primary commodities receiving only a very small portion of the price paid for the finished good. The concept of 'fair trade' has been developed to describe attempts to address this problem. Through a system of accreditation and labelling, started in The Netherlands in 1989 (FLO, 2007), increasing numbers of products are being produced and sold as '**fair trade**' goods. The **fair trade** system guarantees the producers a fair price for their goods, regardless of the market value. There is also an added contribution to be spent on community activities, such as health or education facilities (Blowfield, 1999; Fairtrade Foundation, 2007). Most of the accreditation for '**fair trade**' status is done by European-based organizations travelling to visit producers in the Global South. However, there is a Southern-based accreditation organization in Mexico called *Comercio Justo* (Comercio Justo México, 2008).

Fair trade products, particularly coffee and bananas, have become an increasingly visible part of the market for some goods in some Northern countries (see Table 10.2), but they still represent a very small share of the overall market. Their contribution to social well-being in the communities involved has often been significant. The Kuapa Kokoo Cooperative in Ghana (see Plate 10.2) is one of the most well-known producer organizations within the fair trade movement. It was set up in 1992 by a group of cocoa farmers who were seeking to improve the prices they received for cocoa. Under Ghana's **SAPs** (Konadu-Agyemang, 2000) the state-run Cocoa Marketing Board was abolished and farmers were left to sell their own cocoa. The marketing board had purchased farmers' cocoa at a set price which provided security, but there had also been problems including allegations of corruption. Two international NGOs provided support to the Kuapa Kokoo Cooperative in terms of loans, financial advice and training, for example in bookkeeping. They also helped with access to fair trading opportunities in the Global North. In 1998 the Day Chocolate Company was set up with representation from the Cooperative, thus making a direct link between the producers and the consumers of chocolate. The Farmers' Trust set up by the Cooperative and run by elected farmers makes decisions about how to spend the social premium element of the fair trade price (Tiffen, 2002). This has enabled the communities involved to invest in projects which will have long-term benefits for residents (Purvis, 2003).

As outlined earlier, the consumption of **fair trade** goods has increased greatly in the UK (and in other parts of the Global North) in the past decade or so as consumers have used their purchasing power to contribute to a better quality of life for producers. However, some commentators (see e.g. Lindsey, 2004; Sidwell, 2008) suggest that by interfering in the workings of the free market, **fair trade**

Table 10.2 Growth in UK certified Fairtrade sales, 1998–2007

	Annual retail sales, £ million				
	1998	**2000**	**2002**	**2005**	**2007**
Coffee	13.7	15.5	23.1	65.8	117.0
Tea	2.0	5.1	7.2	16.6	30.0
Chocolate/cocoa	1.0	3.6	7.0	21.9	34.0
Honey products	n/a	0.9	4.9	3.5	5.0
Bananas	n/a	7.8	17.3	47.7	150.0
Flowers	n/a	n/a	n/a	5.7	24.0
Wine	n/a	n/a	n/a	3.3	8.2
Cotton	n/a	n/a	n/a	0.2	34.8
Other	n/a	n/a	7.2	30.3	90.0
TOTAL	**16.7**	**32.9**	**92.3**	**195.0**	**493.0**

Source: Adapted from Fairtrade Foundation, 2008a.

Plate 10.2 Kuapa Kokoo Cooperative, Ghana.
Credit: Kim Naylor/Divine Chocolate.

prevents more positive, long-term developments. For Mark Sidwell, **fair trade** is in fact 'unfair trade' (see Box 10.5). The debate about Fairtrade coffee highlights a key point of disagreement between free trade supporters and opponents; free trade arguments are premised on the operation of a perfect free market, but opponents make the case that the global economic system is not set up as a perfect free market, meaning that particular countries or corporations have the **power** to work the system to their own advantage.

The role of consumers in using their purchasing power to support companies whose working practices they approve of, or alternatively boycotting companies with undesirable practices, has become part of what can be termed 'ethical consumerism' (Blowfield, 1999) or 'ethical **consumption**'. Such decisions may be based on prices paid to producers, as with '**fair trade**', but can also relate to environmental activities, working conditions (for example in **MNC** factories) and involvement in broader social programmes. In most cases, although not all, the consumers are based in the Global North and the producers and workers in the Global South. Such market-based interventions by consumers have led to shifts in the ways in which many companies, most notably **MNCs**, claim to operate. Such companies have increasingly developed **corporate social responsibility** (CSR) strategies (see Concept Box 10.2) to inform the consumer of their operating practices (Conroy, 2007).

Using consumer power to promote social justice and **sustainable development** in the Global South has become an increasingly important and sophisticated

BOX 10.5

Debates around fair trade coffee

Coffee was the first product which was widely sold in the UK as a certified Fairtrade product and, in 2007, it was second only to bananas in the value of Fairtrade retail sales (see Table 10.2). The focus on coffee came about in a context of plummeting global coffee prices in the aftermath of the collapse of the International Coffee Agreement in 1989 which had set coffee prices through negotiations between main producing and consuming countries. For the Fairtrade Foundation and consumers who purchase Fairtrade-certified coffee, the **fair trade** system helps support producers who are suffering from the low free market prices.

For **free market** proponents, such as Mark Sidwell (2008) and Brink Lindsey (2004), **fair trade** does not help coffee producers overall and other strategies should be adopted to support small-scale coffee farmers and labourers. The first main argument they make is that by interfering in the operation of the free market by guaranteeing a set price, **fair trade** organizations are trapping coffee producers in low-income situations, rather than encouraging them to diversify into other crops or activities. Sidwell (2008: 13) states: 'The Fairtrade model assumes that poor farmers must always remain farmers, and it seeks to subsidize their agrarian niche, denying them the possibility of dreams of a better life.' In relation to the demand and supply curves in Figure 10.1, the price paid for **fair trade** coffee is not determined by where the curves crossed, but by the **fair trade** organizations. This means that producers will stay as coffee farmers when they could be better off doing something else.

Sidwell also argues that the certification requirements for Fairtrade recognition means that only farmers who own their land can benefit; labourers who work on the coffee farms are not able to participate even though they are often poorer. Similarly, strict certification requirements make it harder for producers in the very poorest countries, such as Ethiopia and Rwanda, to participate; there are far more certified producers in Mexico and Colombia, for example. The Fairtrade system therefore benefits poor farmers, but not necessarily the very poorest agricultural workers in the world. The solution for Sidwell and Lindsey is for coffee to be sold at free market prices. This may require some farmers to move out of coffee production, but appropriate support in the form of training could be provided. Alternatively, farmers could

BOX 10.5 (*CONTINUED*)

BOX 10.5 (*CONTINUED*)

seek to improve the quality of their produce as this would command a higher price on the world market.

The Fairtrade Foundation (2008b) counters such arguments with reference to the lack of a level playing field in international coffee markets. It is not a simple case of coffee producers selling directly to the people who consume the coffee, rather there is a complex **commodity chain** from fields to coffee shop or supermarket. They also argue that many producers have used the money they receive from selling Fairtrade coffee to diversify into new crops such as citrus and bananas in Guatemala, or processing facilities so that they can get involved in more value-added processes. Finally, the Fairtrade Foundation agrees that small farmers would benefit from greater training and technical support.

(Sources: Adapted from Fairtrade Foundation, 2008b; Lindsey, 2004; Sidwell, 2008)

form of action (see e.g. the RED campaign to raise funds for the Global Fund for AIDS, Tuberculosis and Malaria). Consumer **power** is also apparent in the Global South. For example, Plate 10.3 shows a poster campaigning against Coca-Cola in Colombia due to its alleged involvement in the murder of union activists in its Colombian bottling plants (Killercoke, 2008).

POVERTY REDUCTION

Since the end of the twentieth century, international development efforts (as represented by the United Nations, World Bank and most national governments in both Global North and Global South) have focused on **poverty** reduction. The **Millennium Development Goals** (MDGs) (see Table 5.2) have set **poverty** reduction targets by 2015, using the **poverty** line of less than US$1 per day as the measure. Such an economic-based conception of **poverty** clearly fails to recognize the multidimensional nature of **poverty**, which also includes poor health, low levels of education, political marginalization and poor living conditions, among other factors (Narayan, 2000); but the assumption is that these other factors are outcomes of economic **poverty** (see Chapter 2).

To achieve the **poverty** reduction targets, the broad policy framework which the international financial institutions and Northern donor governments have adopted is a **neoliberal** one. Such an approach has been adopted by most Southern governments, although there are some exceptions, including the policies

---

CONCEPT BOX 10.2

Corporate social responsibility

The concept of **corporate social responsibility** (CSR) has received increasing attention since the mid-1990s as companies (particularly large **MNCs** and **TNCs**) have responded to consumer pressure regarding the ethics of their sourcing, labour or environmental practices. High profile campaigns against companies such as Nike, Gap, Shell and Nestlé, for example, caused considerable concern in the corporate world as brand image is a very valuable commodity. At its heart, **corporate social responsibility** is based on the premise that corporations should not just focus on profit maximization, but that the generation of profit should not come at the cost of environmental destruction or poor working conditions, for example. Most large companies now include statements in their annual reports about their socially responsible practices and many have signed up to voluntary codes of conduct.

(Source: Adapted from Dicken, 2007)

Plate 10.3 Anti-Coca-Cola campaign poster, Bogotá, Colombia.
Credit: © Still Pictures.

Table 10.3 Main ways to ensure globalization helps the poor as identified by DFID

Main strategies	
Promote effective government and efficient markets	Combat corruption
	Respect for human rights
	Resolution of conflict
	Enforcement of the rule of law
	State regulation of service provision and market operation
Investing in people, sharing skills and knowledge	Improve access to health and education services for the poor
	Ensure poor people have access to new technologies e.g. the Internet
	Promote pro-poor research
	Protect intellectual property
Harnessing private finance	Promote corporate social responsibility
	Strengthen international financial system
	Encourage private investment in Global South
	Provide access to finance for poor
	Simplify tax regimes
Capturing gains from trade	Create fairer international trading system and reform WTO
	Reduce trade barriers
	Invest in transport infrastructure to help poor people reach markets
Tackling global environmental problems	Integrate environmental sustainability into development planning
	International cooperation for environmental protection
	Improve private sector use of resources
Using development assistance more effectively	Focus development assistance on poverty reduction
	Debt relief for poor countries
	Nationally owned poverty reduction strategies
	Promote local procurement in development assistance
Strengthening the international system	Reform international institutions
	Mobilize civil society
	Give poorer countries a greater voice

Source: Adapted from Department for International Development (DFID), 2000a.

of the Chávez government in Venezuela (see Box 2.5) and that of Evo Morales in Bolivia. As summarized in the WTO quotation above, opening national economies to foreign investment, reducing the role of the state in the national economy, and promoting individual entrepreneurship and innovation is viewed as the way to generate wealth which will benefit all sectors of society. In the DFID (2000) White Paper on *Making Globalisation Work for the Poor* (see Table 10.3), the role of the private sector is stressed, as is the need for greater openness in world trade. DFID does recognize the current inequalities in bargaining **power** for the poorer countries and argues for reform of global institutions, but the overall message is that **poverty** reduction will be achieved through market-led mechanisms.

Opening up the economy for foreign investment and competition has had some benefits, both in terms of job creation in **MNCs** (see Chapter 4) and service provision (see below). However, the gap between rich and poor has risen greatly and conditions for the poorest have often worsened (World Commission on the Social Dimension of Poverty, 2004).

The focus on **poverty** reduction is also seen in the conditions laid down for further borrowing from the World Bank by governments of the Global South. Governments wanting access to such funding are required to produce **Poverty Reduction Strategy Papers (PRSPs)** outlining the ways in which **poverty** alleviation measures will be implemented (Craig and Porter, 2003). The World Bank argues that compared with the earlier **SAP** agreements, **PRSPs** are unique to the country involved and must include consultation with appropriate **stakeholders** (such as NGOs and community groups: see Box 11.7) throughout the country as part of the '**good governance**' agenda (see World Bank, 2008; and Chapter 9). However, these **PRSPs** are still expected to fall within a **neoliberal** framework (see Box 10.6). Despite claims that **PRSPs** must be 'nationally-owned' (to use DFID's wording – see Table 10.3), this national ownership can only operate within certain externally imposed constraints.

BASIC SERVICE PROVISION

According to **neoliberal** ideas, market principles should be adopted not only in the more conventional sale of goods and services, but also in the provision of what are termed 'basic services' such as infrastructure, health and education. It is argued that this will allow for a more efficient and equitable provision of such services. These principles have been put into operation in two main ways in both the Global North and Global South.

First, the principles of 'cost recovery' have been promoted, particularly in the health sector. This means that rather than having free health care provided by the state, patients are expected to pay for their treatment and medicines. In most countries where such policies have been adopted, there are a few basic services which are free, such as some childhood inoculations, but patients are increasingly

BOX 10.6

Mali's Poverty Reduction Strategy Paper

Mali is one of the economically poorest countries in the world (GDP p.c. of US$1,033 PPP in 2005) and in 2007 was ranked 173 out of 177 countries in the UNDP Human Development Index ranking. It is therefore a perfect candidate for inclusion in the PRSP process.

The Government of Mali has produced two PRSPs. The first, in 2002, covered the period 2002–06, while the second PRSP was published in 2006 and runs from 2007–11. In both cases, **stakeholders** were invited to participate in the development of the document, but such a process has significant limitations due to time, expense and access, and who is invited to participate in the consultation events (see also Box 9.6 on Uganda and Box 11.7 on Nicaragua).

The aims and policies outlined in both of Mali's PRSPs resonate with **neoliberal** ideas about efficient and decentralized government, reduced bureaucracy, greater role for the private sector, expansion of foreign investment and trade, enhanced roles for **civil society** actors and **good governance** (see Chapter 9). Barriers to the successful expansion of private sector involvement in the Malian economy are identified as corruption, lack of infrastructure, poor regulatory frameworks, low levels of education and limited access to finance, especially for small and medium enterprises (SMEs). Policies for the second PRSP period therefore address these concerns, focusing, for example, on putting the running of Mali's airports and railways out to private sector tender and implementing stronger anti-corruption measures. There will also be attempts to encourage private sector provision of financing for SMEs and investment in health and education services through the decentralized service provision framework.

The Malian PRSPs are clearly written by Malians with their country at the centre; for example, there are discussions of the problems around drought and locust invasions, as well as the impacts of the conflict in Côte d'Ivoire which were highly detrimental to landlocked Mali as over 70 per cent of its trade travels through that country and it is also Mali's major trading partner in the region. There was also mention of the efforts the Malian Government has made to lobby the WTO regarding the cotton subsidies which US farmers receive and which are viewed as undermining Mali's cotton production. Despite these context-specific examples, most of the policy plans outlined in

BOX 10.6 (*CONTINUED*)

> **BOX 10.6 (*CONTINUED*)**
>
> the Mali PRSP could have been written for many other Sub-Saharan nations. This is not to deny that many of the policies could have beneficial effects, but that the framework within which such documents are developed is highly prescribed by the World Bank and the IMF.
>
> (Sources: Adapted from Government of Mali, 2002; Ministry of the Economy and Finance, 2006; UNDP, 2007)

having to pay for medical attention. As seen earlier in relation to **SAPs**, this can have very detrimental effects on large sections of the population, who are unable to afford medical treatment, or who use their scarce funds for health care, leaving nothing for other expenditures including food and education (Lee, *et al.*, 2002).

The arguments behind cost recovery in the health sector include a recognition of the limits on government funds. By charging patients, the costs can be recovered and then invested in other parts of the health system. In addition, charging for health care may also encourage private sector involvement, either directly as medical care providers or as providers of health insurance (Sen, 2003). However, to participate in the market as a consumer, you need to have money; for the very poorest, lack of finance may lead to exclusion from formal health care. Similarly, private sector providers are not going to enter into a market if the prices they receive for their service are too low. This has meant that while private health providers are booming for middle-class households in some parts of the world, the very poorest are being excluded (see PAHO, 2002, for Latin America).

Second, governments are increasingly being urged to act as 'enablers' rather than 'providers' in services such as water, electricity, telecommunications and housing (Jones and Ward, 1994). Again, the arguments in favour of such policies are that the state has insufficient money or capacity to provide such services to the whole population and that the private sector would be able to provide them more efficiently. The adoption of such policies has led to massive programmes of privatization and a resulting wave of foreign investment.

For some sectors, particularly telecommunications, privatization and foreign investment have resulted in great successes, with a massive expansion of coverage, more reliable services and cheaper calls. In other sectors, such as water and electricity networks, while networks have usually been extended, the cost of accessing these networks has led to the exclusion of the poorest sectors of the population (see Box 10.7). In many parts of the world, most notably in Bolivia and Ghana, protests and street demonstrations against possible water privatization have been widespread. These protesters argue that water is not a commodity, but rather a basic right, and analysts of such protests have viewed water privatization as an example of 'neoliberalizing nature', that is, turning what was previously a 'free' good available to the population into a commodity that can be bought and sold on the market (Bakker, 2007; McCarthy and Prudham, 2004).

As with the health system, some governments have provided a safety net of basic provision, as in the case of access to water for poor households in South Africa (Box 10.7). However, this safety net is only provided for households with a permanent address, so some families lose out. Where such provision is inadequate, it has often been left to individuals and communities to provide support for themselves, or for NGOs to step in as service providers (see Chapter 11). The introduction of free market principles into water provision in the Global South has not led to widespread water access for the poor. However, as the Bolivian and South African examples demonstrate, public ownership of water companies also faces challenges, particularly around financing.

BOX 10.7

'Pro-poor' water policies in Bolivia and South Africa

In both Bolivia and South Africa, governments have attempted to increase access to drinking water in urban areas, focusing on incorporating the very poorest into the water network, in what are termed 'pro-poor policies'. These policies have all had **neoliberal** aspects, but vary from the award of concessions to private companies to reduced cross-subsidization between sectors in publicly run services.

In 1996, a 30-year concession to run the water system in La Paz-El Alto, Bolivia, was awarded by the Bolivian Government to the French-led consortium Aguas del Illimani. As part of the contract, there were agreements about the extension of the water network and provision for the poor. A regulatory structure was set up by the Government to ensure that the company was abiding by the agreements. This was part of the Government's role in the enabling process, compared to its previous water provision role through the nationalized water company.

The La Paz-El Alto project received very public support from the World Bank as a shining example of pro-poor water provision, but Nina Laurie and Carlos Crespo (2007) reveal that while Aguas del Illimani had put processes in place in favour of the poor, water access for these groups was often not

BOX 10.7 (*CONTINUED*)

possible. The company achieved its targets of expanding the water network, but the majority of these new connections were in areas with an existing water service, rather than the unserviced areas on the margins of the city. The pricing structure was organized so customers paid only for the water they used, but this required the installation of water meters which were not provided in many cases. Prices were also elevated due to taxes and the fact that tariffs were calculated against the US dollar. In a survey conducted by Laurie and Crespo (2007), about two-thirds of households said that they could not afford to access water through the network.

Growing public dissatisfaction at the situation led to public demonstrations. These followed the 'water wars' in Cochabamba, where the company Aguas del Tunari withdrew from its 40-year concession following violent protests about high water tariffs in 2000. In both cases water provision was returned to the public sector. However, access to water remains problematic with unreliable services and little investment in the infrastructure.

In South Africa, cities have adopted varying strategies to expand water provision in the post-Apartheid period. In Johannesburg, for example, a public-private partnership was set up to improve the management of the water supply company. The partnership included Suez Lyonnaise des Eaux – one of the partners in the Aguas del Illimani consortium in Bolivia. This partnership achieved efficiency improvements in the management and provision of water, but the company remained in public hands.

In Cape Town, water is still provided by the public water company but, again, **neoliberal** approaches to efficiency and cost recovery have become more common. Water tariffs are banded according to household income; this means that higher-income households pay more than poorer ones. In addition, very poor households (as identified by national criteria) are provided with 6kl water free per month. There are also attempts to expand the network to encompass about 30,000 households, largely in the peripheral settlements, which currently lack access to piped water. While these policies are clearly targeted at supporting the poor, there are concerns about their medium-term sustainability, due to the drain on the municipal budget. There is a lack of public funds for maintenance and in 2004 water tariffs were raised to generate money. Sylvia Jaglin (2008) discusses the challenges faced by the Cape Town Government as it seeks to introduce more **neoliberal** policies to try to improve efficiency and reduce the use of cross-subsidies within the system, while also trying to ensure that poor residents have access to clean water.

(Sources: Adapted from Jaglin, 2008; Laurie and Crespo, 2007; Smith, 2008

CONCLUSIONS

Since the 1980s, market-led approaches to development have become dominant. Proponents of such approaches have argued that these approaches are more efficient and equitable than state-led policies, as they provide technical and rational solutions to problems, rather than being caught up in politics. However, as this Chapter has argued, this very notion of 'technocratic solutions' and rationality is based on a particular political view and the introduction of **neoliberal** policies reflects the exercise of **power** at both international and national levels.

While **neoliberalism** is clearly the dominant **discourse** in international development circles and among governments in the Global South, processes of **neoliberalization** are spatially and temporally distinct. The cases of water provision in Bolivia and South Africa demonstrate how approaches seeking to improve efficiency and cost recovery, while protecting the very poorest, have been implemented in very different ways. The shift from **SAPs** to **PRSPs** also highlights how **neoliberal** policies originating from the IMF and World Bank have been adapted over time, albeit not that significantly.

Finally, as the focus of Part 4 is 'making a difference' it is important to recognize how market-led or **neoliberal** policies have had a positive effect on some groups of people and places. The creation of jobs and new economic opportunities have come about as part of increased market openness in some parts of the world. Greater choice of consumer products and better services can also be observed. However, reliance on market forces excludes those who do not have the money to participate in the market as consumers, or are excluded or marginalized from the labour market. In the next Chapter we consider how grassroots or community-based approaches may help support such groups, as well as possible alternative development models to the market-led or state-led systems discussed in Chapters 9 and 10.

 Review questions/activities

1   Read a PRSP (available from the World Bank website). How far do the policies fit with a neoliberal model of development?

2   What are the arguments for and against fair trade? Use the reports by Lindsey (2004) and Sidwell (2008) and material on the Fairtrade Foundation website as a starting point.

3   'Rising inequalities are inevitable when neoliberal policies are adopted.' Discuss.

SUGGESTED READINGS

Department for International Development (DFID) (2000a) *Making Globalisation Work for the Poor*, London: HMSO. (Available from: http://www.dfid.gov.uk/pubs)
An example of how international development policy has drawn on neoliberal ideas to promote poverty alleviation.

Harvey, D. (2007) *A Brief History of Neoliberalism*, Oxford: Oxford University Press.
A clear overview of neoliberalism and its spatially differentiated impacts.

Geoforum (2007) Themed issue on Pro-Poor Water? The Privatisation and Global Poverty Debate, *Geoforum* 38(5).
A multidisciplinary collection of articles discussing pro-poor water policies throughout the world.

WEBSITES

www.adamsmith.org The Adam Smith Institute
A UK-based free market think tank.

www.comerciojusto.com.mx Comercio Justo México
Mexico's Fairtrade labelling organization.

www.fairtrade.net Fairtrade Labelling Organizations International

www.fairtrade.org.uk The Fairtrade Foundation

www.globalenvision.org
Organization promoting the global free market system as a way of reducing poverty for the world's poorest people.

www.maketradefair.com Oxfam Make Trade Fair Campaign

www.joinred.com Red Campaign
Set up to raise funds through selling clothes, electronic equipment and other products for the Global Fund to Fight AIDS, Tuberculosis and Malaria.

www.worldbank.org/prsp/ World Bank Poverty Reduction Strategies site

www.wto.org World Trade Organization

11 Grassroots development

INTRODUCTION

In Chapters 9 and 10 we have looked at the contrasting plans for state- and market-led development, approaches that have dominated mainstream academic and policy debates about development for much of the postwar era. Because state-led strategies are often associated with left-wing political projects, and calls to free the market with the political right, it can sometimes seem that arguments for 'more state' or 'more market' exhaust the full range of development options in both academic and practical terms. This chapter, however, looks at a third important source of development ideas, that of autonomous development, or development in which communities, **civil society** and non-governmental organizations (NGOs) are key actors alongside (or in place of) states and markets. Although they have managed to become part of the 'mainstream' in recent decades, these ideas have deep historical roots, and raise important questions not only of how development can be achieved, but of what development actually is, and who gets to define it.

The Chapter is split into three parts. First we look at some of the different arguments for community-led and community-scale development that have emerged across the Global South. We then turn to the question of how development academics and practitioners have engaged with these ideas through research and action that aims to place the knowledge and capabilities of local communities at centre stage. Finally, we look at what happens when some of these 'alternative' development ideas have been taken in to the development mainstream, and when community-based and NGO activities are 'scaled up' to wider programmes of change. Community-led development has done much to enliven debates and inspire new forms of practice, but it does not offer easy or universal solutions. As with the state- or market-led approaches addressed earlier, there is a need to reflect critically on its limitations as well as its potential, and a number of questions are important here: Who gets to represent, or speak for, 'the community'? When and how can 'grassroots' views be accommodated within wider institutional systems? Finally, but importantly, when does an emphasis on grassroots empowerment allow states and powerful market actors to pass the risks and responsibilities of development on to communities? We return to these questions in the conclusions.

MORAL ECONOMIES AND HUMAN-SCALE DEVELOPMENT

Although market- and state-led development are frequently presented as political opposites, they have often shared rather disparaging attitudes towards local communities within the South, and the values that they represent. This was perhaps at its most extreme in **modernization** theory (Box 9.2), which explicitly aimed to transform traditional ways of life and labelled entire indigenous value systems as 'non-economic barriers to development' (Hoselitz, 1995 [1952]), but has been present in a range of other development approaches too. **Neoliberalism** stresses the dynamism created by linking the South's 'emergent markets' into a global economy (see Chapter 10), and calls for a **developmental state** stress the benefits of planned intervention in society. The strong implication is that without dramatic intervention whole areas of the South will be left 'behind' by the inefficiency of a subsistence economy, or the irrationality of traditional systems of **governance**. This way of labelling communities or areas as lagging or backward depends, of course, on a clear view of what progress actually is, and the political right and left have often shared ideas of the *goals* of development – such as increasing material wealth, well-being and opportunity – even when their approaches on how to get there have differed.

Many 'anti-development' academics (e.g. Sachs, 1992; Escobar, 1995; and see Chapter 2) argue that these negative perceptions of local communities have dominated mainstream development thinking, but ignore a much more positive side of local cultures and the value systems they contain. At a practical level, there are a range of indigenous cultural practices of sharing work and wealth that draw on close kinship and other ties in order to build up shared assets at a community level, and reduce people's exposure to risk in times of hardship. For example, in Kenya group work parties (usually single-sex) are organized by women or men at a community level to undertake important tasks – such as building a house or clearing an area of bush for farming – that would be difficult to undertake within a single household. These activities, called *harambee* in Swahili (literally 'let's all pull together'), are widespread throughout East Africa, and participation within them not only greatly helps the beneficiary, it also reinforces social bonds among all those doing this work. Similarly, traditional value systems often have ways of sharing wealth at a community level: under the Hindu *jajmani* system, all castes within a village provided goods and services for each other in return for a share of the harvest, and although this has disappeared with the monetization of the economy, other traditions of charitable support remain strong. Alms-giving within Muslim cultures, *zakat*, is seen as a religious duty, and provides socially sanctioned mechanisms by which the poorest households can ask for help from their richer neighbours.

As well as providing material support, these cultural practices are important in embodying alternative ideas about the value of community, or indeed of development itself. Rather than being viewed as barriers to the spread of **modernity**, dense community ties can be the basis of a localized **moral economy**: a set of social relations that places limits on the individual pursuit of profit in favour of a notion of good or just behaviour. More broadly, traditional value systems can provide diverse and local challenges to 'universal' definitions of what development should be, and here a simple life, living in harmony with neighbours, or in accordance with religious practices, may be valued at least as much as the focus on material uplift that has been stressed within 'mainstream' development approaches. Development itself can profoundly disrupt these value systems, as James C. Scott's *The Weapons of the Weak* (Scott, 1985) shows. Scott's study looks at the impact of **Green Revolution** (see Concept Box 7.2) agricultural technology on rural Malaysia in the late 1970s: whilst food security was improved for most people, the poorest rapidly found themselves being excluded from village society as pre-existing social relationships governing the use of labour and sharing of wealth were replaced with a more capitalist outlook.

A sense of the practical and moral worth of community, and the potential for it to be disrupted by social change (whether driven by the market or the state), has been important in promoting alternative development ideas that have explicitly aimed to challenge 'mainstream' approaches. For Mohandas K. Gandhi, the values, skills and knowledge inherent within rural society were to be the centre of both political and economic development (Box 11.1). In many ways, his ideas were about recovering or reinventing a moral economy based around an idealized Indian village, and as such they posed a direct challenge to those of his contemporaries (such as Nehru) who wished to modernize India through industrial development. Gandhi's attempt both to draw on Hindu moral values and to transform elements within them (such as caste-based hierarchy) had its internal contradictions and, as will be discussed further in the following section, there are important critical questions to be asked about any idea of development that assumes that communities are harmonious. Nevertheless, an important part of Gandhi's legacy has been to inspire a range of alternative development practitioners both in India and well beyond.

Another source of inspiration for alternatives to development as **modernization** has come from the work of Paulo Freire and the radical Latin American context of the 1960s in which it emerged (Box 11.2). If there is a sense within Gandhi that communities are unchanging and harmonious, within Freire's work they are to be created through self-awareness and struggle. Freire was active in the radical liberation theology movement within the Roman Catholic Church that was important in Latin America during the 1960s. This organized rural peasants and the urban poor into Christian Base communities, which helped to replace the

BOX 11.1

M. K. Gandhi, *swaraj* and *sarvodaya*

Mohandas Karamchand Gandhi (1869–1948) was an important leader of India's struggle for independence from 1920, and is globally recognized for leading non-violent **resistance** to British **colonialism** through mass civil disobedience. For Gandhi, however, self-rule (*swaraj*) was not merely about political independence, it was also about using principles of non-violence to spread an alternative model of socio-economic development, *sarvodaya* or 'the welfare of all'. *Sarvodaya* is based around the recognition of the dignity of all work, communal self-sufficiency and practical education. In the immediate run-up to Independence, India was faced with mass **poverty** and unemployment: Gandhi saw the solution to these problems not in the rapid expansion of heavy industry, but in a revival of the economic life of India's villages. Gandhi wanted villages to be democratic self-governing communities, and also to be reinstated as central to the economic life of the nation. He advocated the cooperative organization of cottage industries and agriculture at the local scale wherever possible; large-scale industry was to be kept to a minimum, serving only needs identified by village communities, and operated by the state. The wearing of *khadi* (home-spun) cloth and the spinning-wheel (which became the central emblem of India's flag) were important political symbols of **resistance** to British **colonialism**. For Gandhi, they also represented an alternative vision of village-based development.

Following Gandhi's assassination in 1948, these ideas were largely ignored by Prime Minister Jawaharlal Nehru, who placed India firmly on a path of centrally planned industrial growth. Although marginalized within government policy, however, Gandhi's ideas remained influential. The ideals of *sarvodaya* were put into practice by social activists who founded model village communities: these included one of Gandhi's closest followers, Acharya Vinoba Bhave, who launched a national *bhoodan* (land-gift) movement to support this work by promoting voluntary contributions of land to 'untouchable' Scheduled Caste groups. Beyond India, a vibrant *sarvodaya* movement sprang up in Sri Lanka, and Gandhi is still cited as an inspiration for community-focused and environmental development initiatives across the globe.

(Sources: Adapted from Bhatt, 1982; Kantowsky, 1985; Hardiman, 2003)

BOX 11.2

Paulo Freire and conscientization

Like Gandhi, the Brazilian Paulo Freire (1921–97) was born into a middle-class family, trained as a lawyer and dedicated his life to radical social reform. Freire saw education as having radical potential to transform society towards a more socially just and humane path, and his work has had an important influence on both educational theory and social activists, especially within the Global South. His own work was inspired by Marxist analyses of oppression, and also Christian ideas of self-sacrifice and redemption: as such it was closely related to liberation theology, a radical movement within the Roman Catholic Church that was also important in Latin America at the time when he was writing.

His most famous book, the *Pedagogy of the Oppressed* (1968), argues against a 'banking' model of education that sees students as passive recipients of 'deposits' of knowledge made by their teachers, and instead promotes education based around dialogue, respect and mutual learning between teacher and student. Furthermore, within Freire's model, the current injustices in society were to be important *subjects* of education in themselves: education should 'make oppression and its causes objects of reflection by the oppressed, and from that reflection will come their necessary engagement in the struggle for their liberation. And in the struggle this pedagogy will be made and remade' (Freire, 1968: 33, cited in Leeman, 2004). Here, the teacher is acting not only to transfer skills to the student, but also to facilitate their conscientization – awareness of the structures of oppression, and of action they can take to overcome these.

For Freire, these were not simply tenets of educational theory: they had to be linked directly to practice. Soon after completing his PhD on Brazil's education system in 1959, he put his ideas of consciousness-raising into practice through adult literacy in Brazil, teaching 300 sugar-cane workers to read in just 45 days. After being imprisoned and then exiled by Brazil's military government, he went on to work first in Chile, and then to have a significant impact on national literacy efforts in Guinea-Bissau. His ideas inspired others in education and beyond, for example, early work within the Bangladesh Rural

BOX 11.2 (*CONTINUED*)

BOX 11.2 (*CONTINUED*)

Advancement Committee (BRAC), now one of that country's largest developmental NGOs, drew on ideas of conscientization, working with poor farmers to uncover the rural **power** structures that stopped government benefits from reaching them (Chambers, 1983).

(Sources: Adapted from Chambers, 1983; Freire, 1972 [1968]; Leeman, 2004; Smith, 1997)

hierarchical relationships between priest and congregation with relationships of solidarity and mutual support. For poor people, these communities also provided a space where they could develop their own critical awareness of the different forms of oppression they faced on a daily basis, and so closer contact with the Catholic Church went hand in hand with increasing class consciousness.

Despite their many differences, both Gandhi's and Freire's work stress in different ways the oppression that exists within modern society, and see the potential that lies within the community for the empowerment of those oppressed. It is this sense of the community or the 'grassroots' as a space of possibility that has inspired a range of practical development interventions – and the NGO sector has been particularly important in bringing these to fruition. NGOs can be defined as 'self-governing, private, not-for-profit organizations that are geared to improving the quality of life for disadvantaged people' (Vakil, 1997: 2060). The term embraces a wide variety of different organizations, from the large-scale Northern development institutions such as CARE or Oxfam to the equally important range of smaller Southern NGOs based within individual countries. NGOs themselves have risen to become prominent actors within international development since the 1980s: the total number of *international* NGOs alone was estimated at 28,900 in 1993 (Hulme and Edwards, 1997, cited in Desai, 2002), and Southern NGOs are almost innumerable. For example, in Nepal alone the numbers of officially registered NGOs increased from 249 in 1990 to 12,388 in 2001, and up to treble this number may remain unregistered (Whitehand, 2003: 106–7). By the late 1980s, both state- and market-based development approaches had had a range of practical failures, leading some commentators in the 1980s to hope that a rapidly expanding NGO sector – possessing the organizational flexibility to experiment with new ideas and approaches – might provide a 'magic bullet' for development (see Edwards and Hulme, 1995). Although there is no automatic connection between NGOs and empowerment-based approaches to development, Northern and Southern NGOs alike often publicly express these values, and this in turn implies a greater degree of responsiveness to local communities and their knowledge.

RESPECTING COMMUNITIES AND LOCAL KNOWLEDGE

Implicit within the approach of many organizations and movements inspired by the 'alternative' development philosophies outlined previously is the idea that the knowledge and capabilities of communities themselves should be central to development practice. Throughout much of history, however, it has been the knowledge of professionals (economists, planners, scientists, administrators and others) that has dominated development practice, and there have been implicit assumptions that local capabilities need to be transformed by new technologies, forms of organization or training that are transferred from these experts to lay people. While skilled professionals clearly have very important roles to play in many aspects of life in the Global South, there can be dangers if this transfer of knowledge is all one-way: local understandings of development problems and their potential solutions are entirely replaced by the views of outsiders, a lot of context-specific knowledge is potentially lost and locally valued traditions can be demonized by experts as 'backward', inefficient or just plain wrong.

The control of natural resources is one important area where lay and expert knowledge have often been directly opposed, and attempts by government to enforce their vision of good environmental management has led to conflict in many instances across the Global South. Research conducted by geographers, anthropologists and others has provided evidence that local environmental knowledge deserves to be reassessed in a more positive light, as the example of forest conservation in Kissidougou shows (Box 11.3), but it can be hard to shift official attitudes even in the light of new evidence. Melissa Leach and Robin Mearns (1996b) explain why this is the case: official ideas of good practice in environmental management make up a body of 'received wisdom', a set of standardized and simplified ways of framing environmental policy, that is then held in place through a combination of inertia and the interests of powerful groups. Studies of Africa by early colonial scientists often drew on Europe's own ecological conditions – and heavily emphasized ideas of equilibrium and stability as a result – even though this was not always appropriate to African conditions. Several of these scientists, such as French botanist André Aubréville (1897–1982), went on to develop and lead colonial departments of forestry or the environment, so their ideas continued to inform the training of officials even after subsequent developments in ecology produced more complex analyses of non-equilibrium environments. The state's desire to manage its natural resources and rural populations gave these institutions the force of new (and often draconian) laws, and so the opportunities to counter 'received wisdom' were limited: it was far easier for the local populations to carry on their traditional practices surreptitiously rather than trying to explain their own environmental expertise to disinterested or hostile officials. Even today, forms of received wisdom are reproduced by international **aid** agencies and the media: these often work around established 'storylines' with a global reach such as combating desertification or preserving biodiversity, and local environmental knowledge

BOX 11.3

Misreading the forest-savannah mosaic

Seen from the air, the landscape of Kissidougou, Guinea, is a patchwork of savannah grassland interspersed with ribbons and islands of forest (Plate 11.1); it is also an environment that appears to have been repeatedly misread by officials to the detriment of the local population. Early European travellers to the area thought that forest islands were the remnants of earlier complete tree cover, and this understanding that Kissidougou was suffering from deforestation became central to colonial and post-independence administrators' approach to environmental management. The assumption was that local land-use practices, such as the shifting cultivation of rice, were to blame for tree loss and as a result local inhabitants were prohibited by law from activities which had been important for their farming and hunting: tree felling was heavily restricted, and fire-setting even carried the death penalty.

Research by James Fairhead and Melissa Leach (1996) has challenged this picture. A time series of historical records, air photos and satellite images suggests that while some forest islands have disappeared, others have sprung up in their place, and that the total amount of tree cover may be increasing. Not only this, but it is traditional land-use practices that are responsible for the forest regeneration: grazing cattle on grass nearest to the village provides a natural firebreak, gardening and the spreading of ash and manure improve the soil, controlled burning of grasses prevents devastating dry-season fires, and tree species are encouraged or actively planted as they are valued for a range of purposes. From the testimony of village elders, it appears that a range of settlements now surrounded by forest were founded in grassland – the trees have grown up with and because of the villages. Despite having a wealth of knowledge of techniques that have been important in maintaining

Plate 11.1 The forest-savannah mosaic, Kissidougou Province, Guinea.
Credit: Google images.

BOX 11.3 (*CONTINUED*)

tree cover, these lay perspectives were still being ignored by the forest
administrators in the 1990s. Their professional training and the flow of
international **aid** for environmental conservation – both based around
wider concerns about forest loss at a West African regional scale – support
a resistant body of 'received wisdom' about a landscape in danger of
desertification, however ill-matched this is to local environmental conditions
and indigenous stewardship of the land.

(Source: Adapted from Fairhead and Leach, 1996)

is often too complex, inaccessible or place-specific to be easily fitted in to their
agendas.

When the knowledge of local populations is marginalized, it is perhaps no
surprise that their own technological achievements are often ignored in favour of
expert-designed systems. From the 1960s, large quantities of **aid** money (from the
Ford and Rockefeller Foundations, the Food and Agriculture Organization (FAO),
United Nations Development Programme (UNDP) and others) were spent on the
development and dispersal of **Green Revolution** technology (see Concept Box 7.2),
with the Consultative Group for International Agricultural Research (CGIAR)
being set up in 1971 to coordinate work in 16 major international research centres.
Following the initial work of Dr Norman Borlaug in Mexico, scientifically cross-
bred high-yielding varieties (HYVs) of staple crops such as rice, wheat and maize
were developed that gave vastly enhanced yields in response to the addition of
nitrogenous fertilizer. These 'miracle seeds' did indeed dramatically improve
harvests, allowing a step-change in cereal production especially across Latin
American and Asia, but at a price: achieving higher output relied upon high-cost
inputs of chemical fertilizer, tightly controlled irrigation and (in the initial decades
of the revolution at least) widespread use of artificial pesticides (Conway, 1997).
The social and ecological impacts of the 'improved' technology were therefore
mixed: local farming systems and crop strains were displaced by standardized
monocropping of HYVs, and although richer farmers in particular profited from
technological change this wealth often did not trickle down to the poorest people in
the countryside, particularly agricultural labourers (Lipton and Longhurst, 1989).
The technology also exposed farmers to new forms of risk, both natural (such as
increased susceptibility to pest attack) and man-made (increased dependency on
chemical inputs – and hence vulnerability to increases in oil prices).

Later research, both within and outside the CGIAR institutes, has uncovered
a less dramatic but equally important history of community-based agricultural
innovation, stretching from improved methods of potato cultivation in the Andes

BOX 11.4

Indigenous and modern irrigation in Kenya's Kerio Valley

In Kenya's Kerio Valley, along the Marakwet Escarpment and the northern slopes of the Cherangani Plateau, the indigenous irrigation systems of the Marakwet and Pokot people have provided a key source of food security since pre-colonial times. Rainfall in the area is highly seasonal, and irrigation is vital to supporting agriculture during the April/May rains, allowing a range of crops (including cassava, bananas, maize and vegetables) to be grown in areas that would otherwise only support millet and livestock. Rock and brushwood dams divert water from fast-flowing rivers into irrigation furrows, which carry water to fields on the hill slopes and valley floor (Plate 11.2). The technology involved is relatively simple – irrigation is entirely by gravity, and the furrows themselves are earth channels, sometimes strengthened with rocks – but the operation of the system requires careful social organization to ensure that labour is provided for vital tasks of channel maintenance, and that water is shared between all users along furrows that may be as much as 14km long. Irrigation rights are divided up between all the villages that have land served by a furrow at public meetings. These allocate times at which different groups can take water, and ensure some degree of agreement among users, although this does not mean equal shares for all (women farmers have no direct right to water, and those who have contributed more to maintenance work may get more water). Compliance is ensured by the *kokwa* – a meeting of circumcised men – which may punish those caught stealing water with a fine of a goat (physical beatings also occur), and at a more day-to-day level by the fact that most farmers have a detailed knowledge of their neighbour's crops and water use.

Pokot and Marakwet furrows contrast strongly with the area's Wei-Wei irrigation project (Plate 11.3), with its modern sprinklers, and dependency on Italian aid-funding and a small core of professional and technical staff. Traditional irrigation does, however, face its own challenges to sustainability: as wider social change occurs in the Kerio Valley, the authority of the *kokwa* is diminishing, migration for work is becoming common among the younger population, and the links between furrow maintenance and gaining status and standing within the community are eroding. Despite its use of simple technology and resources from the local environment, furrow irrigation may be undermined as the community's skills, aspirations and organizational forms change.

(Sources: Adams, *et al.*, 1997; Adams and Watson, 2003)

BOX 11.4 (*CONTINUED*)

Plate 11.2 A traditionally-irrigated *shamba* (field), Kenya.
Credit: Glyn Williams.

Plate 11.3 Modern sprinkler irrigation, Wei-Wei Valley, Kenya.
Credit: Glyn Williams.

to community-managed systems of irrigation in South India and East Africa (Box 11.4). Inspired by these and other findings, some authors have championed **indigenous technical knowledge**, arguing that it should become an accepted part of agro-ecosystem research, providing low-impact 'alternative technology' solutions that can work in parallel with high-cost (and centralized) plant-breeding and genetic engineering programmes (Chambers, *et al.*, 1989; Conway and Barbier, 1990; Conway, 1997).

Drawing on these experiences, alternative development practitioners have explicitly tried to disrupt some of the 'expert bias' that has led to the marginalization of local knowledge and indigenous technologies. From the 1980s, the experiences of applied anthropologists, field researchers working on farming systems, Freire-inspired activist participatory researchers and others came together to create a new set of methods for accessing local knowledge. Collectively, these became known as RRA (Rapid Rural Appraisal) techniques, with the rural of the title reflecting both the contexts within which they were initially used as well as a commitment to include areas and people of the South marginalized by mainstream development (Chambers, 1983). These methods seemed to have significant advantages over their competitors: they appeared to offer far more speedy insights into rural life than the extended field studies of anthropologists, and a far more direct representation of people's own thoughts and ideas than was possible within the constraints of a formal questionnaire (Box 11.5). It is perhaps unsurprising therefore that they spread rapidly initially within the NGO community. In situations where official data were absent or misleading, they offered NGOs working with severe time and financial constraints the opportunity to gain practical information that could shape priorities for action, or even be a vital resource in itself (see also Box 11.6). Plate 11.4 shows a participatory research activity allowing group discussion around a ranking exercise in Sigor market, northwest Kenya.

Participatory techniques are often conducted with groups of respondents in settings deliberately designed to facilitate social exchange and mutual learning, and as such they not only aim at the rapid production of useful information, but also to blur the distinctions between 'expert' researchers and 'lay' participants. Listening to local opinions, co-experimentation with new techniques and giving people control over the knowledge that is shared, all require different attitudes and behaviour on the part of researchers, and these in turn are deliberately intended to distance participatory work from more 'extractive' modes of data collection. For the proponents of participatory techniques, these methods are not simply about making local knowledge visible or accessible to outsider 'experts', they are also about challenging the entire set of **power** relationships that marginalized that knowledge in the first place. They are intended to demonstrate publicly the value of participants' own understanding and analysis (and the limitations of that of outside experts) and by doing so to have two important effects: first, participatory working at a grassroots level can raise self-awareness of people's own capacities and foster new networks and alliances within and beyond the community; and second,

BOX 11.5

Participatory techniques in action

The range of participatory techniques used by development practitioners is vast and varied, and the names used to describe these approaches have also changed over time, from RRA (Rapid Rural Appraisal) to PRA (*Participatory* Rural Appraisal) and now PLA (Participatory *Learning and Action*). Regardless of the acronym, some common elements have underpinned participatory approaches from the outset: rapid, flexible, field-based work which involves the *active* involvement of research participants (Conway, 1988; Chambers, 1994). The journal *PLA Notes* (available via the International Institute for Environment and Development (IIED) website: www.iied.org) provides an important space in which users of these techniques share ideas and innovations, and reflect on using these methods in practice.

The techniques encourage participants to discuss, debate, draw, map, rank and compare important aspects of their lives or experiences – and are often (but not always) highly visual, as Figure 11.1 shows. This is a copy of an historical matrix, tracing the coping strategies of households in Senegal in times of crisis, produced by a group of villagers who placed stones within each box to show the relative importance of different actions. It demonstrates a number of important features of participatory methods.

BOX 11.5 (*CONTINUED*)

Crisis / Strategy	WWII 39/45	Locust invasion 1950	Fire in village 1967	Drought 1973	Rat invasion 1976	Locust invasion 1988
Eat Neow tree fruit	•••	••	••	••	••	•
Eat wild leaves	•••	•••	•	•	•	
Eat Manioc	•••	•••				
Eat Dugoor tree fruits	••					
Food Aid	•••	••	••	••	••	•
Cultivate and weave cotton	••	••				
Eat Millet Bran	•••	••				
Hunting	•••	••	••		•••	
Eat Cowpeas	•••	•••	••	••	••	••
Dig trenches against locusts	••	•••				
Trade Neow fruit for millet	•••	•••	••	••	•	
Sell chickens	•••	•••	•••	•••	•••	•••
Migration — Rural / Rural — Urban				•••	•••	••
International migration				•••	••••	•••••
Sell weak animals to buy food for strong			••	•••	••	••
Buy flour				•••		
Cut branches for animal feed				•••		
Eat own animals	•••			•••		

Figure 11.1 Coping matrix: village strategies in times of crisis.
Source: Adapted from Schoonmaker Freudenberger and Schoonmaker Freudenberger (1993: 32).

> **BOX 11.5 (*CONTINUED*)**
>
> The historical events and coping strategies that define the matrix were produced through discussion among the participants: they therefore include categories outsiders would have been unable to predict in advance. The matrix is also a product that is directly accessible to villagers: they can see the results of pooling their knowledge, and interpret these for themselves. At their best, participatory techniques also explicitly note the constraints or limitations of the knowledge produced: in this case the facilitators noted that the matrix was produced 'by a large group of men (with a few women on the periphery)' (Schoonmaker Freudenberger and Schoonmaker Freudenberger, 1993: 32), and so may not represent different **gender** perspectives equally.
>
> (Sources: Adapted from Chambers, 1994; Schoonmaker Freudenberger and
> Schoonmaker Freudenberger, 1993)

for outside professionals, it can challenge both their 'received wisdom', and also hierarchical ways of working. As Robert Chambers, one of its key proponents, claims:

> In an evolving paradigm of development there is a new high ground, a paradigm of people as people. On the new high ground, decentralization, democracy, diversity and dynamism combine. Multiple local and individual realities are recognized, accepted, enhanced and celebrated. Truth, trust, and diversity link. Baskets of choice replace packages of practices. Doubt, self-critical self-awareness and acknowledgement of error are valued. ... For the realities of lowers [lay people] to count more, and for the new high ground to prevail, it is uppers [development experts] who have to change.
>
> (Chambers, 1997: 188)

There is much to be admired in this approach, and the hope it holds out for a reformed development practice, but questions need to be raised about its expectations of community empowerment and institutional change. As has already been indicated within the discussion of indigenous irrigation (Box 11.4), and the operation of 'traditional authority' in South Africa (Box 6.3), community empowerment can also have its darker side. Traditional moral economies are often based around severe **gender** and other social inequalities. At their best, participatory methods can help to uncover these, but this can place participatory development activists in a difficult position. Should they be working hard to transform forms of oppression *within* the community, or stressing the community's positive aspects to help protect it from unsympathetic outsiders, or both? If the first of these is important, whose values should dominate in this process of transformation? The inspirational figures of participatory development had their own particular answers to these questions, drawing on values selectively chosen from Christianity and Marx (Freire), or Hinduism and the work of Ruskin (Gandhi). Equally, development activists today cannot claim that they are simply representing the consensual views of holistic communities: they need to be clear about how they negotiate internal community tensions and differences, and their own role in filtering or influencing the picture of 'the community' that is presented to a wider world.

Hard questions also need to be asked about Chambers's upbeat account of the potential for reforming development professionals. In his writing there is a strong sense of voluntarism and almost inevitability: once their eyes can be opened up to the possibilities of participation, the 'uppers' will be willing and

Plate 11.4 Participatory Rural Appraisal, Sigor, NW Kenya.
Credit: Vandana Desai.

able agents of change. There is little mention here of the institutional constraints, personal jealousies or internal political battles that will need to be confronted if the 'new high ground' is to become dominant within the development industry itself, let alone implemented in practice. However, what undoubtedly *has* happened since the explosion of interest in participatory techniques in the late 1980s is that key elements of 'alternative development' thinking – participation, decentralization and community empowerment – have become part of the rhetoric and **discourse** of 'mainstream' development. We turn to the issues raised by this 'scaling up' of alternative development now.

FROM MARGINS TO MAINSTREAM? SCALING UP

Just as the best participatory development practitioners recognize that there are problems inherent in any vision of 'the community', they also realize that there is limited scope for development interventions that begin and end at the grassroots level. The effects of changes in world markets are felt even in the most remote rural communities in the Global South, and few if any places are untouched by national and international processes of **governance**. Alternative development practitioners today cannot hope to act only at the scale of Gandhi's 'village republics' – more than ever they need to think and act globally as well as locally.

The international NGO community has long recognized this, with David Korten (1987) talking of the major Northern international NGOs (such as Oxfam or CARE) as exhibiting three generations or modes of operation. At their foundation, many of these organizations were primarily engaged in disaster relief work, but over time this evolved into a concern with ongoing development grassroots programmes and projects, and it is here that some of them aimed to mark themselves out as an alternative to official (national or international) development programmes through their community-based and participatory approaches. The 'third-generation' approach, however, has been to play a leadership and coordination role, working in partnerships that fund and support indigenous NGOs and CBOs (community-based organizations) that are often better placed to undertake grassroots projects directly. This gives Northern NGOs greater scope to use their international connections to play advocacy roles, and engage with policy debates at the national or global scale.

Importantly, however, 'scaling up' is by no means dependent upon these large, Northern-based institutions: as Box 11.6 shows, the problems facing slum dwellers in Mumbai have meant that 'the Alliance', a group of NGOs based in the city, needed to engage with government at the metropolitan, state and national levels, and in doing so have developed their own international networks. The Alliance has been successful in building links 'upwards' to powerful international bodies such as the World Bank and UNDP, and these have helped its views be heard in some quarters of the Indian state. What is particularly significant about the Alliance, however, is that it has not been content simply to rely on these 'vertical' links, it also has 'horizontal' links to equivalent NGOs elsewhere in the South. The support provided to the Alliance by its international partner NGOs in Shack-Slum Dwellers International (SDI) is a clear reminder that there is considerable scope for Southern-to-Southern exchange and linkages (see also Box 7.6 on SEWU's activities in Durban), and that these can help to develop alternative international development and political agendas that do not have to be routed through the cities and institutions of the North.

The fact that the World Bank has entered partnerships and funding arrangements with NGOs such as the Alliance indicates that some important changes have also occurred on the part of major international development organizations since the 1980s. First, and significantly, NGOs have increasingly become a conduit for official development **aid**. For example, less than 10 per cent of World Bank projects between 1973 and 1988 involved NGOs, but over the 1990s this figure increased to 33 per cent (World Bank data cited in Nelson, 2002: 500). This shift in development finance itself helps to explain part of the rapid growth of the NGOs

BOX 11.6

Shack-Slum Dwellers International: Mumbai's NGOs go global

The population of Mumbai today exceeds 12 million people, of whom less than half have access to formal sector housing (Appadurai, 2001). It is therefore perhaps unsurprising that Mumbai is also home to 'the Alliance', a network of NGOs campaigning on housing issues. The key partners in the Alliance are SPARC (an NGO of social work professionals), the National Slum Dwellers Federation (a CBO with its historical base in Mumbai) and Mahila Milan (which focuses on women's self-organized savings groups). Arun Appadurai (2001) describes three key elements of their operation. First, daily savings schemes recruit slum dwellers into the NGOs' federated structure, building the Alliance's own organizational strength. Second, participatory slum surveys give the Alliance the detailed knowledge of living conditions in the informal sector that government agencies often lack – and mean that the NGOs can argue with authority on slum dwellers' behalf. Finally and importantly, the Alliance organizes local people in slum upgrading activities that range from building and exhibiting community-maintained public toilets (Plate 11.5) to managing the resettlement of rail-side shack dwellers within alternative housing they themselves helped to design and build. By providing practical solutions to the city's massive housing problems the Alliance has become a sought-after partner, working with UN-Habitat and the World Bank in their programmes within Mumbai, and influencing all-India policy on the relocation of slum dwellers (Burra, 2005).

To ensure that the Alliance remains an independent and community-based voice within these partnerships, in 1996 it set up an international network of its own, Shack-Slum Dwellers International (SDI). SDI consists of precisely the groups that are going to be most critically constructive of the Alliance's work: equivalent NGOs in other Southern countries (which include Nepal and South Africa). Networking with these groups at a distance and through face-to-face visits (often funded by the World Bank or Northern governments) not only provides a wealth of comparative experience for the organizations involved, it also sends an important message to local politicians: 'the poor themselves have cosmopolitan links' (Appadurai, 2001: 42). The importance of this message has never been higher in Mumbai, as powerful business interests have controversial plans to level the city's largest slum,

BOX 11.6 (*CONTINUED*)

BOX 11.6 (*CONTINUED*)

Dharavi, replacing it with commercial property and high-spec apartment blocks (Whitehead and More, 2007). At the time of writing, it is uncertain whether Dharavi's residents will be forcibly relocated to other parts of the city; what is clear is that the Alliance is using its strong community base and external connections to oppose these plans in the national and international media, and will not allow Dharavi to be redeveloped without a fight (Patel and Arputham, 2007).

(Sources: Adapted from Appadurai, 2001; Burra, 2005; McFarlane, 2008a, 2008b; Patel and Arputham, 2007; Whitehead and More, 2007)

Plate 11.5 An Alliance-built toilet block, Khotiwadi, Mumbai, India. Credit: Colin McFarlane.

discussed earlier in this chapter: with more official development assistance being routed through 'third sector' organizations, the opportunities for local NGOs have increased dramatically.

This is, however, a shift that holds rather ambivalent promise for the NGOs involved. Within World Bank projects, NGO involvement is often relatively small in financial terms, and limited to specific tasks that do not challenge overall policy direction or project design. As Paul Nelson (2002) notes, these typically involve NGOs acting as service deliverers within 'social fund' components of projects, a limited role that is hardly the radical refocusing of development goals around the needs and aspirations of marginalized communities that Gandhi or Freire would have called for. This is also an opportunity for NGO involvement that has arisen in part because **structural adjustment policies** (see Chapter 10) have eroded the practical effectiveness of the state in many countries in the South. Prior to spending and staffing cutbacks, Southern governments may well have been in a position to deliver these social services directly themselves, and so increasing NGO involvement may well be substituting for their activities within a wider 'rolling back' of the state. This is clearly a far from ideal position: it leaves social services administered by NGOs which are accountable to the donors rather than to communities or local government. It also associates the delivery of social services with 'charity' (for which the recipients should presumably be grateful) rather than with the rights they should have as *citizens* in relation to the local state.

To balance this, not all NGOs have been 'bought off' by such partnerships, and there is evidence that they can exert some influence over official development agencies. Sustained pressure from environmental NGOs opposing major dam-building projects, including the Sadar Sarovar dam in India's Narmada Valley (see Chapter 9) and Nepal's Arun River, have caused the World Bank not only to withdraw its funding from these individual projects but also to add new forms of environmental and social conditions on the loans it provides for all major infrastructure projects. Indeed, the Bank's own strengthened rules on the resettlement of communities affected by infrastructure projects have become an important resource for the Alliance as it seeks to ensure fair treatment of the informal settlement dwellers displaced by the upgrading of Mumbai's transport linkages and the extension of its airport (Patel and Arputham, 2007).

The second, and perhaps even more important, change within major development institutions is that they have themselves increasingly adopted the **discourse** of participation and community empowerment that was seen as 'alternative' in the 1980s. This is undoubtedly resulting in changes in the ways in which 'mainstream' development interventions are undertaken. Since the early 1990s, there has been an explosion of mechanisms to ensure public participation and community representation within a vast array of different development activities across the South. Community-based natural resource management schemes, which hand over some degree of control over key natural resources to the local community, would be one such activity which has been supported by a broad range of international donor agencies. Examples include village forest committees in India (supported by the World Bank, among others), water-user associations to manage local irrigation needs in Egypt (supported by German **aid** agency GTZ) and community-based wildlife protection in Zimbabwe's CAMPFIRE programme (supported by the World Wide Fund for Nature (WWF)). In all of these schemes, the local community or 'user groups' are invited to participate within new grassroots institutions, which often include safeguards against the more negative aspects of local **power** structures. For example, most Indian states have ensured that women have guaranteed representation in both the general body and the management committees of their village forest committees (Jewitt and Kumar, 2000).

To some extent, this uptake is exactly what some supporters of participatory development were hoping for in the 1990s. Jeremy Holland and James Blackburn discussed the scope for experimentation with participation not only being extended 'outwards' into ever more areas of development intervention, but also 'upwards' through institutionalization in the internal practices of international development agencies (Holland and Blackburn, 1998). While there is evidence that this has happened to some degree, important questions still remain about the practical impacts this change is actually having on the ground. The World Bank increasingly stresses the importance of participation in its public statements, and *Attacking Poverty* (World Bank, 2000) – its third decadal World Development Report to focus on **poverty** – was supposed to incorporate in its analysis the results of 'The Voices of the Poor', a multi-country participatory research project examining the experience of **poverty** across the globe. More fundamentally, the process whereby Southern countries produce **Poverty Reduction Strategy Papers** to qualify for the Bank's debt relief and **poverty** alleviation funding (see Chapters 9 and 10) is supposed to be inherently participatory. Governments are required to facilitate widespread consultation with **civil society** groups in the production of this document which is of key national importance. The outcomes in both cases have been less impressive than the rhetoric. In some countries, such as Thailand, the participatory research exercises that contributed to *Attacking Poverty* were superficial at best (Parnwell, 2003), and although many direct illustrations of people's experiences of **poverty** did make it into the final report, this research did not dramatically change the Bank's underlying **neoliberal** approach to **poverty** alleviation (see Mawdsley and Rigg, 2002). Nicaragua's experience of the **PRSP** process has been that even with active and mobilized **civil society** groups wanting to ensure national government fulfils its duty to facilitate participation, their actual opportunities for meaningful input can be limited (Box 11.7).

Bill Cooke and Uma Kothari's edited book *Participation: The New Tyranny?* (2001) expresses wider concerns that community engagement and participation itself are being co-opted through these processes of institutional change, with some negative consequences. The contributors to this volume raise the possibility that institutionalized participation is incorporating grassroots voices within the development process without giving them real **power**. The fact that members of the public have 'had their say' through their involvement in participatory learning activities, or their membership of a village forest committee, means that their opposition to development interventions may be rather more muted. At a national level, this was the situation the Civil Coordinator of Emergency and Reconstruction (CCER) felt it was being forced into within the Nicaraguan **PRSP**

BOX 11.7

Nicaragua's PRSP process

In October 1998, Nicaragua was hit by Hurricane Mitch, one of Central America's worst natural disasters for over 200 years. The hurricane highlighted and deepened the country's growing **poverty** – those living in fragile housing and flood-prone areas were particularly badly affected – but it also galvanized an impressive reconstruction effort from the country's active **civil society**. The Civil Coordinator of Emergency and Reconstruction (CCER) emerged as a national umbrella organization for over 350 local NGOs, producers' associations, unions, social movements and others, and was active in trying to make national policy responses to reconstruction more people-centred.

During the post-hurricane recovery period, Nicaragua entered the World Bank's HIPC process for debt relief (see Chapter 9), as a result of which it had to produce a national **Poverty Reduction Strategy Paper**. The Government of Nicaragua quickly produced a **PRSP** with very limited **civil society** consultation (the interim paper was not even translated into Spanish): this worked within a narrow approach to **poverty** and **poverty** alleviation, focusing on economic growth. The CCER criticized this, and used its network quickly to implement an impressive national **civil society** consultation process on **poverty** alleviation. The result of this was a parallel consultation document *La Nicaragua que Queremos* (The Nicaragua we Want), produced within a mere nine months, that called for a different approach to **poverty** alleviation, centring on sustainable human development. This was not met within the final **PRSP**, but some elements from the CCER's consultation were taken up within the report, and the World Bank acknowledged that greater public participation was necessary.

The conflicting roles of the Government of Nicaragua, the CCER and the World Bank show the difficulties of embedding participation and **civil society** engagement into formal **governance** structures. The Government of Nicaragua argued it did not have time to produce a thorough national participatory exercise on **poverty** alleviation, given that the speedy approval of a **PRSP** was necessary to start the process of debt relief. The World Bank's own guidelines on what constitutes adequate participation within a **PRSP** were also not clear. For CCER, the entire process of engagement posed a difficult dilemma. If it stood back, the government's neoliberal approach to **poverty** alleviation would

BOX 11.7 (*CONTINUED*)

BOX 11.7 (CONTINUED)

be implemented unopposed; if it took part, it risked lending legitimacy to a process which it fundamentally questioned, and thereby possibly losing its integrity as a campaigning organization.

(Source: Adapted from Bradshaw and Linneker, 2003)

process (Box 11.7). Worse still, by promising to 'empower' grassroots participants, 'participatory' development can offload the responsibility for project success on to the communities themselves, allowing the state and international development agencies to limit their own responsibilities. The fear here is that participatory development's focus on engaging with 'the community' at the micro-scale is a very useful way of taking the spotlight away from the operation of development agencies themselves, and from wider structures of **power** and inequality (Mohan and Stokke, 2000); in short, it is taking the politics out of development.

These are powerful arguments, and provide a useful warning about the 'scaling up' of alternative development. Just because ideas such as community engagement or participation that have entered the development mainstream can trace their roots back to radical thinkers such as Freire or Gandhi, does not mean that their actual effects are always as benign as their promoters might hope. It is, however, a warning that should make those searching for community-based alternative development reflect carefully on their own practice rather than give up hope altogether. Some contributors to *Participation: The New Tyranny?* go as far as to suggest that participatory development is an attempt mentally to control communities in the Global South today (Henkel and Stirrat, 2001; see also Rahnema, 1992), describing this in similar terms to Franz Fanon's account of **colonialism** (see Chapter 3). This needs to be balanced against the possibilities for change that participation opens up: it can provide new rules and expectations for the ways in which governments and other development agencies *should* interact with local communities, and these can offer new opportunities for people to press their political claims, or express their rights as citizens (Williams, 2004; Mohan and Hickey, 2004). Examples such as the ongoing rights to information struggle being fought by the Rajasthani NGO the Mazdoor Kisan Shakti Sangathan in India (Jenkins and Goertz, 1999) or the activities of SDI in and beyond Mumbai (Box 11.6) are valuable reminders that people's participation can be highly political – and empowering.

CONCLUSIONS

There is no doubt that community-based development is an idea that has great appeal. Far too often in the past, development interventions have assumed that rational scientific knowledge and large-scale projects automatically represent 'progress', and this has meant the devaluing of both more local-scale development alternatives, and the expertise and skills that exist within communities of the Global South. The work of thinkers such as Gandhi and Freire helps us to question assumptions that the 'modern' should always be preferred to the 'traditional', or that there is a clear hierarchy of knowledge between 'experts' and lay people. As such, both continue to be important figures not only within their native India and Brazil, respectively, but in their wider impact on environmentalists, educators and social activists across the Global South (and North).

The attractiveness of community-based development does, however, need to be balanced by careful reflection if it is not to descend into dewy-eyed idealism, and this Chapter has highlighted three central areas of concern. First, we need to think very carefully about the nature of 'the community' itself. Alternative development practices cannot base themselves on romanticized images of the community as eternal, fixed and harmonious bodies: communities are and always have been in flux, and include within themselves **power** differences that can mean some of their members face exclusion on the grounds of **gender**, ethnicity or other forms of difference. The example of indigenous irrigation in Kenya (Box 11.4) is a good case in point: this is an ingenious low-technology solution to improving agricultural output, but it is dependent on the social control of the *kokwa*, an institution that is male-only, and may no longer be meeting the changing needs of the local population. A related second point is that we need to think carefully about how 'the community' gets represented: a rapidly growing NGO community has been influential in the spread of alternative and community-based development ideas, but questions still remain over quite who and what these groups represent. CCER (Box 11.7) and Mumbai's Alliance (Box 11.6) may have quite clearly articulated views and maintain strong links with the communities they are involved with, but not all NGOs are as open or democratic in their practices. Finally, we need to look very carefully at the **power** relationships involved in integrating community-based approaches with mainstream development. Here, 'scaling up' can result in the watering down of core ideas and values, and community-based organizations need to be very careful if they are not to lend legitimacy to more powerful institutions with different agendas to their own.

As the examples in this Chapter have shown, there can be no final or objective resolution of some of these concerns around community-based development. As a result, it is especially important for alternative development practitioners to be explicit about the values and the politics (both formal and informal) that guide their work. Similar unresolved tensions are, however, equally present within the state- or market-led visions of development examined in Chapters 9 and 10. What is certain is that by raising fundamental questions about the ultimate aims of development and the role local voices and ideas can and should play within it, community-led approaches have proved themselves an equally vital part of the Global South's development today.

Review questions/activities

1 Look through recent issues of *PLA Notes* (see website below): How far do you think participatory methods are useful in representing community voices within development practice?

2 Use the empirical examples presented in *The Lie of the Land* (see below) or Richard Peet and Michael Watts's *Liberation Ecologies* (2004) to examine when and why indigenous environmental knowledge gets overlooked within development practice.

3 Look back at Box 11.6 and the website of SDI (see below). To what extent are the activities of the Alliance and its international partners managing to 'scale up' community-level voices within urban development practice?

SUGGESTED READINGS

Chambers, R. (1983) *Rural Development: Putting the Last First*, Harlow: Pearson Education.

Chambers, R. (1997) *Whose Reality Counts: Putting the First Last*, London: Intermediate Technology Publications.
Although it needs to be read with a critical eye, Robert Chambers's work is an accessible and enthusiastic introduction to alternative development approaches.

Leach, M. and Mearns, R. (eds) (1996a) *The Lie of the Land: Challenging Received Wisdom on the African Environment*, Oxford: James Currey.
One of the best collections of examples of the conflict between 'expert' and 'traditional' knowledge systems about the environment.

Cooke, B. and Kothari, U. (eds) (2001) *Participation: The New Tyranny?*, London: Zed Books.

Hickey, S. and Mohan, G. (eds) (2004) *Participation: From Tyranny to Transformation: Exploring New Approaches to Participation in Development*, London: Zed Books.
Two more recent volumes on the tensions inherent within participatory approaches to development.

WEBSITES

http://www.iied.org/NR/agbioliv/pla_notes/ *PLA Notes* online
  A free-access archive of all back issues of a publication that has been influential
  in developing and disseminating knowledge about participatory learning and
  action (PLA) techniques.

http://www.sdinet.org/ Shack-Slum Dwellers International
  Provides one example of a NGO that is attempting to 'scale up' and build
  networks of South–South interaction to represent the activities and interests of
  informal settlement dwellers within urban development debates internationally.

12 Conclusion

We hope that you have found the previous chapters engaging and that they have challenged you to think about the peoples and places of the Global South in new ways. While the book has covered a vast range of material and ideas, we have made four main arguments throughout which we outlined in Chapter 1 and which we review in this chapter. The first argument is that **representations** of the Global South matter – they can have important effects in the real world which we illustrate in various ways through the book. Chapter 2 explored this argument at some length, and did so deliberately to disrupt from the very outset some prominent Northern images of the Global South as a place of backwardness, **poverty** or danger. While historically, such negative **representations** of the South may have been an important element of psychological control by colonizers (as was argued by Franz Fanon, see Chapter 3), today the **representation** of the Global South is not wholly dominated by the narrow concerns of Northern development agencies, media outlets and others. Ways of imagining the Global South *from* the South are both important and varied, and can draw on or undermine representations from the North. Some actively aim to build a sense of the South as a unified or united group, such as within the **G77** or the **Non-Aligned Movement**, whereas others may selectively adopt aspirations of **modernity** and difference within the South, such as within the '**place marketing**' of particular cities. Exploring this array of **representations** is important because of their material effects which are felt at a range of spatial scales, from Southern countries bargaining for environmental or trade concessions through the United Nations, to the personal impacts of Southern governments presenting themselves as guardians of 'traditional' and moral values. It also helps to place 'development' as one way – but only one among many – in which the Global South is imagined.

Parts 2 and 3 of the book have particularly emphasized the importance of two of our other main arguments, that the Global South is a vital and active part of processes of **globalization**, and that to understand the South better requires a detailed engagement with its particular histories and geographies. Part 2 approached these arguments from a 'macro' perspective, looking at the emergence of the Global South, and its location within and impact upon broader political, economic and social processes. Importantly, the role of the South in shaping these processes was stressed. China's growing role in Africa (see Boxes 3.6 and 4.5) is clearly one example here, but it would be wrong to leave the reader with impression that the Global South only impacts upon processes of **globalization** when it poses such an obvious geopolitical challenge to the interests of the North. Singapore's status as both an emergent 'global city' and an alternative model of urban development (Box 4.4), or the reshaping of global Christianity through the influence of African churches (Box 5.6), are both powerful reminders that within **globalization** more widely the traffic in ideas and values does not simply run from North to South.

In presenting the material in Part 2 in this way, we deliberately wanted to move away from those accounts that describe the place of the Global South as simply a periphery within a 'world system' (Wallerstein, 1984). While there are massive inequalities in global structures of wealth and **power** today, the experiences of the Global South cannot simply be 'read off' from their places in global social, economic or political structures. Thus the fact that much of the Global South underwent colonization and **decolonization** should not blind the reader to the great variation in the ways in which these processes have been experienced – and their very different legacies today (compare the discussion of Hugo Chávez in Box 2.5 with the contemporary challenges of nation-building in India in Box 3.3). Similarly, today, much of the Global South may be experiencing processes of market integration (Chapter 4), **migration** and **urbanization** (Chapter 5), but these are not simply homogenizing processes: they affect the South in different ways, and are important in reproducing rather than eliminating the particularity of places within it.

The third part of the book again brings both of these ideas to the fore by looking at people's everyday lives. Chapters 6, 7 and 8 all highlight the diversity of life in the South, and the differential impacts of **globalization** as it is experienced on the ground, by looking at people's everyday practices. Through our varied snapshots of people's changing working practices (Chapter 7), patterns of **consumption** (Chapter 8) or modes of political engagement (Chapter 6), we hope that we have presented some of the contradictory effects of **globalization** – and also have illustrated people's roles in actively reworking these at the grassroots level.

Engaging with the histories and geographies of the South at both a micro- and a macro-scale thus illustrates that **globalization** is not experienced homogenously, and that in many contexts it contributes to increased differentiation. More than that, however, it can also lead us to reflect critically on some of the assumptions that implicitly underlie much social science theory based in the experiences of the Global North. Although industrialization and **modernization** may have made the North overthrow many of its traditional social ties and become increasingly secular from the nineteenth century onwards, there is no guarantee that these trends will be universal or irreversible. The experience of the South indicates that instead the significance of religion, and its intimate connections to moral, cultural and economic practices, is constantly being renewed (Chapter 5), and that 'traditional' forms of authority and **power** are important constituent parts of modern political systems (Chapter 6).

Our final argument is that the Global South should not be read through the lens of development alone. Hence we introduced our discussion of deliberate attempts to 'develop' the South in the final part of the book. It was also our intention here to show that development is not a singular entity, but a constantly contested set of aspirations. This follows from our earlier arguments about the Global South's active role within **globalization,** and the importance of engaging with the histories and geographies of the South. The key institutions and agencies of the international development industry may well be important in setting part of the context in which Southern states, markets and civil societies press their contrasting claims as to what development is, and how it should be promoted, but they by no means have a monopoly here. Southern development experiences have often been of equal importance in driving forwards debate within international development, whether this has been in the economic transformations created within the East Asian 'Tiger Economies' (Box 10.2), the participatory **governance** experiences of Porto Alegre (Box 9.7) or the innovative forms of social action undertaken by groups such as the Alliance in Mumbai (Box 11.6).

In presenting this critical approach to development, we should stress that we don't wholly agree with the 'anti-development' or 'post-development' perspectives popular since the 1990s. There *is* value in deconstructing some of the representational tricks and **tropes** present within the development industry (see Box 1.4 and Chapter 2), but it would be wrong to stop at that point and leave the reader with the perception that 'Development' is a singular set of alien values and practices imposed upon an unsuspecting Global South. We feel that critical engagements with development have an equal responsibility to consider seriously the important aspirations within different – and often inherently hybridized – local understandings of the term. As a result, our intention has been to arm the reader with important questions that should be asked of *all* intentional development initiatives: Who is governing change and how (Chapter 9)? At what or whose cost is economic growth being achieved (Chapter 10)? How and to what ends are ideas of community ownership of development processes being mobilized? We hope that this leaves the reader healthily sceptical of some of the overblown 'success stories' of international development, but not so cynical that the search for better forms of development intervention seems like a hopeless exercise.

By structuring this book around these main arguments, there are, inevitably, other important aspects of the Global South that may have been downplayed. Questions of sustainable development, environmental degradation and climate change are clearly pressing at a global scale, and have great significance for the futures of many people in the South, but are not explicitly brought out within this structure. Environmental aspects are, however, threaded through our discussion at various points – such as in looking at global environmental governance (Box 3.8), or the struggles over whose environmental knowledge is seen as valid (Chapter 11). These threads provide an indication of the relationships between human and natural environmental processes in the Global South, but a comprehensive consideration of environmental debates in Africa, Asia, Latin America and the Caribbean requires a whole book in itself (see Adams, 2008, or Robbins, 2004, for different, but very good and accessible examples).

A critical discussion of the importance of space and scale is often curiously absent from much writing in Development Studies, and we hope we have gone some way to redressing this here. We look at contemporary changes in the South from contrasting global and local scales in Parts 2 and 3, and we address spatiality (and the complex realities of space and place) particularly in relation to **urbanization** (Chapter 5), informal living and also the home (Chapter 8). We are aware, however, that there is still much more that could be said through a thorough theoretical investigation of spatiality in the Global South, and would strongly recommend Jenny Robinson's *Ordinary Cities* (2006) to those who would like to explore these ideas further.

There are always difficult balancing acts to perform, of both subject matter and approach. For example, we hope that we have challenged negative stereotypes of the South as being universally 'backward' and poverty-stricken, but at the same time have not overlooked the extreme inequalities in wealth and **power** that exist within the South. Thus Chapter 5 deliberately highlighted some of the realities of global inequality through the lens of health concerns. In other chapters, people facing political marginalization in rural India (Box 6.6), making a living in Durban's informal settlements (Box 7.4) or coping with the social outcomes of SAPs in Peru (Box 10.4) are clearly doing so with a vastly different set of opportunities from most people reading this book. Our hope is that by emphasizing examples such as these, rather than foregrounding technical debates over the measurement of **poverty**, we've also given our readers a richer sense of the sources – and at least some insights into the lived experience – of these inequalities. In a similar vein, we have chosen to illustrate the importance of **gender** throughout the book, instead of bracketing it off as an isolated set of practical problems or theoretical concerns (as was often the case in earlier approaches to gender and development, see Momsen, 2004). Our approach to **gender** throughout the book also aims to signal our view that other axes of difference are also central to understanding life in the Global South. Ethnicity, caste, class, age, religion and sexuality are, along with **gender**, key to shaping people's experiences throughout the Global South.

What we hope we've gained by this approach is a way of presenting some of the complexity of the Global South that does map reasonably easily onto the current concerns of 'mainstream' geography and other social science. Our primary political themes within the book have been the South's position within international **governance** structures, its processes of nation-building, and people's experiences of and resistance to state **power**. These themes raise critical questions about contemporary processes of state-led development in the South (Chapter 9), but equally well each of these themes is important within political geography in its own right – even if much of political geography has taken the Global North as the arena in which they are explored. Similarly, our look at social and cultural issues in the South raises issues about the role of **material culture** in **identity** formation and performance and highlights how global flows of commodities, ideas and people from the Global South contribute to the social and cultural geographies of places in the Global North. The spatialities of economic processes are as important in the Global South as they are in studies of economic geography in the Global North and contribute to debates around the need to challenge the centring of Global North experiences in economic geography (Yeung, 2007). Recognizing the diversity of economic processes in the Global South also signals alternative ways of conceptualizing 'the economic' (*Geoforum*, 2008) and considering economic spaces beyond the formal market economy (Gibson-Graham, 2005).

WHERE NEXT FOR THE GLOBAL SOUTH?

The Global South has changed dramatically in the two decades since we as authors first engaged with 'development geography' as university students. The intervening years been significant in the parts of the South with which we have had our strongest personal ties (see Boxes 1.1–1.3): Mexico has moved from **debt crisis** to membership of the Organization for Economic Cooperation and Development (OECD); India has become a nuclear power with a globally significant economy; and Apartheid has thankfully been replaced with multiracial democracy in South Africa. In the late 1980s, it would have been very difficult to foresee the impact these and other global changes – such as the rapid rise of **neoliberalization** (Chapter 10), the search for international responses to environmental change (Chapter 3) and the growing **transnationalization** of lives and lifestyles (Chapters 5 and 8) – would have on the Global South we see today. As such, it would be rash here to predict what the Global South will look like by the late 2020s. We are submitting this book to press in summer 2008, a point in time when some of the greatest issues of the day are global economic uncertainty, around credit markets, and rapidly increasing oil and food prices. It is impossible to predict how the widely anticipated global economic slowdown will play out over the next two years (Will this accelerate or slow down China's rise vis-à-vis the USA? What will its impacts be on people – as producers and consumers – across the Global South? Will it deflect attention from another issue of crucial importance to the South, that is, tackling climate change?), let alone whether it will seem significant, or even be remembered, from the perspective of anyone writing in 2028.

Two things do, however, seem certain. The first is that the subject of this book – the Global South – will become a still more contested term. As we noted in Chapter 1, the Brandt Line dividing 'South' from 'North' is already looking like a boundary that belongs in history books rather than contemporary geography texts. It is a line that does not neatly divide rich from poor, and fails to capture the diversity – and extreme inequality – that exists *within* the Global South, let alone the complexity of its links with the Global North. We sincerely hope that by 2028, some of the structural inequalities that divide the world will have been ameliorated to some degree: that the current imbalance within institutions such as the World Bank, the IMF or the UN Security Council will have been corrected to allow forms of **global governance** more inclusive of the Global South, and also that what are currently seemingly irresistible processes of **neoliberalization** will at least be tempered by greater concern with issues of democratization, and social and environmental justice. If they are, there is hope that the effects of nineteenth- and twentieth-century histories of **colonialism** will begin to be overtaken by more inclusive forms of **globalization**, and that the world of 2028 will be one in which the countries, regions and people of the Global South are treated as important in their own terms, rather than as 'development problems'. If not, the alternative is an increasingly dystopian world of growing inequality, but this is also likely to be one where global cities, islands of wealth, and surrounding seas of **poverty** will continue to map less and less neatly on to Brandt's division.

Whichever way processes of **globalization** proceed, the other point that seems certain to us writing now is that those living in 2028 will need more, rather than less, knowledge of the Global South. Throughout this volume we've repeatedly stressed the diversity that exists within the South, and we have also pointed out at the very outset our intention of disrupting unthinking assumptions that the experience of people living in those parts of the world fit neatly within the categories and representation of 'development'. We hope that by 2028, the statistics we quoted in Chapter 1 about the current marginalization of the Global South within Anglo-American Geography will seem like a distant memory, and that there will be no need to write a textbook such as this that deliberately sets out to 'restore the balance'. The 80 per cent of the world's population that live in the Global South are far too important to all our futures for the study of 'developing areas' to be treated as a narrow academic specialism. Rather, we would argue, it is the 'mainstream' that needs to change. Geography and other social sciences as taught in the Global North must become more engaged with the scholarship and concerns of Southern countries if they are to avoid becoming parochial and irrelevant subjects. In the meantime, there is a fascinating and rapidly changing world that lies to the South of Brandt's line, and our intention in this volume was to communicate some of this dynamism and interest. Our own lives have been greatly enriched by engaging – both personally and professionally – with that world, and we hope that this book will encourage our readers to do likewise.

Glyn Williams, Paula Meth and Katie Willis,
July 2008

Glossary

Agency The power of individuals to act. Often used in contrast to 'structure' which refers to abstract economic, social and political structures (such as capitalism) which some argue shape and control our actions.

Aid The transfer of resources on concessional terms to promote economic development and improvements in social welfare (see Concept Box 10.1).

Bretton Woods Conference held in the USA in 1944 to establish post-World War II international economic institutions.

Capitalism System of economic organization which gives markets a central role in the exchange and valuing of goods and services.

Civil society Formal and non-formal organizations outside of government, usually excluding private sector organizations and household/kinship groups.

Cold War Period of political tension between the capitalist world led by the USA and the Communist world led by the USSR beginning in the late 1940s until 1989.

Colonialism Formal political control of territory and peoples by another state.

Commodity chain Series of economic interactions linking raw materials through production to final point of sale (see Concept Box 4.1)

Communism Political and economic system based on the works of Karl Marx and Friedrich Engels. Property is owned by the state and the economy is planned centrally.

Complex political emergency A humanitarian crisis that has political instability as one of its main causes (see Concept Box 3.2)

Consumption The use of goods and services.

Corporate social responsibility The need for corporations to be concerned with the ethics of their activities as well as profit maximization (see Concept Box 10.2).

Cronyism A system of political and/or economic relationships which favours personal (and often hidden/opaque) relationships over merit.

Cultural capital Skills and knowledge that allow successful performance in particular cultural settings or navigation through particular cultural settings.

Cultural change Changes, e.g. in practices of religion, consumption or nationalism, that have an impact on people's cultural norms or identities (see Concept Box 5.2).

Cultural globalization The spread of cultural norms and practices internationally. Often used to refer to processes of cultural homogenization.

Debt crisis (The) The default of a series of countries in the Global South on the servicing of their debts to Northern financial institutions in the 1980s.

Debt servicing Payments made by national governments related to their international borrowing. Usually includes interest and the partial repayment of original loan.

Decolonization Transfer of formal political control of a territory and people from foreign to indigenous rule.

Deconstruct(ion) The in-depth analysis of texts, language, images and other forms of representation. This normally involves an investigation of underlying and implicit ideas.

Democracy A system of government in which citizens hold political power directly, or through their elected representatives.

Demography The study of human populations.

Deregulation The removal of regulations or controls particularly associated with reduced direct role of the state.

Developmental state A state that takes an active role in directing (rather than merely regulating) the drive towards economic growth, particularly through industrialization.

Diaspora Population which has been dispersed from their original home (see Concept Box 5.4).

Discourse A form of representation through writing, visual images or other means (see Concept Box 2.2).

Division of labour The division of labour processes into smaller tasks which are done by different people, possibly in different places.

Domestic labour Unpaid work that is performed in or around the home (see Concept Box 7.3).

Eurocentricism Way of thinking which sees the European (or Northern) experience as the only possible way of doing things, or sees Europe/the Global North as superior (see Concept Box 2.3).

Eurodollars Reserves of US dollars outside the USA (see Concept Box 4.4).

Export processing zones (EPZs) 'Industrial zones with special incentives set up to attract foreign investors, in which imported materials undergo some degree of processing before being exported again' (ILO, 2004: 37).

Fair trade System of trade where producers are guaranteed a price for their goods regardless of the market price. The price paid to producers includes an additional payment for social programmes.

Floating exchange rate System whereby exchange rates are determined by market forces, rather than fixed by governments or pegged to gold or another currency (usually US dollar).

Fordism System of production geared around mass production, often using assembly lines and significant divisions of labour (see Concept Box 4.2).

Foreign direct investment (FDI) Investment by governments or private companies into foreign countries.

Formal economy Economic activities that operate within legal frameworks and are regulated.

Formal politics 'The operation of the constitutional system of government and its publicly defined institutions and procedures' (Painter, 1995: 8) (see Box 6.1).

Free trade Trade that is unencumbered by barriers such as tariffs and quotas.

G77 Grouping of 77 countries from the Global South formed in 1964 at UNCTAD conference. Now contains 134 members.

Gender Socially produced differences between men and women (see Concept Box 5.3).

Geopolitics An approach to studying the actions of states with reference to geographical concerns, such as territory and resources.

Global governance Practices of governance at a global scale (see Concept Box 3.3).

Global political economy The interrelated system of economic and political processes at a global scale.

Globalization Economic, political, social and cultural processes that are bringing peoples and places of the world closer together. Often summarized in the concept of 'a shrinking world'.

Good governance A developmental agenda, associated particularly with the World Bank, aimed at improving the quality of governance within Southern countries (see Chapter 9).

Governance Forms of rule-making which encompass government and non-government actors such as private companies and civil society organizations.

Governmentality Ways in which a government and the population it governs interact (see Concept Box 3.1).

Green Revolution Introduction of new forms of agricultural production in the 1960s and 1970s. Included the use of high-yielding seed varieties, irrigation and fertilizers (see Concept Box 7.2).

Gross Domestic Product (GDP) The value of all goods and services produced in a country (see Concept Box 2.7).

Gross National Income (GNI) The measure of the value of all goods and services claimed by the residents of a particular country. It does not matter where those goods or services were produced (see Concept Box 2.7).

Hegemonic State of domination of one idea, way of doing things or government, which may be so embedded that it seems natural.

Home-based work Income-generating activities that take place within the home.

Human Development Index (HDI) Measure devised by UNDP to provide an indicator of levels of human development, including indicators of education, health and standard of living (see Concept Box 2.7).

Hybridization A process of cultural mixing (see Concept Box 8.2).

Identity Ascription as belonging to a particular group which can be either self-identified, or ascribed by others (see Concept Box 5.5).

Imperialism System of domination by one geographical area over another, which may include formal political control.

Import-substitution industrialization (ISI) Economic strategy that aimed to foster domestic industrial development through the erection of tariff barriers.

Indigenous technical knowledge Knowledge possessed by indigenous populations based on their long-standing interactions with their environments.

Informal economy Economic activities that take place outside of formal controls and legislation or through non-formalized relations of employment (see Concept Box 7.1).

Informal politics 'Forming alliances, exercising power, getting other people to do things, developing influence and protecting and advancing particular goals and interests' (Painter, 1995: 9) (see Box 6.1).

Just-in-time production A form of production whereby inputs are employed and outputs are delivered on demand (see Concept Box 4.2).

Laissez-faire A form of capitalism entailing minimal state intervention and maximum market freedom.

Lifestyle The way in which a person lives, or ways of living, which includes the social relations they conduct and their consumption practices. Lifestyle also reflects the cultural values a person has.

Livelihood 'The capabilities, assets (including both material and social resources) and activities required for a means of living' (DFID, 1999).

Malthusian Interpretation of the relationship between population growth and resources which argues that unchecked population growth will lead to massive resource shortages, civil unrest and deaths (see Box 2.2).

Material culture The relationships developed between artefacts and people (see Concept Box 8.4).

Mercantile A form of capitalism whereby the state actively protects a country's trading interests through the use of tariffs.

Migration The temporary or permanent movement of people within or across international borders.

Millennium Development Goals (MDGs) Series of development targets internationally adopted in 2000 that focus on improving living conditions for the poorest.

Modernity The state of being modern or of the time. Often used to refer to Northern ideas of being modern, although alternative forms of modernity exist (see Concept Box 2.5).

Modernization Process of evolution from simpler to more complex, and allegedly more advanced, societies.

Moral economy System of economic relations that is overseen by social sanctions dependent on a close-knit community.

Multinational corporation (MNC) Corporation that has the power to coordinate the production of goods or services in more than one country.

Multiplier effect The additional economic activity produced as a consequence of a given round of investment.

Nation-state A state in which the territory over which it claims sovereignty ideally corresponds to the geographical boundaries of a clearly defined nation of people that are its citizens.

Neoliberalism Approaches to economic and political development that stress the role of the market rather than the state (see Concept Box 4.3).

Neoliberalization The implementation of neoliberal policies.

New International Division of Labour (NIDL) A global division of labour associated with the movement of manufacturing to NICs.

New Public Management Public sector organizational change aimed at creating greater financial efficiency through the introduction of market-like mechanisms (see Concept Box 9.3).

Newly industrializing country (NIC) Country whose rapid industrial growth is delivering levels of economic development near to that of the OECD countries.

Non-Aligned Movement (NAM) An organization primarily of former colonial countries which aims to ensure the self-determination of its members in international affairs.

Normative How things *should* or *ought* to be, rather than simply describing how things actually are.

Official Development Assistance (ODA) Financial aid from members of the OECD's Development Assistance Committee that is loaned to developing countries at concessional rates for developmental objectives.

Offshore banking Financial transactions that take place within lightly regulated territories primarily serving the business of non-residents.

Old age dependency ratio Ratio of older persons to working age persons.

Organization of Petroleum Exporting Countries (OPEC) Organization comprising 12 key oil-producing states which aims to secure their common economic interests.

Orientalism A set of institutional practices that have the effect of defining the non-Western world as different and inferior relative to the West's own standards (see Concept Box 2.4).

Othering A process of description which establishes categories of difference in which 'the Other' is normally defined by and in the terms of the dominant category (e.g. Western/Oriental – see Orientalism).

Paradigm shift Dramatic shift in a previously stable set of ways of understanding a phenomenon, sometimes offering a new worldview.

Passive resistance Resistance that seeks to undermine power through inactivity or non-compliance rather than direct confrontation.

Performativity Speech or other action that performs or 'acts out' elements of an individual's identity, or her/his relationship to wider social norms (see Concept Box 6.1).

Petrodollars Dollar reserves in international banks coming from sales of petroleum (see Concept Box 4.4).

Place marketing The marketing of a particular location (such as a city) with the intention of providing it with economic benefit, such as promoting inward investment or tourism.

Population growth rates The change in population over a particular period of time due to natural increase and/or migration.

Post-colonialism Can be used in two senses: first, the time period after independence, and second, a theoretical approach that challenges Northern-centric interpretations and instead starts from the viewpoint of the South.

Post-development Set of ideas that highlights the ideological nature of development discourse and development interventions (see Concept Box 2.6).

Post-structuralism Theoretical approach that recognizes that knowledge is always partial and situated, rather than adopting singular interpretations of the world.

Poverty Often used to mean economically unable to survive or participate in society in a meaningful way. Non-economic interpretations highlight the multifaceted nature of being poor.

Poverty Reduction Strategy Paper (PRSP) Document prepared by an indebted Southern country in partnership with the World Bank that outlines an agreed national approach to tackling poverty.

Power The capability to act (power *to*), to control others (power *over*) or something that operates *within* everyday techniques, strategies and practices (see Concept Box 1.1).

Protectionism The imposition of trade tariffs and other measures to protect a national economy from outside competitors.

Public works programmes State-directed programmes aimed at providing employment through the development of public infrastructure.

Relational Construction or identity of one person or thing is dependent on its contrasts to or similarities with someone/something else.

Remittances Money or goods sent home by migrant workers.

Representation The ways in which language, symbols, signs and images stand for objects, people, events, processes or things (see Concept Box 2.1).

Resistance The process of opposing power either actively or passively.

Sati Traditional Hindu practice of wife self-sacrifice on her husband's funeral pyre.

Secularism The removal of religious values from the conduct of public affairs.

Social change Processes of change that relate specifically to society and people (see Concept Box 5.1).

Social construction The idea that categories or ways of seeing the world are not objective or natural, but rather are constructed in and through social interaction.

Social reproduction The ways in which social life is reproduced through both material and social means.

Socialism A form of political and economic organization in which there is public ownership and control of the economy.

Sovereignty The absolute right of control over people, a territory or an area of governance.

Stakeholder A person or organization that has a legitimate interest in a particular process or situation.

Structural adjustment policies/programmes (SAPs) Policies imposed by international donor organizations on indebted countries aimed at scaling back public expenditure and achieving balance of payments.

Subsistence agriculture Agriculture that provides for the needs of the household rather than for commercial exchange.

Sustainable development Development that meets the needs of the current generation without compromising those of future generations (see Concept Box 3.4).

Terms of trade The ratio of the value of a nation's exports to its imports.

Transnational corporation (TNC) A company that operates in more than one country.

Transnationalism Economic, social and political processes that take place across international boundaries (see Concept Box 8.3).

Trickle-down The idea that benefits of economic growth will diffuse outwards from their area of main concentration to other sectors of the economy and society.

Trope A common or dominant theme within descriptions or narratives.

Urbanization Increases in the proportion of the population living in urban areas.

Westernization The spread of Western values and practices to non-Western societies.

Bibliography

Abaza, M. (2001) 'Shopping malls, consumer culture and the reshaping of public space in Egypt', *Theory, Culture and Society* 18(5): 97–122.

Abers, R. (1998) 'From clientalism to cooperation: local government, participatory policy, and civic organizing in Port Alegre, Brazil', *Politics and Society* 26: 511–37.

ACTSA, Christian Aid and SCIAF (2007) 'Undermining development: copper mining in Zambia', report published by Action for Southern Africa, Christian Aid and Scotland's aid agency.

Adams, W. M. (2008) *Green Development: Environment and Sustainability in a Developing World* (third edition), London: Routledge.

Adams, W. M. and Watson, E. E. (2003) 'Soil erosion, indigenous irrigation and environmental sustainability, Marakwet, Kenya', *Land Degradation & Development* 14(1): 109–22.

Adams, W. M., Watson, E. E. and Mutiso, S. (1997) 'Water, rules and gender: water rights in an indigenous irrigation system, Marakwet, Kenya', *Development and Change* 28(4): 707–30.

Adelkhah, F. (2000) *Being Modern in Iran*, New York: Columbia University Press.

Afshar, H. (1999) 'The impact of global and the reconstruction of local Islamic ideology, and an assessment of its role in shaping Feminist politics in post-revolutionary Iran', in H. Afshar and S. Barrientos (eds) *Women, Globalization and Fragmentation in the Developing World*, London: Macmillan, pp. 54–76.

Akbar, A. and Donnan, H. (1994) *Islam, Globalization and Postmodernity*, London: Routledge.

Allen, J. (1997) 'Economies of power and space' in R. Lee and J. Wills (eds) *Geographies of Economies*, London: Arnold, pp. 59–70.

Allman, J. (2004) 'Fashioning Africa: power and the politics of dress', in J. Allman (ed.) *Fashioning Africa: Power and the Politics of Dress*, Bloomington and Indianapolis: Indiana University Press, pp. 1–10.

Amin, S. (1989) *Eurocentrism*, London: Zed Books.

Anders, G. (2005) 'Civil servants in Malawi: cultural dualism, moonlighting and corruption in the shadow of good governance', Erasmus Universiteit Rotterdam, PhD dissertation, School of Law. http://hdl.handle.net/1765/1944

Anderson, A. (1997) 'Pluriformity and contextuality in African Initiated Churches', paper presented at Conference on Christianity in the African Diaspora, University of Leeds, September. http://artsweb.bham.ac.uk/aanderson/index.htm

Anderson, B. and Tolia-Kelly, D. (2004) 'Matter(s) in social and cultural geography', Geoforum 35(6): 669–74.

Andresen, S. (2007) 'The effectiveness of UN environmental institutions', International Environmental Agreements: Politics, Law and Economics 7(4): 317–36.

Appadurai, A. (2001) 'Deep democracy: urban governmentality and the horizon of politics', Environment and Urbanization 13(2): 23–43.

Appe, S. (2007a) 'From state intervention to cultural synthesizism in Bogotá, Colombia', Culture Work: A Periodic Broadside for Arts and Culture Workers 11(2), Centre for Community Arts and Cultural Policy, Arts and Administration, University of Oregon.

Appe, S. (2007b) 'Cultura para todos: Colombian Community Arts... and politics', community arts network, Saxapahaw, NC. http://www.communityarts.net/

Atkinson, P. (1999) 'Representations of conflict in the Western media: the manufacture of a barbaric periphery', in T. Skelton and J. Allen (eds) Culture and Global Change, London: Routledge, pp. 102–8.

Baiocchi, G. (2001) 'Participation, activism and politics: the Porto Alegre experiment and deliberative democratic theory', Politics and Society 29: 43–72.

Bakker, K. (2007) 'The "commons" versus the "commodity": alter-globalization, anti-privatization and the human right to water in the Global South', Antipode 39(3): 430–55.

Balassa, B. (1971) 'Trade policies in developing countries', American Economic Review 61: 178–87.

Bale, J. (1999) 'Foreign bodies: representing the African and the European in an early twentieth century "contact zone"', Geography 84(362): 25–33.

Barnett, C. (2003) 'Media transformation and new practices of citizenship: the example of environmental activism in post-apartheid Durban', Transformation: Critical Perspectives on South Africa 51: 1–24.

Barnett, C. (2005) 'The consolations of "neoliberalism"', Geoforum 36(1): 7–12.

Baviskar, A. (1997) In the Belly of the River; Tribal Conflicts Over Development in the Narmada Valley, Delhi: Oxford University Press.

Bayat, A. (1997) 'Un-civil society: the politics of "informal people"', Third World Quarterly 18(1): 53–72.

Baylies, C. and Bujra, J. (2000) AIDS, Sexuality and Gender in Africa: Collective Strategies and Struggles in Tanzania and Zambia, London: Routledge.

Beaverstock, J. (2002) 'Transnational elites in global cities: British expatriates in Singapore's financial district', Geoforum, 33(4): 525–38.

Beck, U. (1992) *Risk Society: Towards a New Modernity*, London: Sage.

Bell, D. and Valentine, G. (1997) *Consuming Geographies: We Are Where We Eat*, London: Routledge.

Bell, M. (1994) 'Images, myths and alternative geographies of the Third World', in D. Gregory, R. Martin and G. Smith (eds) *Human Geography: Society, Space and Social Science*, London: Macmillan, pp. 174–99.

Bennett, T.; Grossberg, L. and Morris, M. (eds) (2005) *New Keywords: A Revised Vocabulary of Culture and Society*, Oxford: Blackwell.

Bhabha, H. (1994) *Locations of Culture*, London: Routledge.

Bhatt, V. V. (1982) 'Development problem, strategy and technology choice: *Sarvodaya* and socialist approaches in India', *Economic Development and Cultural Change* 31(1): 85–99.

Bhattachrayya, D. (1999) 'Politics of middleness: the changing character of the Communist Party of India (Marxist) in rural West Bengal (1977–90)', in B. Rogaly, B. Harriss-White and S. Bose (eds) *Sonar Bangla? Agricultural Growth and Agrarian Change in West Bengal and Bangladesh*. London: Sage.

Bickham Mendez, J. (2002) 'Creating alternatives from a gender perspective: transnational organizing for maquila workers' rights in Central America', in N. A. Naples and M. Desai (eds) *Women's Activism and Globalization: Linking Local Struggle and Transnational Politics*, London: Routledge, pp. 121–41.

Bird, G. and Milne, A. (1999) 'Miracle to meltdown: a pathology of the East Asian financial crisis', *Third World Quarterly*, 20(2): 421–37.

Blackman, A. (2000) 'Informal sector pollution control: what policy options do we have?', *World Development* 28(12): 2067–82.

Blaut, J. M. (1993) *The Colonizer's Model of the World*, London: The Guilford Press.

Blom Hansen, T. and Stepputat, F. (eds) (2001) *States of Imagination: Ethnographic Explorations of the Postcolonial State*, Durham, NC: Duke University Press.

Blowfield, M. (1999) 'Ethical trade: a review of developments and issues', *Third World Quarterly* 20(4): 753–70.

Blunt, A. and Dowling, R. (2006) *Home*, London: Routledge.

Bonnett, A. (2004) *The Idea of the West*, Basingstoke: Palgrave Macmillan.

Bose, M. (2007) 'Women's home-centred work in India – the gendered politics of space', *International Development Planning Review* 29(3): 271–98.

Bourdieu, P. (1984) *Distinction: The Social Judgement of Taste*, London: Routledge.

Bourdillon, M. (2006) 'Children and work: a review of current literature and debates', *Development and Change*, 37: 1201–26.

Brabant, M. (2001) 'Bush's brother to face vote inquiry', *BBC News Online*, 5 January, 2001, 19:25 GMT: http://news.bbc.co.uk/1/hi/world/americas/1102806.stm

Bradshaw, S. and Linneker, B. (2003) 'Civil society responses to poverty reduction strategies in Nicaragua', *Progress in Development Studies* 3(2): 147–58.

Brandt, W. (Chair) (1980) *North–South: A Programme for Survival*, Report of the Independent Commission on International Development Issues, Cambridge, MA: MIT Press.

Brickell, K. (2008) '"Fire in the house": gendered experiences of drunkenness and violence in Siem Reap, Cambodia', *Geoforum* 39(5): 1637–1798.

Brunn, S. D. (ed.) (2006) *Wal-Mart World: The World's Biggest Corporation in the Global Economy*, London: Routledge.

Bryceson, D. (2002) 'The scramble in Africa: reorienting rural livelihoods', *World Development* 30(5): 725–39.

Bulatao, R. and Ross, J. (2002) 'Rating maternal and neonatal health services in developing countries', *Bulletin of the World Health Organization* 80(9): 721–7.

Bulbeck, C. (1998) *Re-Orienting Western Feminisms*, Cambridge: Cambridge University Press.

Burra, S. (2005) 'Towards a pro-poor framework for slum upgrading in Mumbai, India', *Environment and Urbanization* 17(1): 67–88.

Butler, J. (1999) *Gender Trouble: Feminism and the Subversion of Identity*, London: Routledge.

Buxton, J. (2005) 'Venezuela's contemporary political crisis in historical context', *Bulletin of Latin American Research* 24(3): 328–47.

Campbell, C. (2003) *'Letting Them Die': How HIV/AIDS Prevention Programmes Often Fail*, International Africa Institute, in association with James Currey, Oxford.

Campbell, D. (2007) 'Geopolitics and visuality: sighting the Darfur conflict', *Political Geography*, 26: 357–82.

CARICOM (Caribbean Community) (2008) CARICOM homepage: www.caricom.org (accessed 16 July 2008).

Carmody, P.R. and Owusu, F.Y. (2007) 'Competing hegemons? Chinese versus American geo-economic strategies in Africa', *Political Geography*, 26: 504–24.

Castles, S. and.Miller, M. J. (2003) *The Age of Migration: International Population Movements in the Modern World* (third edition), Basingstoke: Macmillan.

Cellular-news (2006) 'Coltan, gorillas and cellphones' at www.cellular-news.com/coltan/ (accessed 21 July 2006).

Chambers, R. (1983) *Rural Development: Putting the Last First*, Harlow: Pearson Education.

Chambers, R. (1994) 'The origins and practice of participatory rural appraisal', *World Development* 22(7): 953–69.

Chambers, R. (1997) *Whose Reality Counts: Putting the First Last*, London: Intermediate Technology Publications.

Chambers, R. and Conway, G. (1992) 'Sustainable rural livelihoods: practical concepts for the twenty-first century', IDS Discussion Paper 296, Brighton: Institute of Development Studies.

Chambers, R., Pacey, A. and Thrupp, L. (eds) (1989) *Farmer First: Farmer Innovation and Agricultural Research*, London: Intermediate Technology Publications.

Chang, H.-J. (2002) *Kicking Away the Ladder: Development Strategy in Historical Perspective*, London: Anthem.

Chant, S. (1991) *Women and Survival in Mexican Cities: Perspectives on Gender, Labour Markets and Low-Income Households*, Manchester: Manchester University Press.

Chant, S. (1997) *Women-Headed Households: Diversity and Dynamics in the Developing World*, Basingstoke: Macmillan.

Chant, S. (2007) *Gender, Generation and Poverty: Exploring the 'Feminisation of Poverty' in Africa, Asia and Latin America*, Cheltenham: Edward Elgar.

Chant, S. and McIlwaine, C. (1995) *Women of a Lesser Cost: Female Labour, Foreign Exchange and Philippine Development*, London: Pluto Press.

Chapman, G. (2002) 'Changing places: the roles of science and social science in the development of large-scale irrigation in South Asia', in R. Bradnock and G. Williams (eds) *South Asia in a Globalising World*, Harlow: Pearson Education, pp. 78–99.

Chatterjee, P. (2004) *The Politics of the Governed: Reflections on Popular Politics in Most of the World*, New York: Columbia University Press.

Chen, M. A. (2001) 'Women in the informal sector: a global picture, the global movement', *SAIS Review* 21(1): 71–82.

China Development Bank (2008) *The Three Gorges Dam*: http://www.cdb.com.cn/english/NewsInfo.asp?NewsId=280 (accessed 5 July 2008).

Chong, A. (2004) 'Singaporean foreign policy and the Asian values debate, 1992–2000: reflections on an experiment in soft power', *Pacific Review* 17: 95–133.

CIA (2008) *CIA World Factbook 2008*: https://www.cia.gov/library/publications/the-world-factbook/index.html (accessed 5 July 2008.)

Clapham, C. (1985) *Third World Politics: An Introduction*, London: Croom Helm.

Cleasby, A. (1995) *What in the World is Going On?*, London: 3WE.

Cohen, R. (1997) *Global Diasporas: An Introduction*, London: UCL Press.

Cole, S. (2007) 'Beyond authenticity and commodification', *Annals of Tourism Research* 34(4): 943–60.

Coleman, S. and Crang, M. (2002) 'Between place and performance', in S. Coleman and M. Crang (eds) *Tourism: Between Place and Performance*, Oxford: Berghahn.

Comercio Justo México (2008) Comercio Justo México homepage: www.comerciojusto.com.mx (accessed 5 June 2008).

Conroy, M. (2007) *Branded: How the 'Certification Revolution' is Transforming Global Corporations*, Gabriola, BC: New Society Publishers.

Constable, N. (ed.) (2004) *Cross-Border Marriages: Gender and Mobility in Transnational Asia*, Philadelphia: University of Pennsylvania Press.

Conway, G. (1988) Editorial, *RRA Notes* 1(1): 1.

Conway, G. (1997) *The Doubly Green Revolution: Food For All in the 21st Century*, London: Penguin.

Conway, G. and Barbier, E. (1990) *After the Green Revolution: Sustainable Agriculture for Development*, London: Earthscan.

Cooke, B. and Kothari, U. (eds) (2001) *Participation: The New Tyranny?* London: Zed Books.

Corbridge, S. (1993a) 'Colonialism, post-colonialism and the political geography of the Third World', in P. J. Taylor (ed.) *Political Geography of the Twentieth Century: A Global Analysis*, London: Belhaven Press.

Corbridge, S. (1993b) *Debt and Development*, Oxford: Blackwell.

Corbridge, S. (1998) 'Beneath the pavement only soil: the poverty of post-development', *Journal of Development Studies*, 34(6): 138–48.

Corbridge, S. (2007) 'The (im)possibility of development studies', *Economy and Society* 36(2): 179–211.

Corbridge, S. and Harriss, J. (2002) *Reinventing India: Liberalization, Hindu Nationalism and Popular Democracy*, Cambridge: Polity.

Corbridge, S. and Kumar, S. (2002) 'Community, corruption, landscape: tales from the tree trade', *Political Geography* 21: 765–88.

Corbridge, S., Williams, G., Srivastava, M. and Véron, R. (2005) *Seeing the State: Governance and Governmentality in India*, Cambridge: Cambridge University Press.

Cornia, G. A., Jolly, R. and Stewart, F. (1987) *Adjustment with a Human Face*, Oxford: Oxford University Press.

Craig, D. and Porter, D. (2003) 'Poverty Reduction Strategy Papers: a new convergence', *World Development* 31(1): 53–69.

Crang, P. (1997) 'Cultural turns and the (re)constitution of economic geography', in R. Lee and J. Wills (eds) *Geographies of Economies*, London: Arnold, pp. 3–15.

Crang, P. (2005) 'The geographies of material culture', in P. Cloke, P. Crang and M. Goodwin (eds) *Introducing Human Geographies* (second edition), London: Hodder Arnold, pp. 168–81.

Crewe, E. and Fernando, P. (2006) 'The elephant in the room: racism in representations, relationships and rituals', *Progress in Development Studies* 6(1): 40–54.

Crush, J. (ed.) (1995) *Power of Development*, London: Routledge.

Cumings, B. (1998) 'The Korean crisis and the end of "late" development', *New Left Review* 231: 43–72.

Dalton, M. (2008) 'My wave is that of David: civil society, women, and other political actors in Oaxaca, May–December 2006', *Antipode* 40(2): 216–20.

Dalzel, A. (1793) *The History of Dahomy, an Inland Kingdom of Africa, Compiled from Authentic Memoirs*, London: G & W Nicol.

Davis, M. (2006) *Planet of Slums*, London: Verso.

de Haan, L. and Zoomers, A. (2005) 'Exploring the frontier of livelihoods research', *Development and Change* 36(1): 27–47.

de Haas, H. (2006) 'Migration, remittances and regional development in Southern Morocco', *Geoforum* 37: 565–80.

de Soto, H. (1989) *The Other Path: The Invisible Revolution in the Third World*, New York: Harper and Row.

de Soto, H. (2000) *The Mystery of Capitalism: Why Capitalism Triumphs in the West and Fails Everywhere Else*, London: Black Swan.

de Sousa Santos, B. (1998) 'Participatory budgeting in Port Alegre: toward a redistributive democracy', *Politics and Society* 26: 461–510.

Deeb, L. (2006) *An Enchanted Modern: Gender and Public Piety in Shi'i Lebanon*, Princeton: Princeton University Press.

Department for International Development (DFID) (1999) Sustainable Livelihoods Guidance Sheets, Overview, DFID.

Department for International Development (DFID) (2000a) *Making Globalisation Work for the Poor*, London: HMSO. (Available from: http://www.dfid.gov.uk/pubs)

Department for International Development (DFID) (2000b) *Viewing the World: A Study of British Television Coverage of Developing Countries*, London: DFID. (Available from: www.dfid.gov.uk)

Department for International Development (DFID) (2006) *Eliminating World Poverty: Making Governance Work for the Poor*, London: The Stationery Office.

Desai, V. (2002) 'The role of non-governmental organisations (NGOs)', in V .Desai and R. Potter (eds) *The Companion to Development Studies*, London: Arnold, pp. 495–9.

Desai, V. and Potter, R. (eds) (2002) *The Companion to Development Studies*, London: Arnold.

Deurenberg, P. (2001) 'Universal cut-off BMI points for obesity are not appropriate', *British Journal of Nutrition* 85: 135–6.

Devasahayam, T. W. (2005) 'Power and pleasure around the stove: the construction of gendered identity in middle-class South Indian Hindu households in urban Malaysia', *Women's Studies International Forum* 28(1): 1–20.

Devenish, A. and Skinner, C. (2004) 'Organising workers in the informal economy: the experience of the Self Employed Women's Union, 1994–2004', School of Development Studies, University of KwaZulu-Natal. http://www.streetnet.org.za/english/DevenishandSkinner.pdf

Díaz del Castillo, B. (1956 [1632]) *The Discovery and Conquest of Mexico, 1517–21*, New York: Grove Press.

Dicken, P. (2007) *Global Shift: Mapping the Changing Contours of the World Economy* (fifth edition), London: Sage.

Dodds, K. (2005) *Global Geopolitics: A Critical Introduction*, Harlow: Pearson.

Douglass, M. (2006) 'Global householding in Pacific Asia', *International Development Planning Review* 28(4): 421–45.

Dowdney, L. (2003) *Children of the Drug Trade: A Case Study of Children in Organized Armed Violence in Rio de Janeiro*, Rio de Janeiro: 7Letras.

Drakakis-Smith, D. (2000) *Third World Cities* (second edition), London: Routledge.

Driver, F. (2001) *Geography Militant: Cultures of Exploration and Empire*, Oxford: Blackwell.

Dwyer, A. (1995) *On the Line: Life on the US–Mexican Border*, London: Latin American Bureau.

Dwyer, C. (1999) 'Migrations and diaspora', in P. Crang, P. Cloke and M. Goodwin (eds) *Introducing Human Geographies*, London: Edward Arnold, pp. 287–95.

Dwyer, D. and Bruce, J. (eds) (1998) *A Home Divided: Women and Income in the Third World*, Stanford, CA: Stanford University Press.

Economist, The (2005) 'Breaking up is hard to do', *The Economist*, 12 October.

Edwards, M. and Hulme, D. (eds) (1995) *Beyond the Magic Bullet: NGO Performance and Accountability in the Post-Cold War World*, London: Macmillan.

Ehrenreich, B. and Hochschild, A. R. (eds) (2002) *Global Woman: Nannies, Maids and Sex Workers in the New Economy*, London: Granta Books.

Ehrlich, P. R. (1968) *The Population Bomb*, New York: Ballantine Books.

Elson, D. (ed.) (1991) *Male Bias in the Development Process*, Manchester: Manchester University Press.

Encyclopaedia Britannica (2005) *The New Encyclopaedia Britannica, Knowledge in Depth*, Vol. 26 (fifteenth edition), London: Encyclopaedia Britannica.

Encyclopaedia Britannica (2007) *Britannica Book of the Year: Events of 2006*, London: Encyclopaedia Britannica.

Energy Information Administration (EIA) (2008) 'Country energy profiles': http://tonto.eia.doe.gov/country/index.cfm (accessed 20 May 2008).

England, K. and Ward, K. (eds) (2007) *Neoliberalization: States, Networks, Peoples*, Oxford: Blackwell.

Escobar, A. (1995) *Encountering Development: The Making and Unmaking of the Third World*, Princeton, NJ: Princeton University Press.

Evans, P. B (1989) 'Predatory, developmental, and other apparatuses: a comparative political economy perspective on the Third World state', *Sociological Forum* 4(4): 561–87.

Expomuseum (2008) Home page of the World's Fair Museum: www.expomuseum.com (accessed 8 June 2008).

Fairhead, J. and Leach, M. (1996) 'Rethinking the forest-savanna mosaic: colonial science and its relics in West Africa', in M. Leach and R. Mearns (eds) *The Lie of the Land: Challenging Received Wisdom on the African Environment*, Oxford: James Currey.

Fairtrade Foundation (2007) Fairtrade Foundation homepage: www.fairtrade.org.uk (accessed 4 October 2007).

Fairtrade Foundation (2008a) 'Facts and figures on Fairtrade': www.fairtrade.org.uk/what_is_fairtrade/facts_and_figures.aspx (accessed 29 July 2008).

Fairtrade Foundation (2008b) 'Fair trade and oversupply': http://www.fairtrade.org.uk/includes/documents/cm_docs/2008/F/1_Fairtrade_and_oversupply.pdf (accessed 29 July 2008).

Fairtrade Labelling Organizations International (FLO) (2007) FLO homepage: www.fairtrade.net (accessed 4 October 2007).

Fanon, F. (2001 [1961]) *The Wretched of the Earth*, London: Penguin.

Farmer, P. (2006) *AIDS and Accusation: Haiti and the Geography of Blame* (second edition), Berkeley: University of California Press.

Featherstone, D. (2003) 'Spatialities of transnational resistance to globalization: the maps of grievance of the inter-continental caravan', *Transactions of the Institute of British Geographers* 28(4): 404–21.

Federal Writers' Project, Tennessee (1939) *Tennessee: A Guide to the State*, State of Tennessee, Department of Conservation.

Feinsilver, J. M. (1989) 'Cuba as a "world medical power": the politics of symbolism', *Latin American Research Review* 24(2): 1–34.

Ferguson, J. (1990a) 'Mobile workers, modernist narratives: a critique of the historiography of transition on the Zambian copperbelt, Part one', *Journal of Southern African Studies* 16(3): 385–412.

Ferguson, J. (1990b) 'Mobile workers, modernist narratives: a critique of the historiography of transition on the Zambian copperbelt, Part two', *Journal of Southern African Studies* 16(4): 603–21.

Ferguson, J. (1990c) *The Anti-Politics Machine: 'Development', Depoliticization and Bureaucratic Power in Lesotho*, Cambridge: Cambridge University Press.

Ferlie, E., Ashburner, L., Fitzgerald, L. and Pettigrew, A. (1996) *The New Public Management in Action*, Oxford: Oxford University Press.

Fincher, R. and Iveson, K. (2008) *Planning and Diversity in the City: Redistribution, Recognition and Encounter*, Basingstoke: Palgrave Macmillan.

Finweek (2006) 'Banking the Muslim market', *Finweek*, 29 June.

Flint, C. and Taylor, P. (2004) *Political Geography: World-Economy, Nation-State and Locality* (fifth edition), Harlow: Pearson.

Flynn, K. C. (2005) *Food, Culture and Survival in an African City*, Basingstoke: Macmillan.

Food and Agriculture Organization (FAO) (2007) Food security statistics: http://www.fao.org (accessed 12 August 2007).

Foucault, M. (1980) *Power-Knowledge: Selected Interviews and Other Writings, 1972–77*, edited by Colin Gordon, Brighton: Harvester Press.

Frank, A. G. (1967) *Capitalism and Underdevelopment in Latin America*, London: Monthly Review Press.

Freedman, L. P. (2003) 'Strategic advocacy and maternal mortality: Moving targets and the Millennium Development Goals', *Gender and Development* 11(1): 97–108.

Freire, P. (1972 [1968]) *Pedagogy of the Oppressed*, Harmondsworth: Penguin.

Freund, B. (1998) *The Making of Contemporary Africa: The Development of African Society since 1800* (second edition), Boulder, CO: Lynne Rienner.

Friedmann, J. (1992) *Empowerment: The Politics of Alternative Development*, Oxford: Blackwell.

Fursich, E. and Robins, M. B. (2002) 'Africa.com: the self-representation of sub-Saharan nations on the World Wide Web', *Critical Studies in Media Communication* 19(2): 190–211.

Gadgil, M. and Guha, R. (1995) *Ecology and Equity: The Use and Abuse of Nature in Contemporary India*, London: Routledge.

Gaetano, A. (2008) 'Sexuality in diasporic space: Rural-to-urban migrant women negotiating gender and marriage in contemporary China', *Gender, Place and Culture.* 15(6): 629–45.

Gender and Aids, UNIFEM (2007) United Nations Development Fund for Women: www.genderandaids.org

Geoforum (2007) Themed issue on Pro-Poor Water? The Privatisation and Global Poverty Debate, *Geoforum* 38(5).

Geoforum (2008) Themed issue on Rethinking Economy, *Geoforum* 39(3): 1111–69.

Gereffi, G. (1994) 'The organization of buyer-driven global commodity chains: how US retailers shape overseas production networks', in G. Gereffi and M. Korzeniewicz (eds) *Commodity Chains and Global Capitalism*, London: Praeger, pp. 95–122.

Gereffi, G. and Korzeniewicz, M. (eds) *Commodity Chains and Global Capitalism*, London: Praeger.

Ghazvinian, J. (2007) *Untapped: The Scramble for Africa's Oil*, London: Harcourt.

Gibson-Graham, J. K. (2005) 'Surplus possibilities: postdevelopment and community economies', *Singapore Journal of Tropical Geography* 26: 4–26.

Gilbert, A. (1997) 'Employment and poverty during economic restructuring: the case of Bogotá, Colombia', *Urban Studies* 34(7): 1047–70.

Global Health Council: www.globalhealth.org

Gold, J. R. (ed.) (1994) *Place Promotion: The Use of Publicity and Marketing to Sell Towns and Regions*, Oxford: Blackwell.

Gollin, J. (2007) Available at: http://members.tripod.com/~D_Parent/threshold_dap.html#Facts

Goodenough, C. (2002) *Traditional Leaders: A KwaZulu-Natal Study 1999–2001*, Durban: Independent Project Trust.

Gootenberg, P. (1993) *Imagining Development: Economic Ideas in Peru's 'Fictitious Prosperity' of Guano, 1840–80*, Berkeley: University of California Press.

Government of Mali (2002) 'Final PRSP'. (Available at: www.worldbank.org/prsp/)

Greenpeace International (2006) 'Eating up the Amazon'. (Available from: www.greenpeace.org/international/press/reports/eating-up-the-amazon.pdf)

Gregory, D (2004) *The Colonial Present: Afghanistan, Palestine, Iraq*, Oxford: Blackwell.

Griffith-Jones, S. (ed.) (1988) *Managing World Debt*, Hemel Hempstead: Harvester Wheatsheaf.

Grillo, R. (2004) 'Islam and transnationalism', *Journal of Ethnic and Migration Studies* 30(5): 861–78.

Grindle, M. (2004) 'Good enough governance: poverty reduction and reform in developing countries', *Governance* 17(4): 525–48.

Gross International Happiness Project (2008) GIH homepage: www.grossinternationalhappiness.org/ (accessed 4 August 2008).

Grove, R. (1995) *Imperialism: Colonial Expansion, Tropical Island Edens and the Origins of Environmentalism, 1600–1860*, Cambridge: Cambridge University Press.

Guardian, The (2006) 'Soya', *Guardian G2 Supplement*, 25 July.

Guardian, The (2008) 'Tariffs: WTO talks collapse after India and China clash with America over farm products', *The Guardian*, 30 July. (Available at: www. guardian.co.uk).

Guha, R. and Martinez-Alier, J. (1997) *Varieties of Environmentalism: Essays North and South*, London: Earthscan.

Gupta, A. (1995) 'Blurred boundaries: the discourse of corruption, the culture of politics and the imagined state', *American Ethnologist* 22(2): 375–402.

Gutmann, M. (1996) *The Meanings of Macho: Being a Man in Mexico City*, Berkeley: University of California Press.

Hagemann, F., Diallo, Y., Etienne, A. and Mehran, F. (2006) *Global Child Labour Trends 2000–04*, Geneva: ILO.

Haggard, S. (2000) *The Political Economy of the Asian Financial Crisis*, Washington DC: Institute for International Economics.

Hall, C. M. and Lew, A. A. (eds) (1998) *Sustainable Tourism: A Geographical Prespective*, Harlow: Longman.

Hall, S. (1997) 'The work of representation', in S. Hall (ed.) *Representation: Cultural Representations and Signifying Practices*, London: Sage, pp. 13–72.

Halliday, F. (2002) 'West encountering Islam: Islamophobia reconsidered', in A. Mohammadi (ed.) *Islam Encountering Globalisation*, London: Routledge, pp. 14–35.

Hamnett, D. (2007) 'Cuban intervention in South African health care service provision', *Journal of Southern African Studies* 33(1): 63–81.

Hampton, M. P. and Christensen, J. (2007) 'Competing industries in islands: A new tourism approach', *Annals of Tourism Research* 34(4): 998–1020.

Hansen, K. T. (2004) 'Dressing dangerously: miniskirts, gender relations, and sexuality in Zambia', in J. Allman (ed.) *Fashioning Africa: Power and the Politics of Dress*, Bloomington and Indianapolis: Indiana University Press, pp. 166–85.

Hardiman, D. (2003) *Gandhi in His Time and Ours: The Global Legacy of His Ideas*, London: C. Hurst.

Hardt, M. and Negri, A. (2000) *Empire*, Cambridge, MA: Harvard University Press.

Harrison, E. (2000) 'Men, women and work in rural Zambia', *European Journal of Development Research* 12(2): 53–71.

Harrison, G. (2004) 'HIPC and the architecture of governance', *Review of African Political Economy* 31(99): 125–8.

Harrison, G. (2005) 'The World Bank, governance and theories of political action in Africa', *British Journal of Politics and International Relations* 7(2): 240–70.

Harrison, P. and Palmer, R. (1986) *News Out of Africa: Biafra to Band Aid*, London: Hilary Shipman.

Harrison, P., Todes, A. and Watson, V. (2008) *Planning and Transformation: Learning from the Post-Apartheid Experience*, London: Routledge.

Hart, G. (2001) 'Development critiques in the 1990s: *culs de sac* and promising paths', *Progress in Human Geography* 25: 649–58.

Hart, G. (2003) *Disabling Globalization: Places of Power in Post-Apartheid South Africa*, Berkeley: University of California Press.

Hartmann, B. (1995) *Reproductive Rights and Wrongs: The Global Politics of Population Control*, Boston, MA: South End Press.

Harvey, D. (2005) *The New Imperialism*, Oxford: Oxford University Press.

Harvey, D. (2007) *A Brief History of Neoliberalism*, Oxford: Oxford University Press.

Hastings, A. (1997) *Construction of Nationhood: Ethnicity, Religion and Nationalism*, Cambridge: Cambridge University Press.

Hay, M. J. (2004) 'Changes in clothing and struggles over identity in colonial western Kenya', in J. Allman (ed.) *Fashioning Africa: Power and the Politics of Dress*, Bloomington and Indianapolis: Indiana University Press, pp. 67–83.

Haynes, J. (1999) *Religion, Globalization and Political Culture in the Third World*, Basingstoke: Palgrave Macmillan.

Hecht, S. (2005) 'Soybeans, development and conservation on the Amazon frontier', *Development and Change* 36(2): 375–404.

Held, D., McGrew, A., Goldblatt, D. and Perraton, J. (1999) *Global Transformations*, Cambridge: Polity Press.

Henkel, H. and Stirrat, R. (2001) 'Participation as spiritual duty; empowerment as secular subjection', in B. Cooke and U. Kothari (eds) *Participation: The New Tyranny?* London: Zed Books, pp. 168–84.

Hickey, S. and Mohan, G. (eds) (2004) *Participation: From Tyranny to Transformation: Exploring New Approaches to Participation in Development*, London: Zed Books.

Hobsbawm, E. (1983) 'Introduction: inventing tradition', in E. Hobsbawm and T. Ranger (eds) *The Invention of Tradition*, Cambridge: Cambridge University Press, pp. 1–14.

Hodgson, D. L. (2001) 'Of modernity/modernities, gender and ethnography', in D. L. Hodgson (ed.) *Gendered Modernities: Ethnographic Perspectives*, Basingstoke: Palgrave, pp. 1–23.

Holland, J. and Blackburn, J. (1998) *Whose Voice? Participatory Research and Policy Change*, London: Intermediate Technology.

Hondagneu-Sotelo, P. and Avila, E. (1997) '"I'm here but I'm there": the meanings of Latina transnational motherhood', *Gender and Society* 11(5): 548–71.

Hoselitz, B. F. (1995 [1952]) 'Non-economic barriers to economic development', *Economic Development and Cultural Change* 1(1): 8–21. Excerpts reprinted as Reading 1 of S. Corbridge (ed.) *Development Studies: A Reader*, London: Edward Arnold, pp. 17–27.

Hudson, R. (2000) 'Offshoreness, globalization and sovereignty: a postmodern geopolitical economy?', *Transactions of the Institute of British Geographers* 25: 269–83.

Huish, R. and Kirk, J. M. (2007) 'Cuban medical internationalism and the development of the Latin American School of Medicine', *Latin American Perspectives* 34(6): 77–92.

Hulme, D. and Edwards, M. (eds) (1997) *Too Close for Comfort? NGOs, States and Donors*, London: Macmillan.

Huneeus, C. (2000) 'Technocrats and politicians in an authoritarian regime: the "ODEPLAN Boys" and the "Gremialists" in Pinochet's Chile', *Journal of Latin American Studies* 32: 461–501.

Ilife, J. (1995) *Africans: The History of a Continent*, Cambridge: Cambridge University Press.

International Labour Organization (ILO) (2004) 'Organizing for social justice', Report of the director-general, Global Report Under the Follow-Up to the ILO Declaration on Fundamental Principles and Rights at Work, International Labour Conference, 92nd Session, Report I (B), Geneva: ILO. (www.ilo.org/Declaration)

International Labour Organization (ILO) (2007) ILO homepage: www.ilo.org

International Labour Organization (ILO) (2008) Child labour by sector: agriculture: http://www.ilo.org/ipec/areas/Agriculture/index.htm (accessed 10 October 2008).

International Monetary Fund (IMF) (2008a) *IMF World Economic Outlook Database, April 2008*: http://www.imf.org/external/pubs/ft/weo/2008/01/weodata/index.aspx (accessed 5 July 2008).

International Monetary Fund (IMF) (2008b) 'The IMF at a glance'. Available at: www.imf.org/external/np/exr/facts/glance.htm (accessed 23 June 2008).

International Organization for Migration (IOM) (2008) IOM homepage: www.iom.int

Jackson, P. (2000) 'Rematerialising social and cultural geography', *Social and Cultural Geography* 1: 9–14.

Jackson, P. (2004) 'Local consumption cultures in a globalizing world', *Transactions of the Institute of British Geographers* 29: 165–78.

Jackson, P. (2005) 'Identities', in P. Cloke, P. Crang and M. Goodwin (eds) *Introducing Human Geographies*, London: Hodder Arnold, pp. 391–9.

Jackson, P., Thomas, N. and Dwyer, C. (2007) 'Consuming transnational fashion in London and Mumbai', *Geoforum* 38(5): 908–24.

Jaglin, S. (2008) 'Differentiating networked services in Cape Town: echoes of splintering urbanism?', *Geoforum* 39(6): 1897–1906.

Janowski, M. (2007) 'Introduction: feeding the right food: the flow of life and the construction of kinship in Southeast Asia', in M. Janowski and F. Kerlogue (eds) *Kinship and Good in South East Asia*, Copenhagen: NIAS, pp. 1–23.

Jarosz, L. (1992) 'Constructing the Dark Continent: metaphor as geographic representation of Africa', *Geografiska Annaler B* 74(2): 105–15.

Jenkins, R. (2002) 'The emergence of the governance agenda: sovereignty, neo-liberal bias and the politics of international development', in V. Desai and

R. Potter, (eds) *The Companion to Development Studies*, London: Edward Arnold, pp. 485–9.

Jenkins, R. and Goetz, A-M. (1999) 'Accounts and accountability: theoretical implications of the right-to-information movement in India', *Third World Quarterly* 20(3): 603–22.

Jenkins, R. and Edwards, C. (2006) 'The Asian drivers and Sub-Saharan Africa', *IDS Bulletin* 37(1): 23–32.

Jewitt, S. and Kumar, S. (2000) 'A political ecology of forest management: gender and silvicultural knowledge in the Jharkhand, India', in P. Stott and S. Sullivan (eds) *Political Ecology: Science, Myth and Power*, London: Edward Arnold, pp. 91–116.

Johnston, R. J. and Sidaway, J. D. (2004) *Geography and Geographers: Anglo-American Human Geography Since 1945* (sixth edition), London: Arnold.

Joint United Nations Program on HIV/AIDS (2007) www.unaids.org

Jones, A. (ed.) (2006) *Men of the Global South: A Reader*, London: Zed Books.

Jones, G. and Ward, P. (1994) 'The World Bank's "new" Urban Management Programme: paradigm shift or policy continuity', *Habitat International* 18(3): 33–51.

Juergensmeyer, M. (1984) 'The Gandhi revival: a review article', *Journal of Asian Studies* 43(2): 293–8.

Kabakian-Khasholian, T., Campbell, O., Shediac-Rizkallah, M. and Ghorayeb, F. (2000) 'Women's experiences of maternity care: satisfaction or passivity?', *Social Science and Medicine* 51: 103–13.

Kandiyoti, D. (2007) 'Old dilemmas or new challenges? The politics of gender and reconstruction in Afghanistan', *Development and Change* 38(2): 169–99.

Kantowsky, D. (1985) 'Gandhi: from East to West', *Manas* 38(44): 1–7.

Kaplinsky, R. (2005) *Globalization, Poverty and Inequality*, Cambridge: Polity Press.

Katz, C. (2004) *Growing Up Global: Economic Restructuring and Children's Everyday Lives*', Minneapolis: University of Minnesota Press.

Kaufmann, D., Kraay, A. and Mastruzzi, M. (2006) *Governance Matters V: Aggregate and Individual Governance Indicators for 1996–2005*, Washington, DC: The World Bank.

Kaviraj, S. (1991) 'On state, society and discourse in India', in J. Manor (ed.) *Rethinking Third World Politics*, London, Longman.

Kelly, P. (2000) *Landscapes of Globalization: Human Geographies of Economic Change in the Philippines*, London: Routledge.

Khan, M. S. and Mirakhor, A. (1990) 'Islamic banking: experiences in the Islamic Republic of Iran and in Pakistan', *Economic Development and Cultural Change* 38(2): 353–75.

Killercoke (2008) Killercoke homepage: www.killercoke.org (accessed 25 July 2008).

Klare, M. and Volman, D. (2006) 'The African "oil rush" and US national security', *Third World Quarterly* 27(4): 609–27.

Klein, N. (2001) *No Logo*, London: Flamingo Press.

Kohli, A. (1994) 'Where do high growth political economies come from? The Japanese lineage of Korea's "developmental state"', *World Development* 22(9): 1269–93.

Kohli, A. (ed.) (2001) *The Success of India's Democracy*. (Contemporary South Asia Series, Vol. 6), Cambridge: Cambridge University Press.

Konadu-Agyemang, K. (2000) 'The best of times and the worst of times: structural adjustment programs and uneven development in Africa: the case of Ghana', *Professional Geographer* 52(3): 469–83.

Kopinak, K. (1997) *Desert Capitalism*, London: Black Rose Books.

Korf, B. (2007) 'Antinomies of generosity: moral geographies and post-tsunami aid in Southeast Asia', *Geoforum* 38(2): 366–78.

Korten, D. C. (1987) 'Third generation NGO strategies: a key to people-centered development', *World Development* 15(1): 145–59.

Kothari, U. (2006) 'Spatial practices and imaginaries: experiences of colonial officers and development professionals', *Singapore Journal of Tropical Geography* 27(3): 235–53.

Kothari, U. and Laurie, N. (2005) 'Different bodies, same clothes: an agenda for local consumption and global identities', *Area* 37(2): 223–7.

Krueger, A. (1974) 'The political economy of rent-seeking society', *American Economic Review* 64: 291–303.

Lai, K. P. Y. (2006) '"Imagineering" Asian emerging markets: financial knowledge networks in the fund management industry', *Geoforum* 37(4): 627–42.

Langhammer, R. J. (2005) 'China and the G-21: a new North–South divide in the WTO after Cancún?', *Journal of the Asia Pacific Economy* 10(3): 339–58.

Last, M. (1999) 'Understanding health', in T. Skelton and T. Allen (eds) *Culture and Global Change*, London: Routledge, pp. 70–83.

Laurie, N. and Crespo, C. (2007) 'Deconstructing the best case scenario: lessons from water politics in La Paz – El Alto', *Geoforum* 38(5): 841–54.

Leach, M. and Mearns, R. (eds) (1996a) *The Lie of the Land: Challenging Received Wisdom on the African Environment*, Oxford: James Currey.

Leach, M. and Mearns, R. (1996b) 'Environmental change and policy: challenging received wisdom in Africa', in M. Leach and R. Mearns (eds) *The Lie of the Land: Challenging Received Wisdom on the African Environment*, Oxford: James Currey, pp. 1–33.

Lee, K., Buse, K. and Fustukian, S. (eds) (2002) *Health Policy in a Globalising World*, Cambridge: Cambridge University Press.

Lee, Y.-J. and Koo, H. (2006) '"Wild geese fathers" and a globalised family strategy for education in Korea', *International Development Planning Review* 28(4): 533–53.

Leeds, E. (1996) 'Cocaine and parallel polities in the Brazilian urban periphery', *Latin American Research Review* 31(3): 47–83.

Leeman, P. (2004). 'Book review: pedagogy of the oppressed'. (Available from: http://fcis.oise.utoronto.ca/dschugurensky/freire/pl.html)

Leinbach, T. R. (1972) 'The spread of modernization in Malaya: 1895–1969', *Tijdschrift Voor Economische en Sociale Geografie* July/Aug: 262–77.

Leinbach, T. R. (1975) 'Transportation and the development of Malaya', *Annals of the Association of American Geographers* 65(2): 270–82.

Lenin, V. I. (1973) *Imperialism: The Highest Stage of Capitalism*, Peking: Foreign Language Press. (Originally published 1917.)

Lever-Tracey, C. (2002) 'The impact of the Asian crisis on diaspora Chinese tycoons', *Geoforum* 33(4): 509–23.

Lewis, D. (2002) 'Non-governmental organisations: questions of performance and accountability', in V. Desai and R. Potter (eds) *The Companion to Development Studies*, London: Arnold, pp. 519–23.

Lewis, W. A. (1955) *The Theory of Economic Growth*, London: George Allen and Unwin.

Li, T. (2000) 'Articulating indigenous identity in Indonesia: resource politics and the tribal slot', *Comparative Studies in Society and History* 42(1): 149–79.

Lidchi, H. (1999) 'Finding the right image: British development NGOs and the regulation of imagery', in T. Skelton and T. Allen (eds) *Culture and Global Change*, London: Routledge, pp. 87–101.

Lindsey, B. (2004) 'Grounds for complaint? "Fair trade" and the coffee crisis', London: Adam Smith Institute. (Available at: www.adamsmith.org)

Lipton, M. and Longhurst, R. (1989) *New Seeds and Poor People*, London: Unwin and Hyman.

Livingstone, D. (1857) *Missionary Travels and Researches in South Africa*, London: John Murray.

Livingstone, David N. (1992) *The Geographical Tradition: Episodes in the History of a Contested Enterprise*, Oxford: Blackwell.

Lloyd-Evans, S (2002) 'Child labour', in V. Desai and R. Potter (eds) *The Companion to Development Studies*, London: Arnold, pp. 215–19.

Lozada Jr, E. P. (2000) 'Globalized childhood? Kentucky Fried Chicken in Beijing', in J. Jing (ed.) *Feeding China's Little Emperors: Food, Children and Social Change*, Stanford, CA: Stanford University Press, pp. 114–34.

Lutz, C. A. and Collins, J. L. (1993) *Reading National Geographic*, Chicago, IL: University of Chicago Press.

McCarthy, J. and Prudham, S. (2004) 'Neoliberal nature and the nature of neoliberalism', *Geoforum* 35(3): 275–83.

McClenaghan, S. (1997) 'Women, work and empowerment: romanticizing the reality', in E. Dore (ed.) *Gender Politics in Latin America*, New York: Monthly Review Press, pp. 19–35.

McClintock, A. (1995) *Imperial Leather: Race, Gender and Sexuality in Colonial Contest*, London: Routledge.

McCord, A. (2005) 'Win-win or lose-lose? An examination of the use of public works as a social protection instrument in situations of chronic poverty', paper presented at the Conference on Social Protection for Chronic Poverty,

Institute for Development Policy and Management, University of Manchester, February.

McCormick, J. (1989) *The Global Environmental Movement: Reclaiming Paradise*, London: Belhaven.

McDowell, L. (1998) *Gender, Identity and Place*, Cambridge: Polity Press.

McEvedy, C. (1972) *The Penguin Atlas of Modern History to 1815*, London: Penguin.

McFarlane, C. (2008a) 'Postcolonial Bombay: decline of a cosmopolitanism city?', *Environment and Planning D: Society and Space* 26: 480–99.

McFarlane, C. (2008b) 'Sanitation in Mumbai's informal settlements: state, "slum" and infrastructure', *Environment and Planning A* 40: 88–107.

Magadi, M., Zulu, E. and Brockerhoff, M. (2003) 'The inequality of maternal health care in urban sub-Saharan Africa in the 1990s', *Population Studies* 57(3): 347–66.

Malthus, T. (1985 [1798]) *An Essay on the Principle of Population*, London: Penguin.

Mamdani, M. (1996) *Citizen and Subject: Decentralized Despotism and the Legacy of Late Colonialism*, Princeton, NJ: Princeton University Press.

Mangin, W. (1967) 'Latin American squatter settlements: a problem and a solution', *Latin American Research Review* 2: 65–98.

Marquette, C. M. (1997) 'Current poverty, structural adjustment and drought in Zimbabwe', *World Development* 25(7): 1141–49.

Martin, P. and Gática, R. (2008) 'An interview with Raul Gática from the Popular Indigenous Council of Oaxaca–Ricardo Flores Magón (CIPO-RFM)', *Antipode* 40(2): 211–15.

Marx, K. (1909) *Capital*, Vol. 1, London: William Glaisher.

Massey, D. (1996) 'A global sense of place', in S. Daniels and R. Lee (eds) *Exploring Human Geography*, London: Arnold, pp. 237–45. (Originally published in *Marxism Today*, June 1991, pp. 24–9.)

Mawdsley, E. (2002) 'Regionalism, decentralisation and politics: state reorganisation in contemporary India', in R. Bradnock and G. Williams (eds) *South Asia in a Globalising World: A Reconstructed Regional Geography*, Harlow: Prentice Hall, pp. 122–43.

Mawdsley, E. (2007) 'China and Africa: emerging challenges to the geographies of power', *Geography Compass* 1(3): 405–21.

Mawdsley, E. and Rigg, J. (2002) 'A survey of World Development Reports I: discursive strategies', *Progress in Development Studies* 2(2): 93–111.

Mawdsley, E. and Rigg, J. (2003) 'The World Development Report II: continuity and change in development orthodoxies', *Progress in Development Studies* 3(4): 271–86.

Mbendi Information for Africa (2007) www.mbendi.co.za

Meadows, D. H., Meadows, D. L., Randers, J. and Behrens III, W.W. (1972) *The Limits to Growth: A Report for the Club of Rome's Project on the Predicament of Mankind*, London: Earth Island.

Mee, L. (2005) 'The role of UNEP and UNDP in multilateral environmental agreements', *International Environmental Agreements: Politics, Law and Economics* 5(3): 303–21.

MegaCity TaskForce of the International Geographical Union (IGU) (2008) http://www.megacities.uni-koeln.de/index.htm (accessed 1 June 2008).

Menegat, R. (2002) 'Participatory democracy and sustainable development: integrated urban environmental management in Porto Alegre, Brazil', *Environment and Urbanization* 14(2): 181–206.

Meth, P. (2003) 'Rethinking the "domus" in domestic violence: homelessness, space and domestic violence in South Africa', *Geoforum* 34: 317–27.

Meth, P (2007) 'Marginal emotions or emotions from the margin? Male informal residents reflect on their positions', paper for the Royal Geographical Society/Institute of British Geographer's Annual Conference, London, August.

Miller, D. (1998) 'Coca-Cola: a black sweet drink from Trinidad', in D. Miller (ed.) *Material Cultures: Why Some Things Matter*, Chicago, IL: University of Chicago Press, pp. 169–87.

Millennium Challenge Corporation (MCC) (2007) *Report on the Criteria and Methodology for Determining the Eligibility of Candidate Countries for Millennium Challenge Account Assistance in Fiscal Year 2008*. Available from: http://www.mca.gov/documents/mcc-report-fy08-criteriaandmethodology.pdf

Ministry of the Economy and Finance, Mali (2006) 'Growth and poverty reduction strategy paper. second generation PRSP, 2007–11'. (Available at: www.worldbank.org/prsp/)

Mirchandani, K. (2004) 'Practices of global capital: gaps, cracks and ironies in transnational call centre in India', *Global Networks* 4(4): 355–73.

Mitchell, K. (1997) 'Different diasporas and the hype of hybridity', *Environment and Planning D: Society and Space* 15: 533–53.

Mitchell, T. (1995) 'The object of development: America's Egypt', in J. Crush (ed.) *Power of Development*, London: Routledge, pp. 129–57.

Mngadi, P. T., Thembi, I. T., Ransjö-Arvidson, A.–B. and Ahlberg, B. M. (2002) 'Quality of maternity care for adolescent mothers in Mbabane, Swaziland', *International Nursing Review* 49: 38–46.

Mohammadi, A. (ed.) (2002) *Islam Encountering Globalisation*, London: Routledge.

Mohan, G. and Stokke, K. (2000) 'Participatory development and empowerment: the dangers of localism', *Third World Quarterly* 21(2): 247–68.

Mohan, G. and Hickey, S. (2004) 'Relocating participation within a radical politics of development: critical modernism and citizenship', in S. Hickey and G. Mohan (eds) *Participation: From Tyranny to Transformation: Exploring New Approaches to Participation in Development*, London: Zed Books, pp. 59–74.

Mohan, G., Brown, E., Milward, B. and Zack-Williams, A. B. (2000) *Structural Adjustment: Theory, Practice and Impacts*, London: Routledge.

Mohanty, C. T. (1991) 'Under Western eyes: feminist scholarship and colonial discourses', in C. T. Mohanty, A. Russo and L. Torres (eds) *Third World Women and the Politics of Feminism*, Bloomington: Indiana University Press, pp. 51–80.

Momsen, J. H. (2004) *Gender and Development*, London: Routledge.

Morrell, R. and Swart, S. (2005) 'Men in the Third World: postcolonial perspectives on masculinity', in M. Kimmel, J. Hearn and R. W. Connell (eds) *Handbook of Studies on Men and Masculinities*, London: Sage, pp. 90–113.

Morris, M. (2006) 'China's dominance of global clothing and textiles: is preferential trade access an answer for Sub-Saharan Africa?, *IDS Bulletin* 37(1): 89–97.

Morse, S. (2004) *Indices and Indicators of Development: An Unhealthy Obsession with Numbers*, London: Earthscan.

Moser, C. (1992) 'Adjustment from below: low-income women, time and the triple role in Guayaquil, Ecuador', in H. Afshar and C. Dennis (eds) *Women and Adjustment Policies in the Third World*, Basingstoke: Macmillan, pp. 87–116.

Murphy, C. N. (2000) 'Global governance: poorly done and poorly understood', *International Affairs* 76(4): 789–804.

Murray, S. O. (1998) 'Sexual politics in contemporary Southern Africa', in S. O. Murray and W. Roscoe (eds) *Boy-Wives and Female Husbands: Studies of African Homosexualities*, Basingstoke: Macmillan, pp. 243–54.

Murray, S. O. and Roscoe, W. (eds) (1998) *Boy-Wives and Female Husbands: Studies of African Homosexualities*, Basingstoke: Macmillan.

Murray, W. E. (2006) *Geographies of Globalization*, London: Routledge.

Mutersbaugh, T. (2008) 'Oaxaca: terror and non-violent protest in a video age', *Antipode* 40(2): 205–10.

Myers, G. (2001) 'Introducing human geography textbook representations of Africa', *Professional Geographer* 53(4): 522–32.

Nafzinger, E. W. (1993) *The Debt Crisis in Africa*, Oxford: Blackwell.

Nagar, R., Lawson, V., McDowell, L. and Hanson, S. (2002) 'Locating globalization: feminist (re)readings of the subjects and spaces of globalization', *Economic Geography* 78(3): 257–84.

Najam, A. (2005) 'Developing countries and global environmental governance: from contestation to participation to engagement', *International Environmental Agreements: Politics, Law and Economics* 5(3): 303–21.

Nanda, V. P. (2006) 'The "good governance" concept revisited', *Annals of the American Academy of Political and Social Science* 603: 269–83.

Narayan, D. (2000) *Voices of the Poor: Can Anyone Hear Us?* New York: Oxford University Press for World Bank.

Nederveen Pieterse, J. (1992) *White on Black: Images of Africa and Blacks in Western Popular Culture*, London: Yale University Press.

Nelson, P. J. (2002) 'The World Bank and NGOs', in V. Desai and R. Potter (eds) *The Companion to Development Studies*, London: Arnold, pp. 499–504.

New Internationalist (2006) 'The Venezuelan revolution', *New Internationalist* 390. (Available at: www.newint.org)

Nieuwenhuys, O. (2007) 'Embedding the global womb: global child labour and the new policy agenda', *Children's Geographies* 5: 149–63.

Norget, K. (2006) *Days of Death, Days of Life: Ritual in the Popular Culture of Oaxaca*, New York: Columbia University Press.

Norris, R. S. and Kristensen, H. M. (2005) 'India's nuclear forces, 2005', *Bulletin of the Atomic Scientists* 61(5): 73–5.

Ntsebeza, L. (2003) 'Democracy in South Africa's countryside: is there a role for traditional authorities?', *Development Update* 4(1): 55–84.

Ntsebeza, L. (2004) 'Democratic decentralisation and traditional authority: dilemmas of land administration in rural South Africa', *European Journal of Development Research* 16(1): 71–89.

O'Reilly, K. (2000) *The British on the Costa del Sol: Transnational Identities and Local Communities*, London: Routledge.

Ó Tuathail, G. (1996) *Critical Geopolitics: The Politics of Writing Global Space*, London: Routledge.

Ó Tuathail, G (2002) 'Post Cold-War geopolitics: contrasting superpowers in a world of global dangers', in R. J. Johnston, P. J. Taylor and M. Watts (eds) *Geographies of Global Change*, Oxford: Blackwell.

Ojaba, E., Leonardo, A. and Leonardo, M. (2002) 'Food aid in complex emergencies: lessons from Sudan', *Social Policy and Administration* 36(6): 664–84.

Ong, F. S. (2002) 'Ageing in Malaysia: a review of national policies and programmes', in D. R. Phillips and A. C. M. Chan (eds) *Ageing and Long-Term Care: National Policies in the Asia-Pacific*, Ottawa: International Development Research Centre (IDRC).

Organization of Petroleum Exporting Counties (OPEC) (2008) OPEC homepage: www.opec.org (accessed 15 June 2008).

Osorno, D. and Meyer, L. (2007) *Oaxaca Sitiada: La primera insurreccion del siglo XXI* (Beseiged Oaxaca: The First Insurrection of the 21st Century), Mexico City: Grijalbo Mondadori Sa.

Ouzgane, L. (ed.) (2006) *Islamic Masculinities*, London: Zed Books.

Overseas Compatriot Affairs Commission (OCAC) (2007) Statistics: www.ocac.gov.tw

Oxfam (2007) 'Make trade fair', at: www.maketradefair.org (accessed 4 October 2007).

Pan American Health Organization (PAHO) (2002) *Health in the Americas: Volume I. 2002 Edition*, Washington: PAHO.

Pan American Health Organization (PAHO) (2007) Available at: www.paho.org

Painter, J. (1995) *Politics, Geography and Political Geography: A Critical Perspective*, London: Arnold.

Palmer, M. (ed.) (2002) *The Times World Religions*, London: Times Books.

Parajuli, P. (1996) 'Ecological ethnicity in the making: developmentalist hegemonies and emergent identities in India', *Identities* 31–2: 15–59.

Parnwell, M. (1993) *Population Movements and the Third World*, London: Routledge.

Parnwell, M. (2003) 'Consulting the poor in Thailand: enlightenment or delusion?', *Progress in Development Studies* 3(2): 99–112.

Parreñas, R. (2001) *Servants of Globalization: Women, Migration and Domestic Work*, Stanford, CA: Stanford University Press.

Parreñas, R. (2005) *Children of Global Migration: Transnational Families and Gendered Woes*, Stanford, CA: Stanford University Press.

Patel, S. and Arputham, J. (2007) 'An offer of partnership or a promise of conflict in Dharavi, Mumbai?', *Environment and Urbanisation* 19(2): 501–8.

Patrick, E. (2005) 'Intent to destroy: the genocidal impact of forced migration in Darfur, Sudan', *Journal of Refugee Studies* 18(4): 410–29.

Payne, A. (2006) 'The end of green gold? Comparative development options and strategies in the Eastern Caribbean banana-producing islands', *Studies in International Comparative Development* 41(3): 25–46.

Peck, J. and Tickell, A. (2002) 'Neoliberalizing space', *Antipode* 34(3): 380–404.

Peet, R. (2007) *Geography of Power: The Making of Global Economic Policy*, London: Zed Books.

Peet, R. and Watts, M. (eds) (2004) *Liberation Ecologies: Environment, Development and Social Movements* (second edition), London: Routledge.

Perrons, D. (2004) *Globalization and Social Change: People and Places in a Divided World*, London: Routledge.

Perry, M., Kong, L. and Yeoh, B. (1997) *Singapore: A Developmental City State*, Chichester: Wiley.

Phelps, N. (2007) 'Gaining from globalisation? State extraterritoriality and domestic economic impacts: The case of Singapore', *Economic Geography* 83(4): 371–93.

Pollard, J. and Samers, M. (2007) 'Islamic banking and finance: postcolonial political economy and the decentring of economic geography', *Transactions of the Institute of British Geographers* 32(3): 313–30.

Power, M. (2003) *Rethinking Development Geographies*, London: Routledge.

Pratt, M. L. (1992) *Imperial Eyes: Travel Writing and Transculturation*, London: Routledge.

Prieur, A. (1998) *Mema's House, Mexico City: On Transvestites, Queens and Machos*, Chicago, IL: University of Chicago Press.

Prunier, G. (2004) 'Rebel movements and proxy warfare: Uganda, Sudan and the Congo', *African Affairs* 103: 359–83.

Purvis, A. (2003) 'The tribe that survives on chocolate', *The Observer Food Magazine* 9 November. (Available at: www.guardian.co.uk)

Raghuram, P. (1999) 'Religion and development', in T. Skelton and J. Allen (eds) *Culture and Global Change*, London: Routlege, pp. 232–9.

Rahnema, M. (1992) 'Participation', in W. Sachs (ed.) *The Development Dictionary: A Guide to Knowledge as Power*, London: Zed Books, pp. 116–31.

Rahnema, M. (1997) 'Introduction', in M. Rahnema with V. Bawtree (eds) *The Post-Development Reader*, London: Zed Books, pp. ix–xix.

Rangachari, R., Sengupta, N., Iyer, R.R., Banerji, P. and Singh, S. (2000) *Large Dams: India's Experience*, a WCD case study prepared as an input to the World Commission on Dams, Cape Town. www.dams.org

Rangan, H. and Gilmartin, M. (2002) 'Gender, traditional authority and the politics of rural reform in South Africa', *Development and Change* 33(4): 633–58.

Ranger, T. (1983) 'The invention of tradition in colonial Africa', in E. Hobsbawm and T. Ranger (eds) *The Invention of Tradition*, Cambridge: Cambridge University Press, pp. 211–62.

Raynolds, L. (1997) 'Restructuring national agriculture, agro-food trade, and agrarian livelihoods in the Caribbean', in D. Goodman and M. Watts (eds) *Globalising Food: Agrarian Questions and Global Restructuring*, London: Routledge, pp. 86–96.

Redmond, I. (2001) (Available at: www.bornfree.org.uk/coltan/coltan.pdf)

Rigg, J. (2003) *Southeast Asia: The Human Landscape of Modernization and Development* (second edition), London: Routledge.

Rigg, J. (2007) *An Everyday Geography of the Global South*, London: Routledge.

Robbins, P. (2004) *Political Ecology*, London: Blackwell.

Roberts, S. M. (2002) 'Global regulation and trans-state organisation', in R. J. Johnston, P. J. Taylor and M. Watts (eds) *Geographies of Global Change: Remapping the World* (second edition), Oxford: Blackwell, pp. 143–57.

Robinson, J. (2006) *Ordinary Cities: Between Modernity and Development*, London: Routledge.

Rodney, W. (1972) *How Europe Underdeveloped Africa*, London: Bogle-L'Ouverture Publications.

Rogerson, C. (1996) 'Urban poverty and the informal economy in South Africa's economic heartland', *Environment and Urbanization*, 8(1): 167–97.

Rostow, W. W. (1960) *The Stages of Economic Growth: A Non-Communist Manifesto*, Cambridge: Cambridge University Press.

Routledge, P. (2003) 'Voices of the dammed: discursive resistance amidst erasure in the Narmada Valley, India', *Political Geography* 22: 243–70.

Roy, A. (1999) 'The greater common good', in A. Roy, *The Cost of Living*, London: Harper Collins.

Rozario, S. (2006) 'The new burqa in Bangladesh: empowerment or violation of women's rights', *Women's Studies International Forum* 29(4): 368–80.

Ruthven, M. (1997) *Islam: A Very Short Introduction*, Oxford: Oxford University Press.

Sachs, W. (ed.) (1992) *The Development Dictionary: A Guide to Knowledge as Power*, London: Zed Books.

Said, E. (1978) *Orientalism: Western Conceptions of the Orient*, Harmondsworth: Penguin.

Salt, J. and Stein, J. (1997) 'Migration as a business: the case of trafficking', *International Migration* 35(4): 467–94.

Santos, M. (1979) *Shared Space: The Two Circuits of the Urban Economy in Underdeveloped Countries*, London: Methuen.

Sautman, B. and Hairong, Y. (2007) 'Friends and interests: China's distinctive links with Africa', *African Studies Review* 50(3): 75–114.

Sauven, J. (2006) 'The odd couple', *The Guardian*, 2 August. (Available at: www. guardian.co.uk)

Schech, S. and Haggis, J. (2000) *Culture and Development: A Critical Introduction*, Oxford: Blackwell.

Schoonmaker Freudenberger, M. and Schoonmaker Freudenberger, K. (1993) 'Fields, fallow and flexibility: natural resource management in Ndam Mor Fademba, Senegal. Results of a Rapid Rural Appraisal', *IIED Drylands Papers* 05, London: IIED.

Scott, D. and Barnett, C. (forthcoming) 'Something in the air: civic science and contentious environmental politics in post-apartheid South Africa', *Geoforum*.

Scott, J. C. (1985) *The Weapons of the Weak: Everyday Forms of Peasant Resistance*, New Haven, CT: Yale University Press.

Scott, J. C, (1998) *Seeing Like a State: How Certain Schemes to Improve the Human Condition have Failed*, New Haven, CT: Yale University Press.

Sen, K. (2003) 'Restructuring health services and policies of privatisation: an overview of experience', in K. Sen (ed.) *Restructuring Health Services: Changing Contexts and Comparative Perspectives*, London: Zed Books, pp. 1–32.

Short, R. J. (1982) *An Introduction to Political Geography*, London: Routledge and Kegan Paul.

Sidaway, J. D. and Pryke, M. (2000) 'The strange geographies of "Emerging Markets"', *Transactions of the Institute of British Geographers* 25: 187–201.

SIDSNET (2008) SIDSNET homepage: www.sidsnet.org (accessed 1 August 2008).

Sidwell, M. (2008) 'Unfair trade', London: Adam Smith Institute. (Available at: www.adamsmith.org)

Silva, P. (1991) 'Technocrats and politics in Chile: from the Chicago boys to the CIEPLAN monks', *Journal of Latin American Studies*, 23: 385–410.

Simone, A. (2004) *For the City Yet to Come: Changing African Life in Four Cities*, London: Duke University Press.

Skaggs, J. M. (1994) *The Great Guano Rush: Entrepreneurs and American Overseas Expansion*, Basingstoke: Macmillan.

Skelton, T. and Allen, T. (eds) (1999) *Culture and Global Change*, London: Routledge.

Sklair, L. (2000) *The Transnational Capitalist Class*, Oxford: Blackwell.

Slater, C. (1995) 'Amazonia as Edenic narrative', in W. Cronon (ed.) *Uncommon Ground: Rethinking the Human Place in Nature*, New York: W. W. Norton.

Smith, A. (1979) *An Inquiry into the Nature and Causes of the Wealth of Nations*, Books 1–3, London: Penguin. (Originally published in 1776.)

Smith, A., Stenning, A. and Willis, K. (eds) (2008) *Social Justice and Neolibealism: Global Perspectives*, London: Zed Books.

Smith, D. M. (1992) 'Introduction', in D. M. Smith (ed.) *The Apartheid City and Beyond: Urbanization and Social Change in South Africa*, London: Routledge, pp. 1–10.

Smith, L. (2008) 'Power and the hierarchy of knowledge: a review of a decade of the World Bank's relationship with South Africa', *Geoforum* 39(1): 236–51.

Smith, M. (1997) 'Paulo Freire and informal education', *Encyclopaedia of Informal Education* (www.infed.org/thinkers/et-freir.htm).

Solomon, N. (2005) 'Aids 2004, Bangkok: a human rights and development issue', *Reproductive Health Matters* 13(25): 174–81.

South African Customs Union (SACU) (2008) SACU homepage: www.sacu.int. (accessed 16 July 2008).

Souza, C. (2001) 'Participatory budgeting in Brazilian cities: limits and possibilities in building democratic institutions', *Environment and Urbanization* 13: 159–84.

Sparke, M. (2007) 'Geopolitical fears, geoeconomic hopes and the responsibilities of geography', *Annals of the Association of American Geographers* 97(2): 338–49.

Standard & Poor's (2006) 'The world by numbers: global stock market review, June 2006'. (Available at: www.standardandpoors.com)

Standard & Poor's (2007) 'Standard & Poor's Emerging Markets Indices. 31 December 2007'. (Available at: www.standardandpoors.com)

Stewart, F. (1995) *Adjustment and Poverty: Options and Choices*, London: Routledge.

Stivens, M. (2006) '"Family values" and Islamic revival: gender, rights and state moral projects in Malaysia', *Women's Studies International Forum* 29(4): 354–67.

Stockholm International Peace Research Institute (2008) Database of information from *SIPRI Yearbook 2007*: http://first.sipri.org/ (accessed 5 July 2008).

Streetnet International (2007) Available at: www.streetnet.org.za

Structural Adjustment Participatory Review Initiative (SAPRI)/Ghana (2001) The Impact of SAP on Access to and Quality of Tertiary Education, Draft, April, Downloadable from: http://www1.worldbank.org/sp/safetynets/publications/files/ghanahealth.pdf

Subrahmanian, R. (2002) 'Children's work and schooling: a review of the debates', in V. Desai and R. Potter (eds) *The Companion to Development Studies*, London: Arnold, pp. 400–5.

Tanski, J. M. (1994) 'Impact of crisis, stabilization and structural adjustment on women in Lima, Peru', *World Development* 22(11): 1627–42.

Tassi, G. (2004) 'Death looms over Miskito lobster divers', *InterPress Service*, 3 February.

Taylor, P. and Bain, P. (2008) 'United by a common language? Trade union responses in the UK and India to call centre offshoring', *Antipode* 40(1): 131–54.

Telfer, D. J. and Sharpley, R. (2008) *Tourism and Development in the Developing World*, London: Routledge.

Tendler, J. (1997) *Good Governance in the Tropics*, Baltimore, MD: Johns Hopkins University Press.

Thomas, A (with Crow, B., Frenz, P., Hewitt, T., Kassan, S. and Treagust, S.) (1994) *Third World Atlas* (second edition), Buckingham: Oxford University Press.

Tiffen, P. (2002) 'A chocolate-coated case for alternative business models', *Development in Practice* 12(3 and 4): 383–97.

Titu Cusi Yupanqui (2005 [1570]) *An Inca Account of the Conquest of Peru*, translated, introduced and annotated by R. Bauer, Boulder: University Press of Colorado.

Toal, G. (2003) 'Reasserting the regional: political geography and geopolitics in a world thinly known', *Political Geography* 22(6): 653–65.

Tomlinson, J (1999) 'Globalised culture: the triumph of the West?', in T. Skelton and T. Allen (eds) *Culture and Global Change*, London: Routledge, pp. 22–9.

Toyota, M. (2006) 'Ageing and transnational householding', *International Development Planning Review* 28(4): 515–31.

Tull, D. M. (2006) 'China's engagement in Africa: scope, significance and consequences', *Journal of Modern African Studies* 44(3): 459–79.

UNAIDS (2006) www.unaids.org/ Joint United Nations Programme on HIV/AIDS.

UN-Habitat (2003) *Slums of the World: The Face of Urban Poverty in the New Millennium?* Nairobi: UN-Habitat. (Available at: www.unhabitat.org)

UN-Habitat (2007) 'World population prospects'. (Available at: www.unhabitat.org)

United Nations (2004) *Basic Facts about the United Nations*, New York: United Nations.

United Nations (2008) 'About the United Nations: introduction to the structure and work of the UN': http://www.un.org/aboutun/ (accessed 14 June 2008).

United Nations Centre for Human Settlements (HABITAT) (1996) *An Urbanizing World: Global Report on Human Settlements, 1996*, Oxford: Oxford University Press.

United Nations Conference on Trade and Development (UNCTAD) (2007) *World Investment Report 2007*, London and Geneva: UNCTAD. (Available at: www.unctad.org)

United Nations Conference on Trade and Development (UNCTAD) (2008a) *UNCTAD Handbook of Statistics*, New York and Geneva: UNCTAD. (Available at www.unctad.org)

United Nations Conference on Trade and Development (UNCTAD) (2008b) *Development and Globalization: Facts and Figures*, New York and Geneva: UNCTAD. (Available at: www.unctad.org)

United Nations Department of Economic and Social Affairs (UNDESA) (2006) 'World migrant stock: the 2005 revision population database'. http://esa.un.org/migration/ (accessed 24 July 2006).

United Nations Department of Economic and Social Affairs (UNDESA) (2008) UNDESA Division for Sustainable Development, SIDS webpage: http://www.un.org/esa/sustdev/sids/sidslist.htm (accessed 1 August 2008).

United Nations Development Programme (UNDP) (2004) *Human Development Report 2004*, Oxford: Oxford University Press.

United Nations Development Programme (UNDP) (2005) *Human Development Report 2005*, Oxford: Oxford University Press. (Available at: www.undp.org)

United Nations Development Programme (UNDP) (2006) *Human Development Report 2006*, Oxford: Oxford University Press. (Available at: www.undp.org)

United Nations Development Programme (UNDP) (2007) *Human Development Report 2007/8*, Oxford: Oxford University Press. (Available at: www.undp.org)

United Nations Human Settlements Programme (UNHSP) (UN-HABITAT) (2003) *The Challenge of Slums: Global Report on Human Settlements 2003*, London: Earthscan Publications.

United Nations Population Division (2007) 'World population ageing, 2007', UNDESA, Population Division.

United States Department of State (2007) *Why Population Aging Matters: A Global Perspective*, Washington, DC: US Department of State.

Unwin, T. (2002) 'War and development', in V. Desai and R.B. Potter (eds) *The Companion to Development Studies*, London: Arnold, pp. 440–4.

Uppsala Conflict Data Program (UCDP) (2008) 'UCDP battle-deaths dataset v.4.1, 2002–05'. http://www.pcr.uu.se/research/UCDP/index.htm (accessed 5 July 2008).

Urry, J. (2002) *The Tourist Gaze* (second edition), London: Sage.

Vakil, A. (1997) 'Confronting the classification problem: towards a taxonomy of NGOs', *World Development* 25: 2057–71.

Vanderbeck, R. M. and Johnson, H. (2000) '"That's the only place where you can hang out": urban young people and the space of the mall', *Urban Geography* 21(1): 5–25.

Vendergeest, P. (2007) 'Certification and communities: alternatives for regulating the environmental and social impacts of shrimp farming', *World Development* 35(7): 1152–71.

Vertovec, S. (1999) 'Conceiving and researching transnationalism', *Ethnic and Racial Studies*, 22(2): 447–62.

Visser, G. (2008) 'The homonormalisation of white heterosexual leisure spaces in Bloemfontein, South Africa', *Geoforum* 39(3): 1347–61.

Wade, R. (1990) *Governing the Market: Economic Theory and the Role of Government in East Asian Industrialization*, Princeton, NJ: Princeton University Press.

Wade, R. (1999) 'Gestalt shift: from "miracle" to "cronyism" in the Asian crisis', *IDS Bulletin* 30(1).

Wade, R. (2000) 'Wheels within wheels: rethinking the Asian crisis and the Asian model', *Annual Review of Political Science* 3: 85–116.

Wagstaff, M. (1994) 'The development of a modern world system', in T. Unwin (ed.) *Atlas of World Development*, Chichester: Wiley, pp. 10–11.

Wainwright, J. (2008) *Decolonizing Development: Colonial Power and the Maya*, Oxford: Blackwell.

Wallerstein, I. (1984) *The Politics of the World-Economy*, Cambridge: Cambridge University Press.

Walton-Roberts, M. and Pratt, G. (2005) 'Mobile modernities: A South Asian family negotiates immigration, gender and class in Canada', *Gender, Place and Culture* 12: 173–95.

Ward, K. (1999) 'Africa', in A. Hastings (ed.) *A World History of Christianity*, London: Cassell, pp. 192–237.

Waters, J. (2006) 'Geographies of cultural capital: education, international migration and family strategies', *Transactions of the Institute of British Geographers* 31: 179–92.

Watson, J. L. (2000) 'China's Big Mac attack', *Foreign Affairs* 79(3):120–34.

Watts, M. (2004) 'Violent environments: petroleum conflict and the political ecology of rule in the Niger Delta, Nigeria', in R. Peet and M. Watts (eds) *Liberation Ecologies: Environment, Development, Social Movements* (second edition), London: Routledge, pp. 273–98.

Watts, M. and Goodman, D. (1997) 'Agrarian questions: global appetite, local metabolism: nature, culture, and industry in *fin-de-siècle* agro food systems', in D. Goodman and M. Watts (eds) *Globalising Food: Agrarian Questions and Global Restructuring*, London: Routledge, pp. 1–32.

Whitehand, C. (2003) 'NGOs and participatory development in Nepal: national, institutional and community dynamics', unpublished PhD thesis, Geography, Keele University.

Whitehead, A. and Tsikata, D. (2003) 'Policy discourses on women's land rights in Sub-Saharan Africa: the implications of the return to the customary', *Journal of Agrarian Change* 3(1 and 2): 67–112.

Whitehead, J. and More, N. (2007) 'Revanchism in Mumbai? Political economy of rent gaps and urban restructuring in a global city', *Economic and Political Weekly* 23 June: 2428–34.

Wilding, R. (2006) '"Virtual" intimacies? Families communicating across transnational contexts', *Global Networks* 6(2): 125–42.

Williams, G. (2004) 'Evaluating participatory development: tyranny, power and (re)politicisation', *Third World Quarterly* 25(3): 557–78.

Williams, G. (2009) 'Good governance', in N. Thrift and R. Kitchin (eds) *International Encyclopedia of Human Geography*, Oxford: Elsevier (in press)

Williams, M. (2005) 'The Third World and global environmental negotiations: interests, institutions and ideas', *Global Environmental Politics* 5(3): 48–69.

Willis, K. (2005) *Theories and Practices of Development*, London: Routledge.

Wilson, F. (2000) 'Representing the state? School and teacher in post-Sendero Peru', *Bulletin of Latin American Research* 19: 1–16.

Wilson, F. (2001) 'In the name of the state? Schools and teachers in an Andean province', in T. Blom Hansen and F. Stepputat (eds) *States of Imagination: Ethnographic Explorations of the Postcolonial State*, Durham, NC: Duke University Press, pp. 313–44.

Wilson, T. D. (1998) 'Approaches to understanding the position of women workers in the informal sector', *Latin American Perspectives* Issue 99, 25(2): 105–19.

Wolpert, S. (2000) *A New History of India* (sixth edition), Oxford: Oxford University Press.

World Bank (1993) *The East Asian Miracle*, Oxford: Oxford University Press.

World Bank (1997) *World Development Report 1997: The State in a Changing World*, Oxford: Oxford University Press.

World Bank (2000) *World Development Report 2000/1: Attacking Poverty*, Oxford: Oxford University Press.

World Bank (2006a) World Bank webpage: www.worldbank.org (accessed 23 July 2006).

World Bank (2006b) *Global Economic Prospects 2006: Economic Implications of Remittances and Migration*, Washington, DC: World Bank.

World Bank (2007a) 'PovertyNet' at: www.worldbank.org (accessed 4 October 2007).

World Bank (2007b) *World Development Report 2007: Selected Indicators*, Oxford: Oxford University Press. (Available at: www.worldbank.org)

World Bank (2008) *World Governance Indicators* dataset http://info.worldbank.org/governance/wgi/sc_country.asp (accessed 5 July 2008).

World Commission on Environment and Development (WCED) (1987) *Our Common Future* (the Brundtland Report), Oxford: Oxford University Press.

World Commission on the Social Dimension of Poverty (2004) *A Fair Globalization: Creating Opportunities for All*, Geneva: ILO. (Available at: www.ilo.org)

World Health Organization Regional Office for Africa (2005) 'Cardiovascular diseases in the African region: current situation and perspectives' (AFR/RC55/12), Brazzaville, Congo: WHO Regional Office for Africa.

World Health Organization (WHO) Statistics (2006) 'Health status: mortality 2006', Geneva: WHO.

World Health Organization (WHO) (2007) 'Tuberculosis fact sheet, 2007', Fact Sheet No. 104, revised March 2007, Geneva: WHO.

World Health Organization (WHO) (2008a) Cardiovascular diseases fact sheets, http://www.who.int/mediacentre/factsheets/fs317/en/index.html

World Health Organization (WHO) (2008b) Global Database on Body Mass Index, http://www.who.int/bmi/index.jsp

World Health Organization (WHO) (2008c) The Tobacco Atlas, http://www.who.int/tobacco/en/atlas40.pdf, Table A: The Demographics of Tobacco.

World Tourism Organization (2006) 'Tourism market trends'. www.world-tourism.org (accessed 22 July 2006).

World Trade Organization (WTO) (2006) 'What is the WTO?': www.wto.org/english/thewto_e/whatis_e/whatis_e.htm (accessed 23 July 2006).

World Trade Organization (WTO) (2008) 'What is the World Trade Organization?' at: www.wto.org (accessed 5 June 2008).

World Travel and Tourism Council (WTTC) (2008) 'Tourism satellite account: executive summary'. (Available from: http://www.wttc.org/eng/Tourism_Research/Tourism_Satellite_Accounting/)

Wright, A. (2006) *Ripped and Torn: Levi's, Latin America and the Blue Jean Dream*, London: Ebury Press.

Yamashita, S. (2003) *Bali and Beyond: Explorations in the Anthropology of Tourism*, Oxford: Berghahn Books.

Yan, Y. (2000) 'Of hamburgers and social space: consuming McDonald's in Beijing', in D. S. Davis (ed.) *The Consumer Revolution in Urban China*, Berkeley: University of California Press, pp. 201–25.

Yeoh, B. S. A. and Huang, S. (2000) '"Home" and "away": foreign domestic workers and negotiations of diasporic identity in Singapore', *Women's Studies International Forum* 23(4): 413–29.

Yeung, H. W. C. (2007) 'Remaking economic geography: insights from East Asia', *Economic Geography* 83: 339–48.

Young, J. (2005) 'Sudan: a flawed peace process leading to a flawed peace', *Review of African Political Economy* 103: 99–113.

Yuval-Davis, N. (1997) *Gender and Nation*, London: Sage.

Zack-Williams, A. B. (2006) 'Kwame Francis Nkrumah', in D. Simon (ed.) *Fifty Key Thinkers on Development*, London: Routledge, pp. 187–92.

Zack-Williams, A. B. and Mohan, G. (2002) 'Editorial: Africa, the African diaspora and development', *Review of African Political Economy* 92: 205–10.

Index

Note: Where an entry appears in a figure, plate or table, the page reference appears in **bold**.